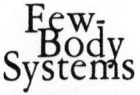

Suppl. 4

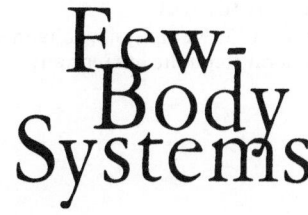

Few-Body Systems

Editor-in-Chief: H. Mitter, Graz   Associate Editor: W. Plessas, Graz

Supplementum 4

K.-M. Schmitt and H. Arenhövel

Complete Atlas of Polarization Observables in Deuteron Photodisintegration Below Pion-Threshold

Springer-Verlag Wien New York

K.-M. Schmitt
H. Arenhövel
Institut für Kernphysik, Johannes-Gutenberg-Universität, Mainz,
Federal Republic of Germany

ISSN 0177-8811
ISBN-13:978-3-7091-9193-4   e-ISBN-13:978-3-7091-9191-0
DOI: 10.1007/978-3-7091-9191-0

Acknowledgement

We would like to thank W. Leidemann for providing us with the computer code for the Nijmegen potential.

This work was partially supported by the Deutsche Forschungsgemeinschaft (SFB 201).

Table of Contents

1	Introductory Notes	1
2	Polarization Observables	2
3	Ingredients of Calculation	4
4	Explanation of Notation	6
	References	8
5	Atlas of Polarization Observables	9
	5.1 Differential Cross Section $d\sigma_0/d\Omega$	11
	5.2 Target asymmetries $P_1^{0,IM}$	12
	5.3 Target asymmetries $P_1^{c,IM}$	16
	5.4 Target asymmetries $P_1^{l,IM}$	20
	5.5 Proton Polarization $P_x^{0,IM}(p)$	29
	5.6 Proton Polarization $P_x^{c,IM}(p)$	33
	5.7 Proton Polarization $P_x^{l,IM}(p)$	38
	5.8 Neutron Polarization $P_x^{0,IM}(n)$	47
	5.9 Neutron Polarization $P_x^{c,IM}(n)$	51
	5.10 Neutron Polarization $P_x^{l,IM}(n)$	56
	5.11 Proton Polarization $P_y^{0,IM}(p)$	65
	5.12 Proton Polarization $P_y^{c,IM}(p)$	70
	5.13 Proton Polarization $P_y^{l,IM}(p)$	74
	5.14 Neutron Polarization $P_y^{0,IM}(n)$	83
	5.15 Neutron Polarization $P_y^{c,IM}(n)$	88
	5.16 Neutron Polarization $P_y^{l,IM}(n)$	92
	5.17 Proton Polarization $P_z^{0,IM}(p)$	101
	5.18 Proton Polarization $P_z^{c,IM}(p)$	105
	5.19 Proton Polarization $P_z^{l,IM}(p)$	110
	5.20 Neutron Polarization $P_z^{0,IM}(n)$	119

VIII

5.21 Neutron Polarization $P_z^{c,IM}(n)$ 123

5.22 Neutron Polarization $P_z^{l,IM}(n)$ 128

5.23 Proton-Neutron Spin Correlation $P_{xx}^{0,IM}$ 137

5.24 Proton-Neutron Spin Correlation $P_{xx}^{c,IM}$ 142

5.25 Proton-Neutron Spin Correlation $P_{xx}^{l,IM}$ 146

5.26 Proton-Neutron Spin Correlation $P_{yy}^{0,IM}$ 155

5.27 Proton-Neutron Spin Correlation $P_{yy}^{c,IM}$ 160

5.28 Proton-Neutron Spin Correlation $P_{yy}^{l,IM}$ 164

5.29 Proton-Neutron Spin Correlation $P_{zz}^{0,IM}$ 173

5.30 Proton-Neutron Spin Correlation $P_{zz}^{c,IM}$ 178

5.31 Proton-Neutron Spin Correlation $P_{zz}^{l,IM}$ 182

5.32 Proton-Neutron Spin Correlation $P_{xz}^{0,IM}$ 191

5.33 Proton-Neutron Spin Correlation $P_{xz}^{c,IM}$ 196

5.34 Proton-Neutron Spin Correlation $P_{xz}^{l,IM}$ 200

5.35 Proton-Neutron Spin Correlation $P_{zx}^{0,IM}$ 209

5.36 Proton-Neutron Spin Correlation $P_{zx}^{c,IM}$ 214

5.37 Proton-Neutron Spin Correlation $P_{zx}^{l,IM}$ 218

5.38 Proton-Neutron Spin Correlation $P_{xy}^{0,IM}$ 227

5.39 Proton-Neutron Spin Correlation $P_{xy}^{c,IM}$ 231

5.40 Proton-Neutron Spin Correlation $P_{xy}^{l,IM}$ 236

5.41 Proton-Neutron Spin Correlation $P_{yx}^{0,IM}$ 245

5.42 Proton-Neutron Spin Correlation $P_{yx}^{c,IM}$ 249

5.43 Proton-Neutron Spin Correlation $P_{yx}^{l,IM}$ 254

5.44 Proton-Neutron Spin Correlation $P_{zy}^{0,IM}$ 263

5.45 Proton-Neutron Spin Correlation $P_{zy}^{c,IM}$ 267

5.46 Proton-Neutron Spin Correlation $P_{zy}^{l,IM}$ 272

5.47 Proton-Neutron Spin Correlation $P_{yz}^{0,IM}$ 281

5.48 Proton-Neutron Spin Correlation $P_{yz}^{c,IM}$ 285

5.49 Proton-Neutron Spin Correlation $P_{yz}^{l,IM}$ 290

1 Introductory Notes

This atlas continues the series of papers concerning systematic investigations of polarization observables in deuteron two-body photodisintegration $d(\gamma, N)N$ [1, 2, 3], in the following called I, II and III, respectively. It completes the most recent paper on a systematic study of polarization observables in deuteron photodisintegration below pion-production threshold [3], where we have presented results on all 59 single and double polarization observables including the unpolarized differential cross section. In this atlas we provide systematic and extensive numerical evaluations of all 288 observables within the conventional framework of nuclear theory including meson and isobar degrees of freedom. For five energies between 4.5 MeV, the maximum of the total cross section, and 140 MeV we show the influence of subnuclear degrees of freedom like meson exchange and isobar currents, of relativistic corrections, the role of the various electric and magnetic multipoles and the dependence on realistic potential models. Our aim is in the same spirit as outlined in III, namely to find out which observables are most sensitive to the various ingredients and which ones are little affected by, for example, the potential model or the different current contributions. In this way we hope to provide guidelines for choosing those observables which are the most interesting ones for the study of different dynamical effects. Hopefully, they will then serve as crucial tests for our present theoretical models allowing to pin down their limits, i.e., to find out where additional degrees of freedom like quark-gluon degrees of freedom have to be introduced.

Since all the details can be found in I to III, we will review only very briefly the essential definition of the polarization observables in Section 2 and the theoretical input in Section 3. In Section 4 we explain the notation for the presentation of our explicit results for all polarization observables for polarized photons and oriented deuterons in Section 5 which constitutes the main part of this atlas.

2 Polarization Observables

We shall briefly review the relevant expressions for the polarization observables as derived in I and II. Note that henceforth equation (n) of I or II will be referred to as (I.n) or (II.n), respectively. In the compact notation of Eq. (II.4) any observable has the general structure

$$P_X \frac{d\sigma}{d\Omega} = \frac{d\sigma_0}{d\Omega} \sum_{I=0}^{2} P_I^d \left\{ \sum_{M \geq 0} \left(P_X^{0,IM}(\theta) \sin\left[M(\phi_d - \phi) + \frac{\pi}{2} \delta_I^X \right] \right. \right.$$
$$\left. + P_c^\gamma P_X^{c,IM}(\theta) \cos\left[M(\phi_d - \phi) - \frac{\pi}{2} \delta_I^X \right] \right) d_{M0}^I(\theta_d)$$
$$\left. + P_l^\gamma \sum_M P_X^{l,IM}(\theta) \sin\left[\psi_M + \frac{\pi}{2} \delta_I^X \right] d_{M0}^I(\theta_d) \right\} \tag{1}$$

where

$$\psi_M = M(\phi_d - \phi) + 2\phi . \tag{2}$$

Here P_l^γ and P_c^γ describe the degree of linear and circular polarization of the incoming photons, respectively, and P_1^d and P_2^d the respective vector and tensor polarization of the deuteron target. Furthermore, $d\sigma_0/d\Omega$ denotes the unpolarized differential cross section and the symbol δ_I^X is defined in (II.12). As explained in detail in II the various observables are labelled by X in the following way:

(i) $X = 1$ with $P_1 = 1$ refers to the differential cross section. In contrast to I, II and III, we have not introduced the notation T_{IM}^α for the beam and target asymmetries $P_1^{\alpha,IM}$ for reasons of conformity of notation.

(ii) $X = x_i(j)$ refers to the polarization component $P_{x_i}(j)$ of an outgoing proton ($j = 1$) or neutron ($j = 2$) as given in Eqs. (I.53, I.54).

(iii) $X = x_i x_j$ refers to the proton-neutron polarization or spin correlation tensor $P_{x_i x_j}$ of the x_i-component of the proton spin and the x_j-component of the neutron spin as given in Eqs. (I.67, I.82).

Like for $X = 1$, any of these observables can be decomposed into beam and target polarization parameters or asymmetries $P_X^{\alpha,IM}$. For further details and in particular for the explicit expressions of $P_X^{\alpha,IM}$ in terms of the T-matrix elements we refer to I and II. For convenience we show in Fig. 1 the geometry and the definition of the angles θ, ϕ, θ_d and ϕ_d which appear in Eq. (1). The reference system for the nucleon polarization components is chosen according to the Madison convention. For the proton it is represented by the primed coordinate system (x', y', z') in Fig. 1b while for the neutron one has to invert the y'- and the z'-axis.

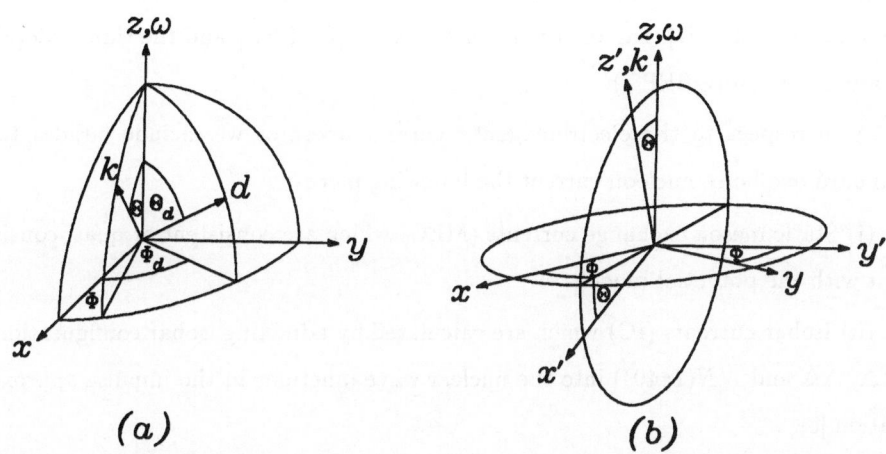

(a)   (b)

Figure 1: (a) Definition of the angles θ_d and ϕ_d of the deuteron orientation axis $\vec{d}$ with respect to the reference frame (x, y, z) associated with the incoming photon momentum $\vec{\omega}$ (see I). The x-axis has been chosen in the direction of maximal linear photon polarization. The relative p-n momentum is indicated by $\vec{k}$ which is characterized by the angles θ and ϕ. (b) The reference frame (x', y', z') for the outgoing proton polarization components.

3 Ingredients of Calculation

The calculation is done in the conventional framework of non-relativistic nuclear physics with nucleon, meson and isobar degrees of freedom but the dominant relativistic corrections of lowest order are included as outlined below.

For the NN interaction we use several realistic potentials: the Nijmegen [4], the Paris [5], the Argonne V_{14} [6] and two one-boson-exchange (OBE) approximations of the Bonn potential [7], namely the r-space version (OBEPR) and the time-ordered, retarded version (OBEPT).

With respect to the electromagnetic current operator we include besides the standard one-body nucleon current the following pieces:

(i) Static meson exchange currents (MEC) which are consistent or quasi consistent with the potential model.

(ii) Isobar currents (IC) which are calculated by admixing isobar configurations ($N\Delta$, $\Delta\Delta$ and $NN(1440)$) into the nuclear wave functions in the impulse approximation [8].

(iii) Relativistic corrections (RC) which are dominated by the relativistic spin-orbit current (SO) [9]. In case of the Bonn OBEPs we include in addition lowest order relativistic corrections to the one-body current beyond the SO-current and for the OBEPT furthermore lowest order retardation corrections to the π- and ρ-MEC [10, 11]. We have not included relativistic corrections in the wave functions, i.e., corrections of dynamic origin nor the boost contributions.

The T-matrix elements are calculated in the standard fashion using a multipole decomposition and including all electric and magnetic multipoles up to the order $L = 4$, if not stated otherwise. In the next section, the notion "one body-current" refers to a calculation where no MEC in the strict sense are included, i.e., no Siegert operators have been used, whereas "normal" (N) is used for a classical calculation with Siegert operators but no explicit MEC, IC and RC. For the Siegert operators we use the Partovi-gauge except in the calculations with the Bonn models where we take the z^l-gauge [12]. This has some relevance only if one considers the normal contribution alone but the total result including explicit MEC beyond the Siegert-MEC is independent of the choice of gauge [12].

4 Explanation of Notation

For the presentation of all 288 polarization observables in the next section we have adopted the following scheme. Each observable is a function of the angle and the photon energy. With respect to the latter we have chosen five energies , namely 4.5 MeV, the maximum of the total cross section, 20 MeV, 60 MeV, 100MeV, and 140 MeV. For each observable and for each of these energies we have studied the following topics:

(i) The influence of meson exchange currents (MEC), isobar configurations (IC) and relativistic corrections (RC). Since the various potential models give qualitatively very similar results, we use in this case the r-space version of the Bonn model (OBEPR).

(ii) The contributions of the different electric and magnetic multipoles to the full calculation including MEC, IC and RC, again for the Bonn OBEPR model.

(iii) The dependence of the full calculation including MEC, IC and RC on the choice of various realistic potential models.

Therefore each observable will be represented by a (5×3)-part figure where the rows refer to the different photon lab energies starting at the top with 4.5 MeV. Each row consists of three parts. The left part shows the subnuclear and relativistic effects of the current: pure one-body (dotted), normal (N) (dashed), $N + MEC$ (dash-dot), $N + MEC + IC$ (dash-double-dot) and $N + MEC + IC + RC$ (full). The middle part shows the multipole contributions starting with E1 (dotted) and adding successively M1 (dashed), E2 (dash-dot), M2 (dash-double-dot), and the full curve shows the total sum up to $L = 4$. The right part shows the potential model dependence: Paris (dotted), Bonn OBEPT (dashed), Nijmegen (dash-dot), Argonne V_{14} (dash-double-dot) and Bonn OBEPR (full). The angle θ refers always to the proton angle in the cm-system, except for the polarization components $P_{x_i}^{\alpha, IM}(n)$ of the neutron where the angle refers to the neutron angle in the cm system. Furthermore, in order to comply with the Madison convention we have changed the signs of $P_y^{\alpha,IM}(n)$, $P_z^{\alpha,IM}(n)$, $P_{x,y}^{\alpha,IM}$ and $P_{x,z}^{\alpha,IM}$ compared to the expressions in I and II. The full curve always refers to a complete calculation with the Bonn OBEPR, i.e., in each row the full curves are identical. At 4.5 MeV, an observable was in three cases less than 10^{-4} and then we have not plotted it.

References

[1] Arenhövel, H.: Few–Body Syst. **4**, 55 (1988)

[2] Arenhövel, H., Schmitt, K.-M.: Few–Body Syst. **8**, 77 (1990)

[3] Schmitt, K.-M., Arenhövel, H.: Few-Body Syst. **11**, in print.

[4] Nagels, M.M., Rijken, T.A., de Swart, J.J.: Phys. Rev. **D17**, 768 (1978)

[5] Lacombe, L., Loiseau, B., Richard, J.M., Vinh Mau, R., Côté, J., Pirés, P., de Tourreil, R.: Phys. Rev. **C21**, 861 (1980)

[6] Wiringa, R.B., Smith, R.A., Ainsworth, T.L.: Phys. Rev. **C29**, 1584 (1984)

[7] Machleidt, R., Holinde, K., Elster, Ch.: Phys. Rep. **149**, 1 (1987)

[8] Leidemann, W., Arenhövel, H.: Nucl. Phys. **A465**, 573 (1987)

[9] Cambi, A., Mosconi, B., Ricci, P.: Phys. Rev. Lett. **48**, 462 (1982)

[10] Schmitt, K.-M., Arenhövel, H.: Few–Body Syst. **7**, 95 (1989)

[11] Schmitt, K.-M., Arenhövel, H.: Few–Body Syst. **6**, 117 (1989)

[12] Schmitt, K.-M., Wilhelm, P., Arenhövel, H., Cambi, A., Mosconi, B., Ricci, P.: Phys. Rev. **C41**, 841 (1990)

5   Atlas of Polarization Observables

5.1 Differential Cross Section $d\sigma_0/d\Omega$

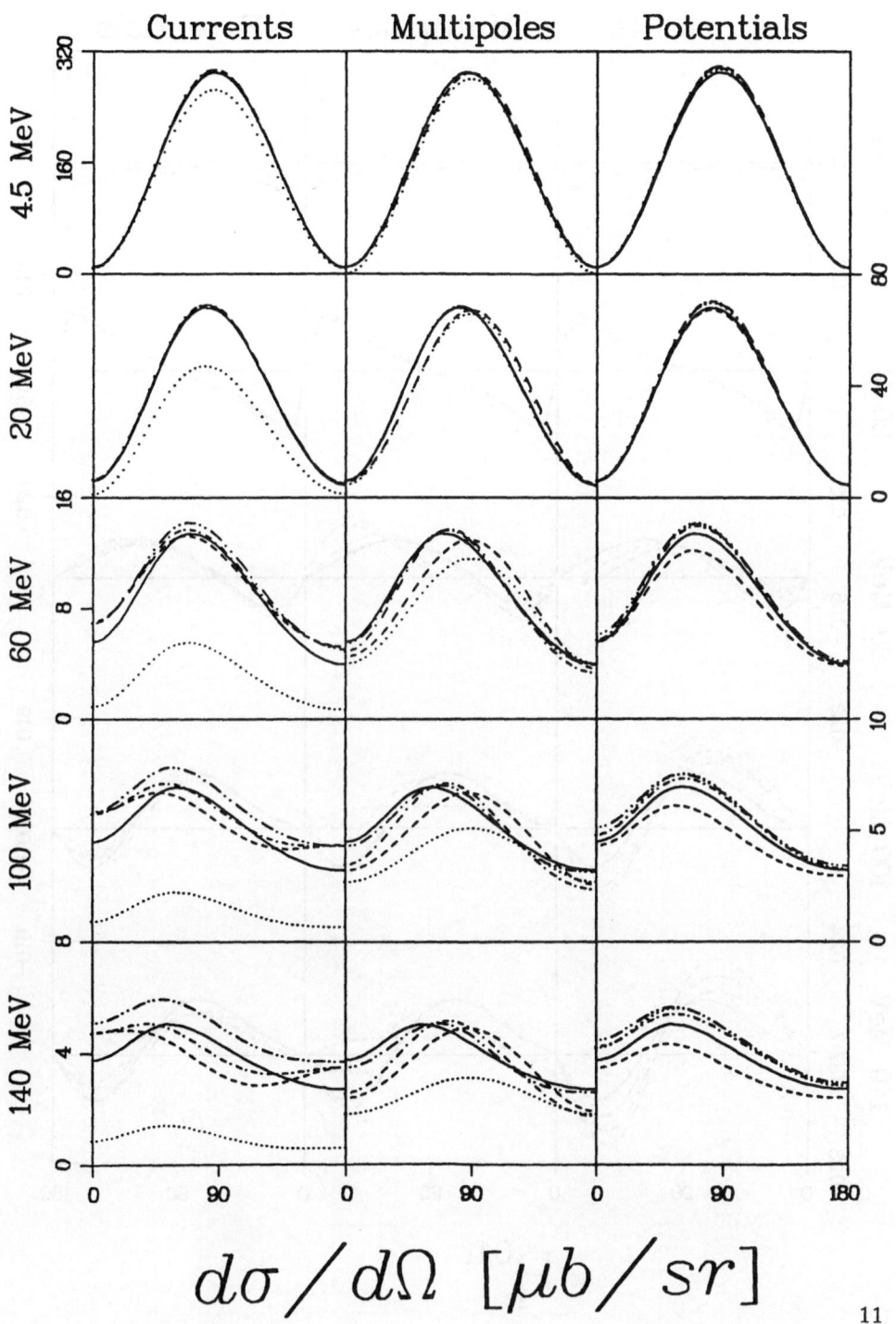

$$d\sigma / d\Omega \; [\mu b / sr]$$

5.2 Target asymmetries $P_1^{0,IM}$

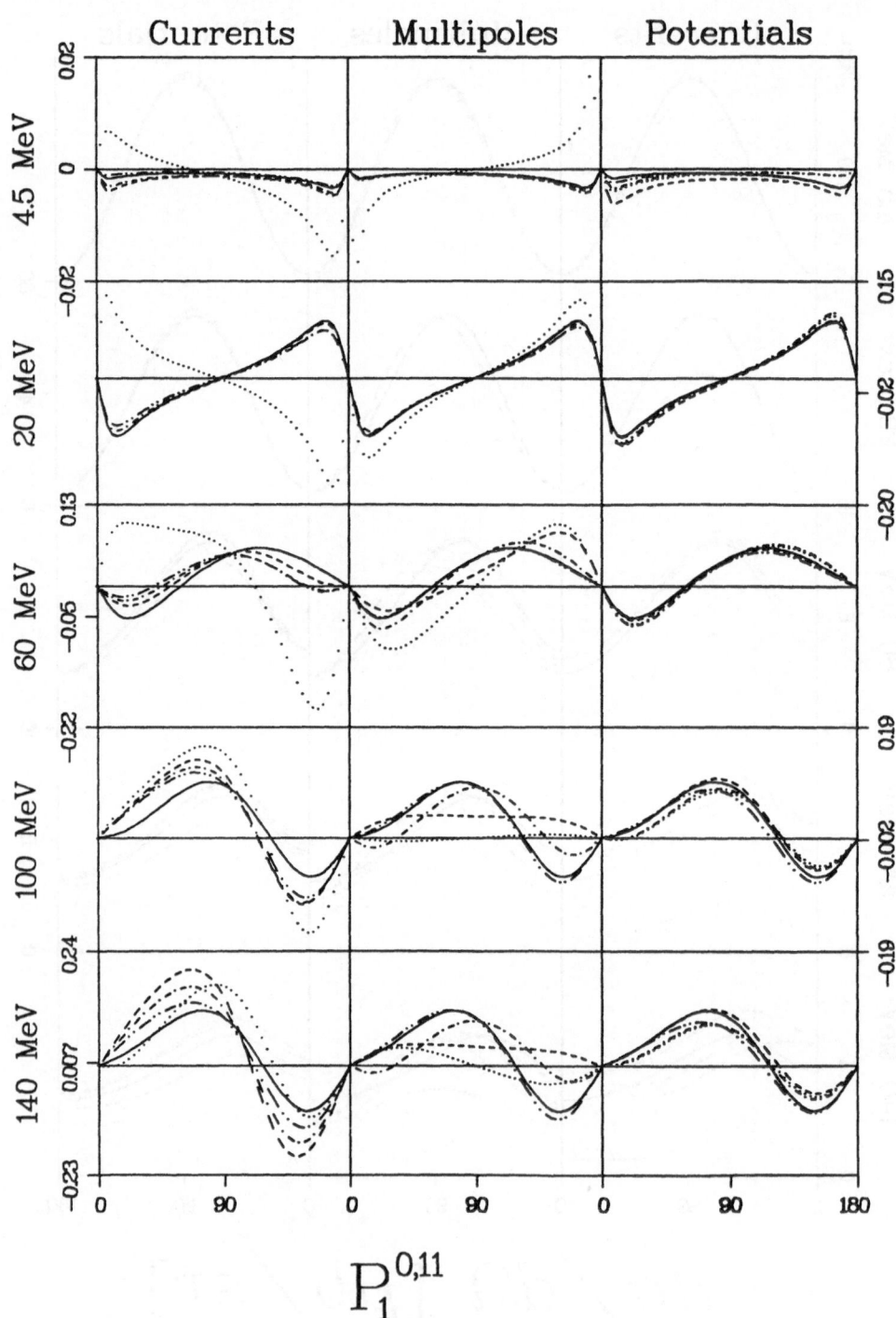

$$P_1^{0,11}$$

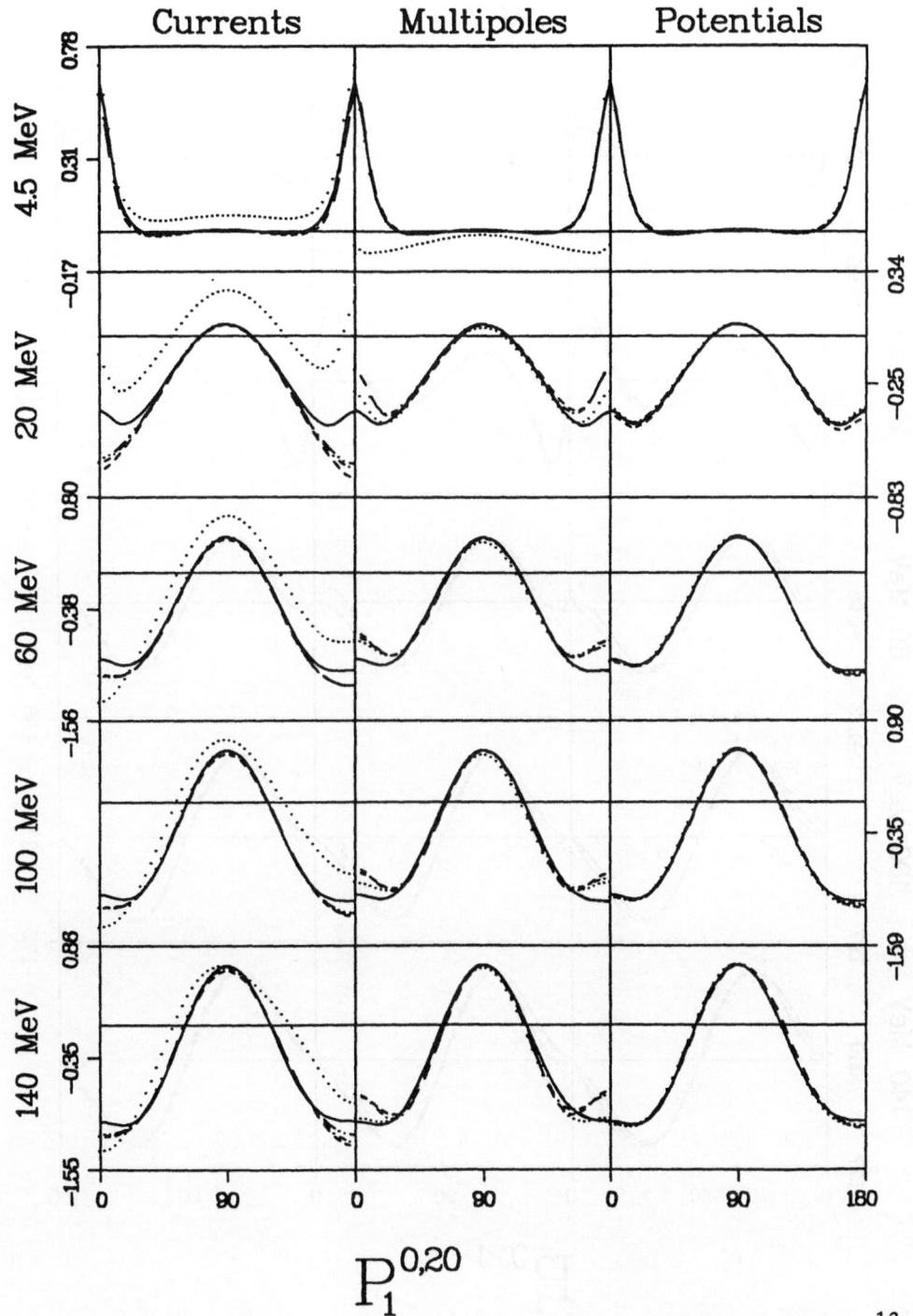

Currents   Multipoles   Potentials

$$\mathrm{P}_1^{0,20}$$

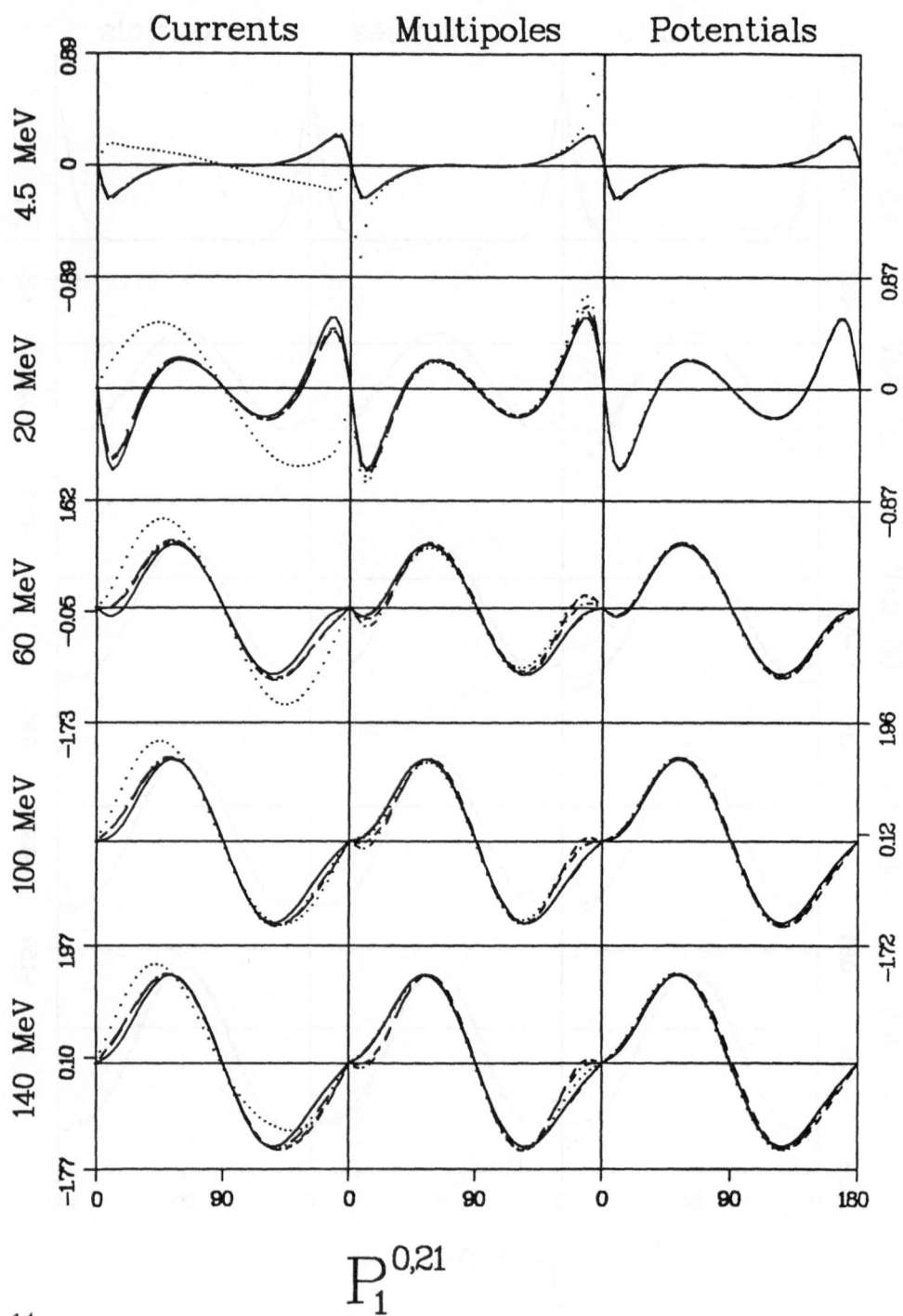

Currents   Multipoles   Potentials

4.5 MeV   20 MeV   60 MeV   100 MeV   140 MeV

$$P_1^{0,21}$$

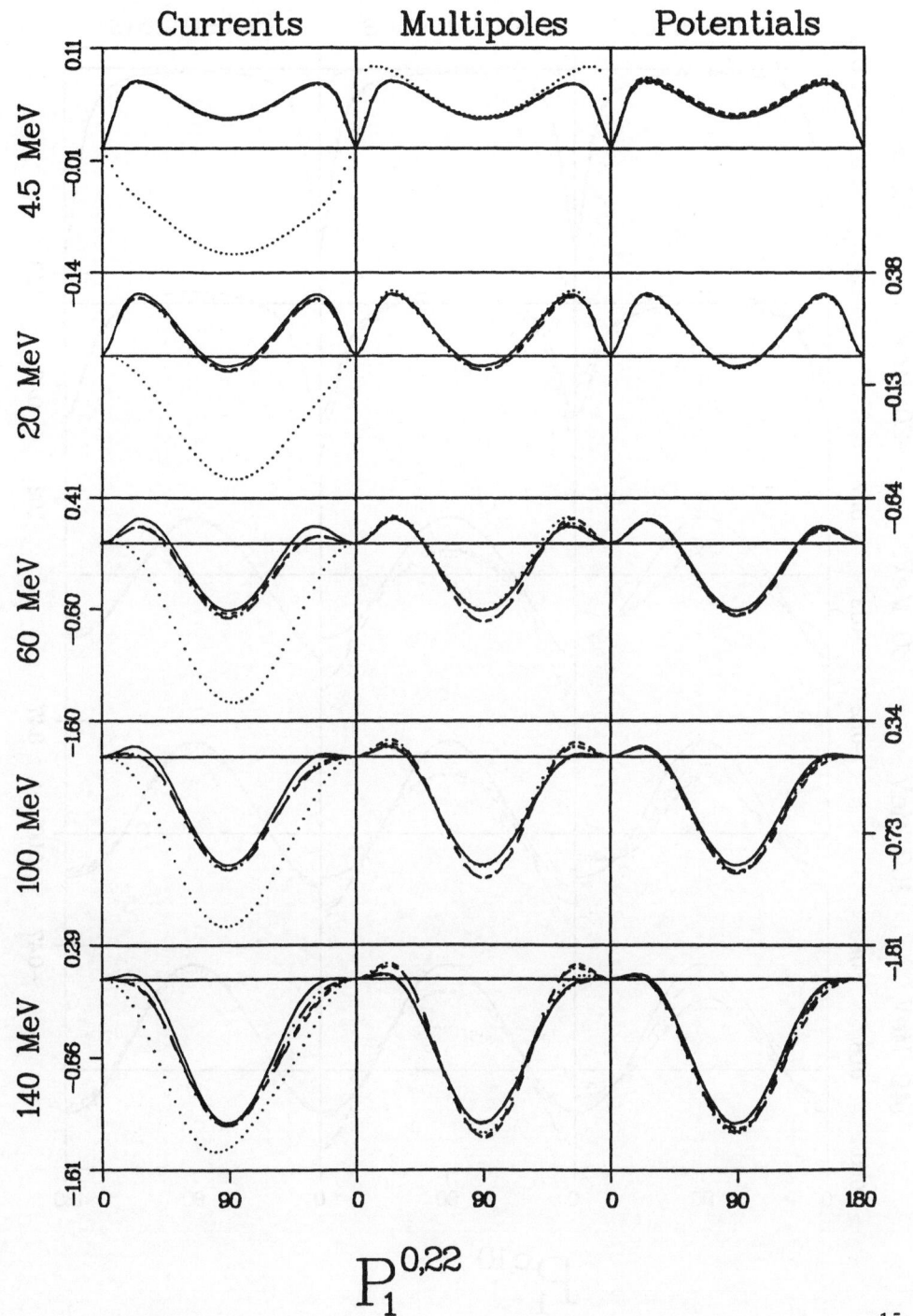

$$P_1^{0,22}$$

15

5.3 Target asymmetries $P_1^{c,IM}$

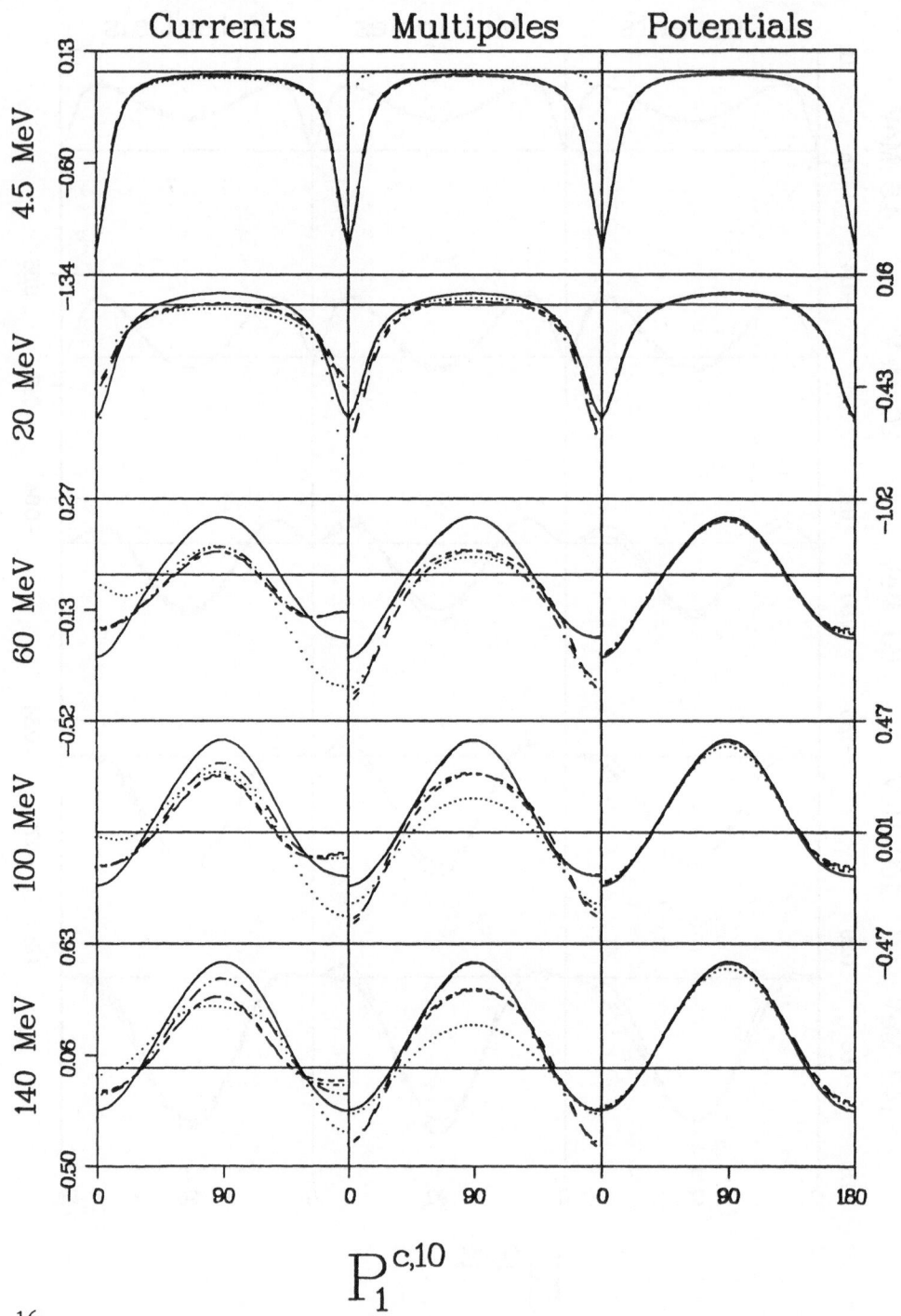

$$P_1^{c,10}$$

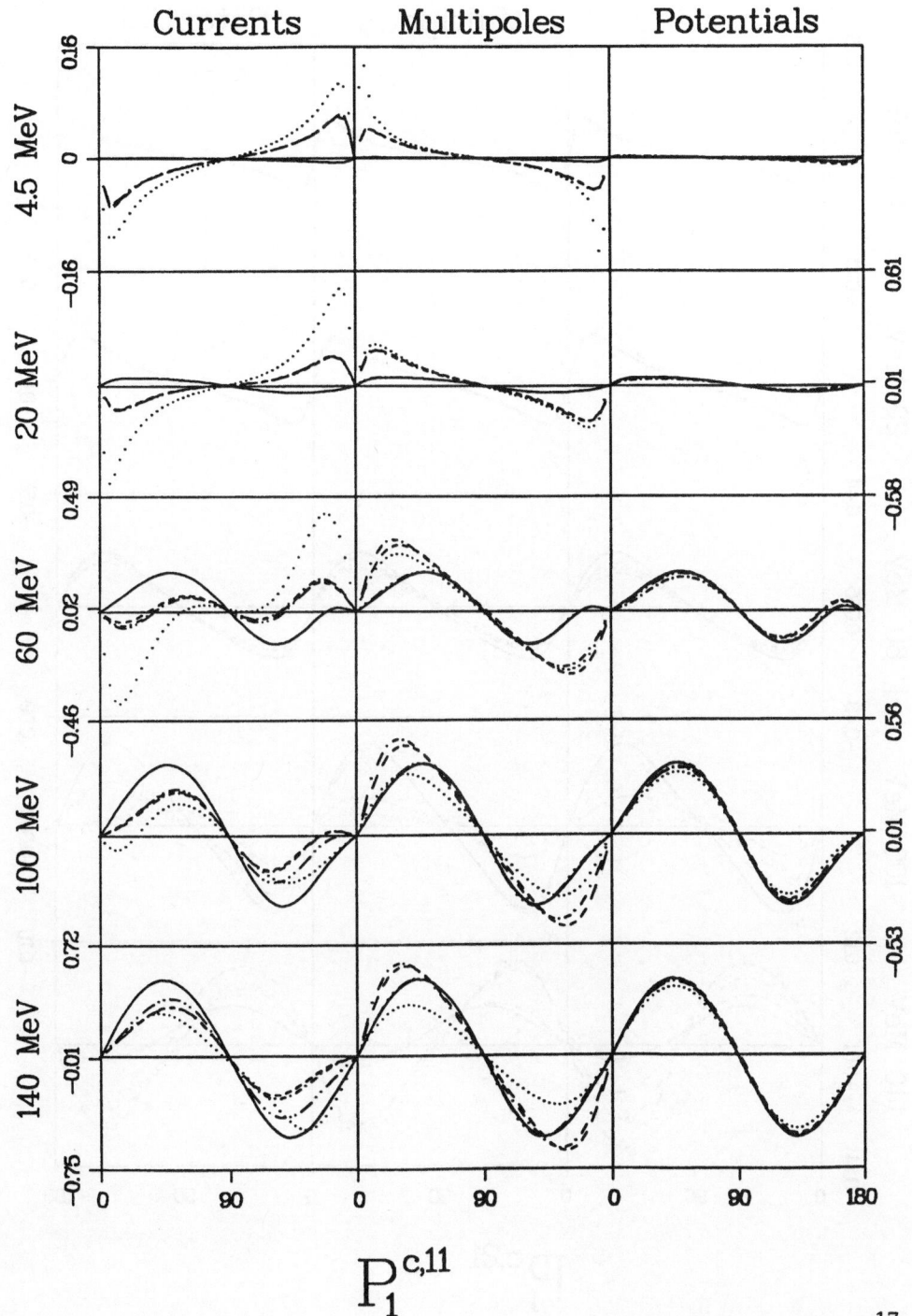

$$P_1^{c,11}$$

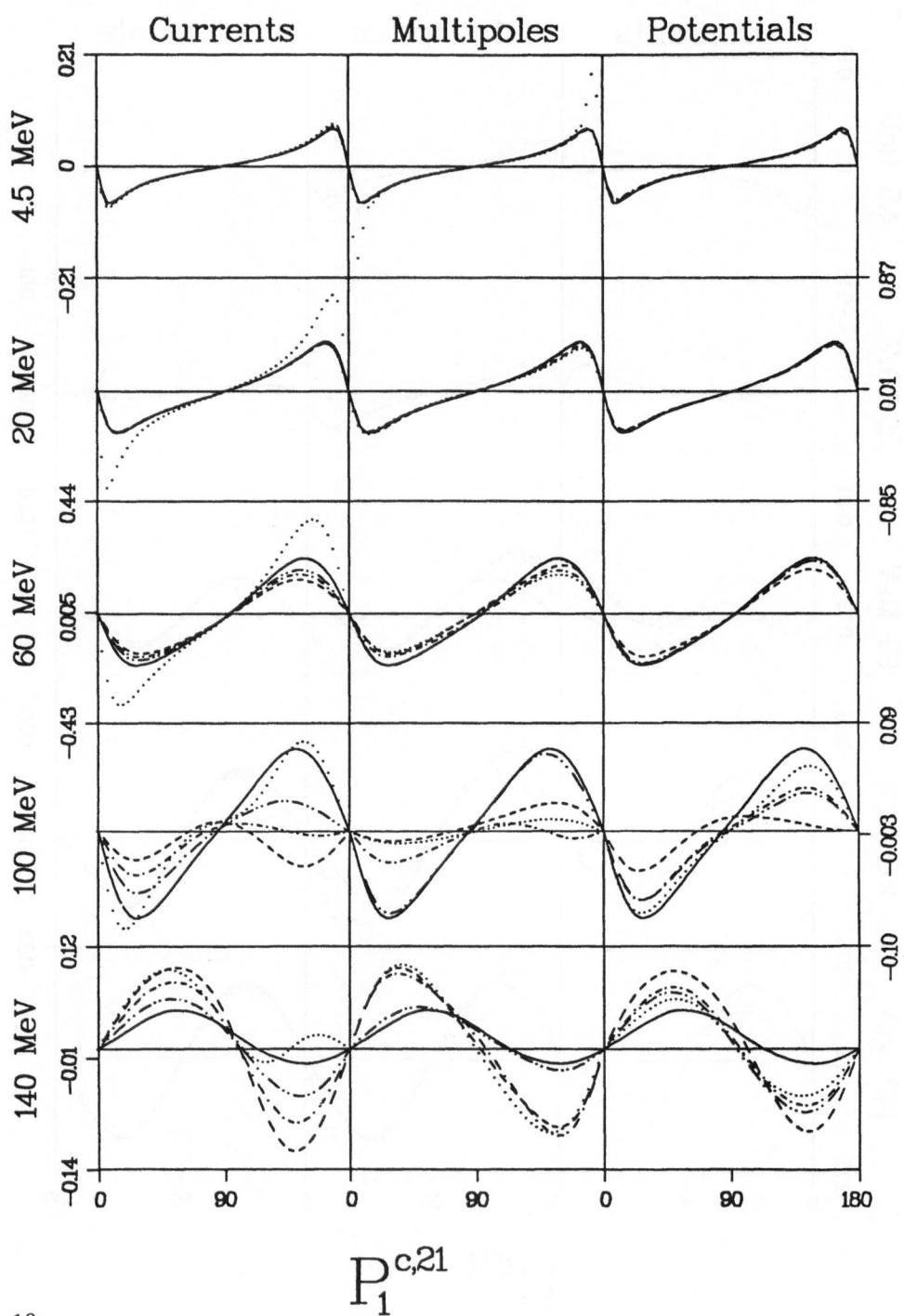

$$P_1^{c,21}$$

Target asymmetries $P_1^{c,IM}$

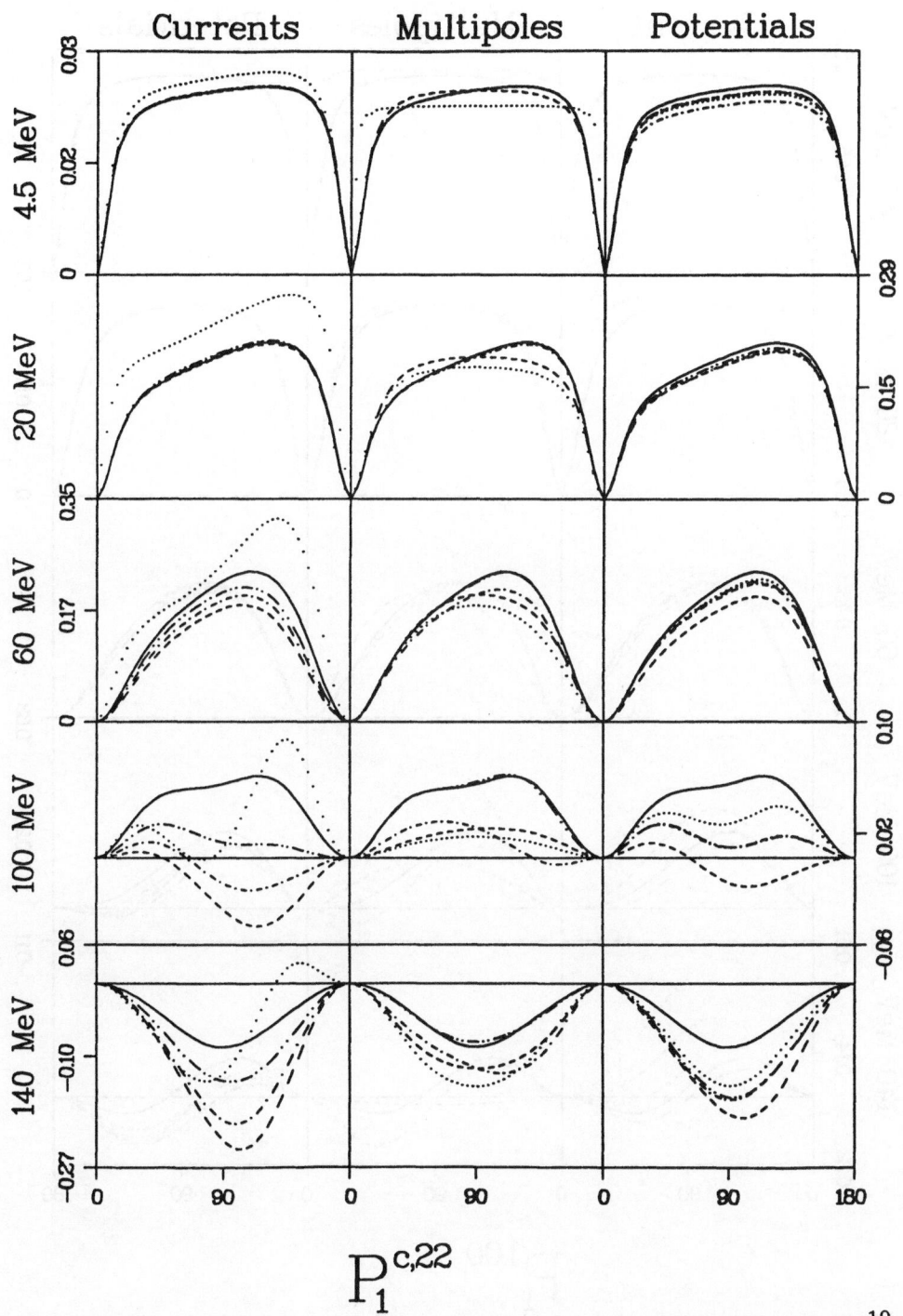

$$P_1^{c,22}$$

5.4 Target asymmetries $P_1^{l,IM}$

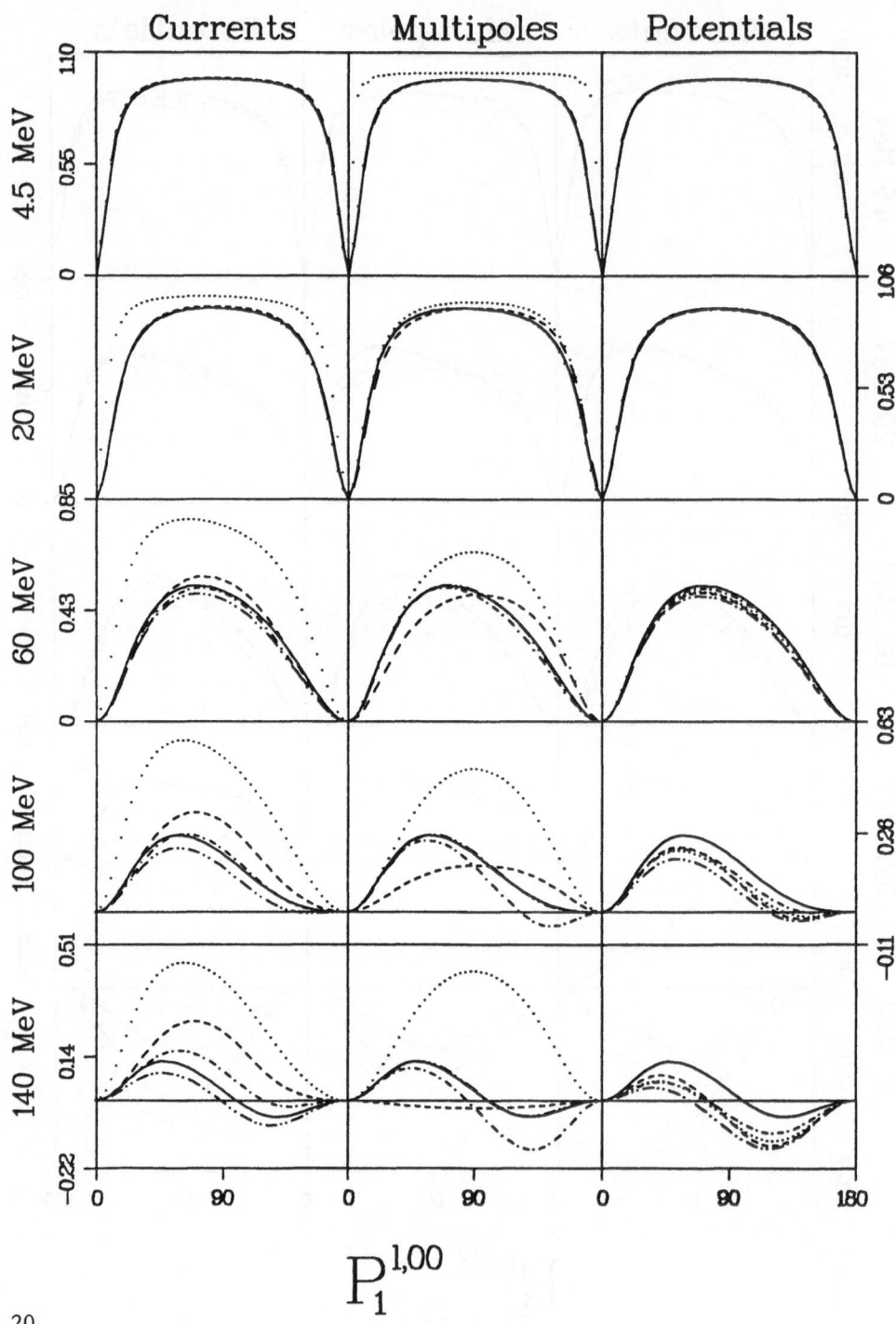

$$P_1^{1,00}$$

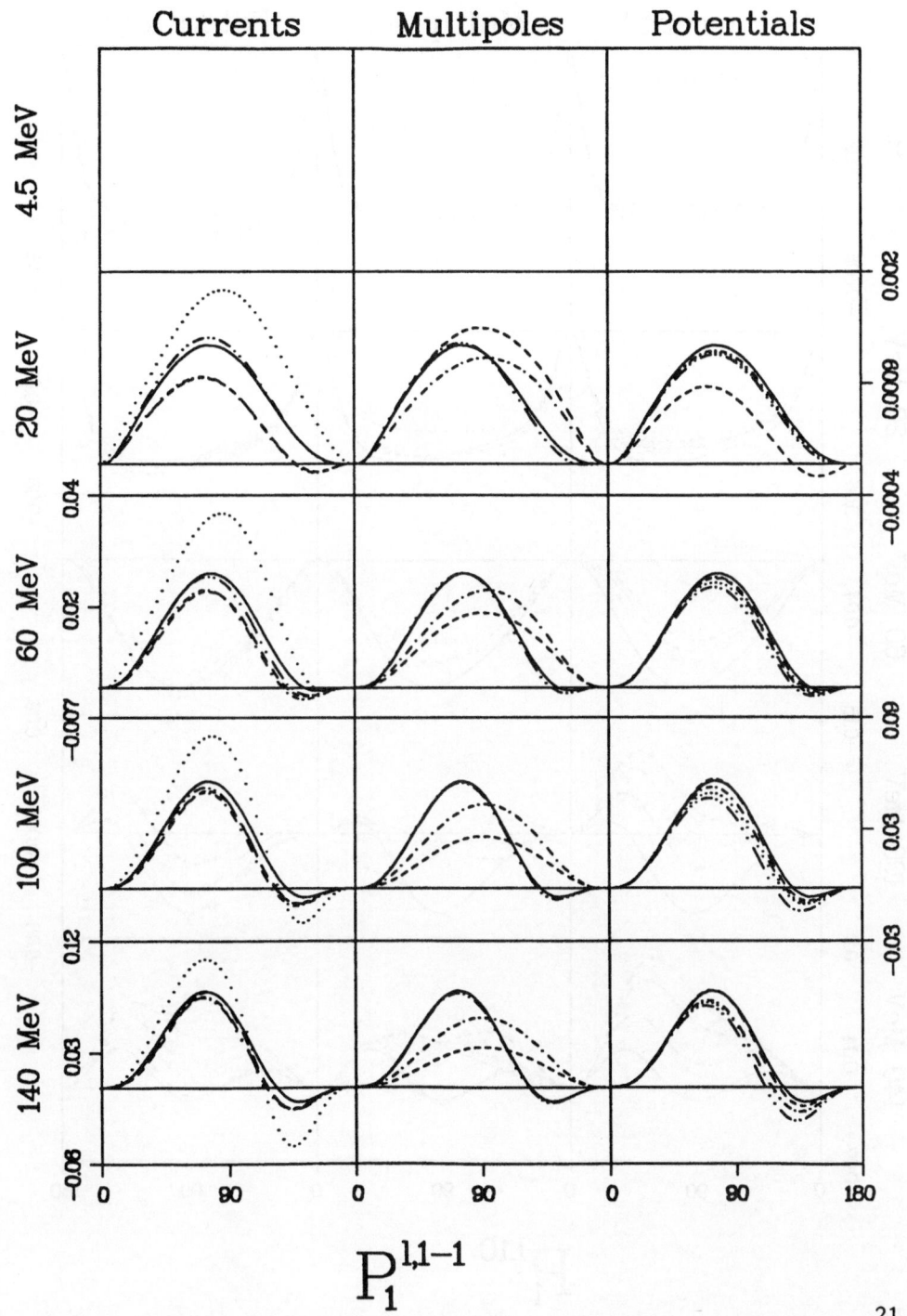

$$P_1^{1,1-1}$$

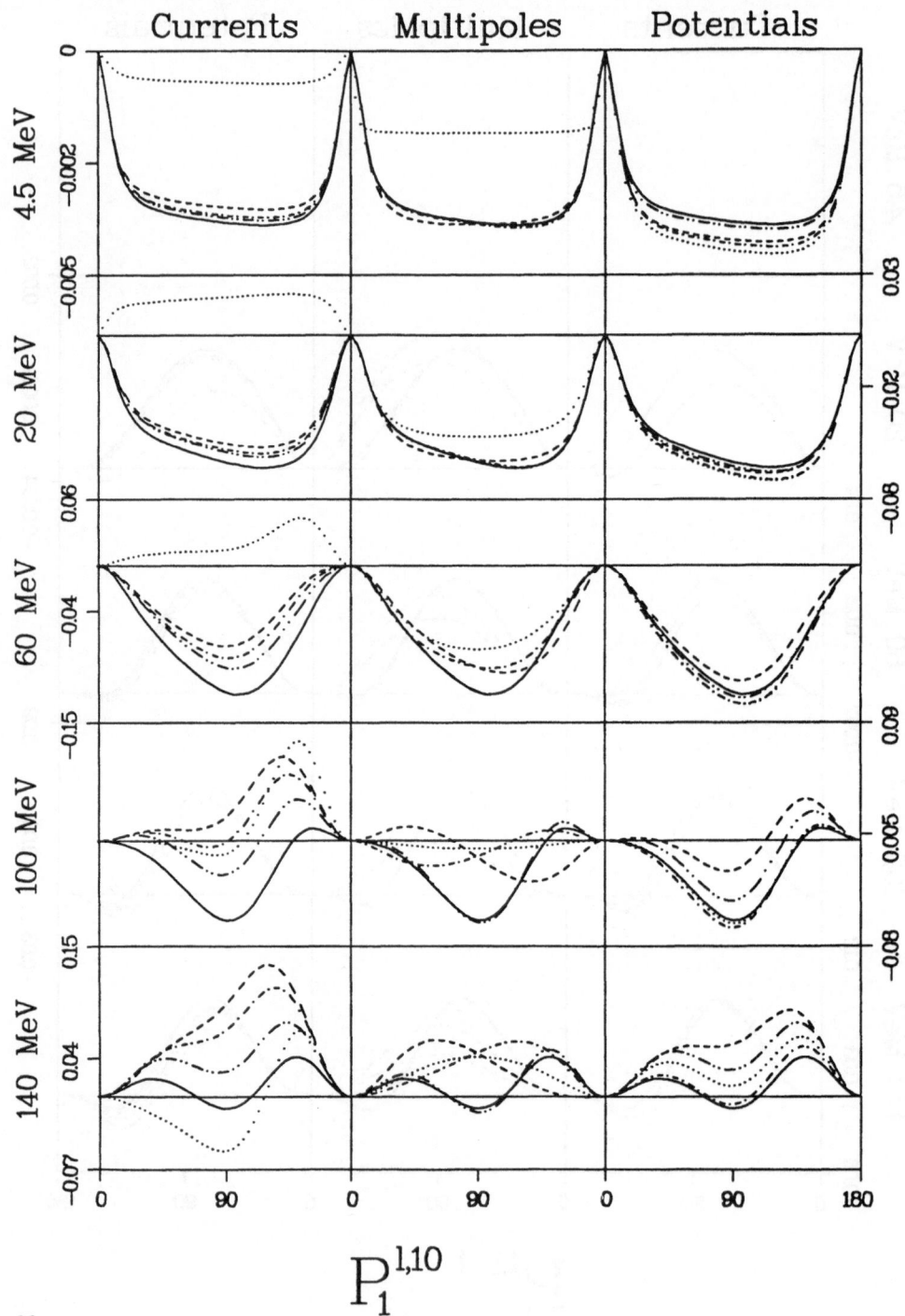

$$P_1^{1,10}$$

Target asymmetries $P_1^{l,IM}$

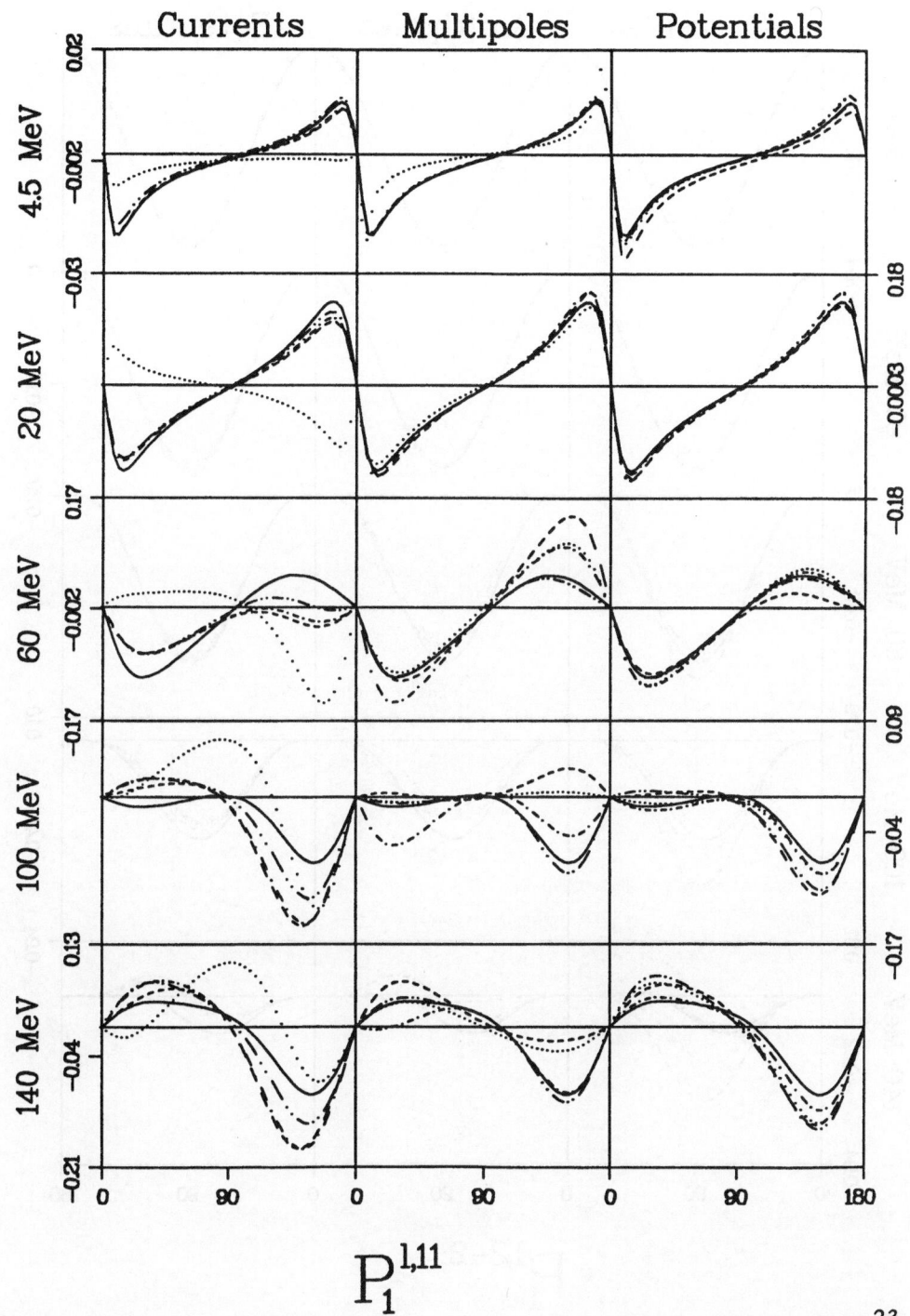

$$P_1^{1,11}$$

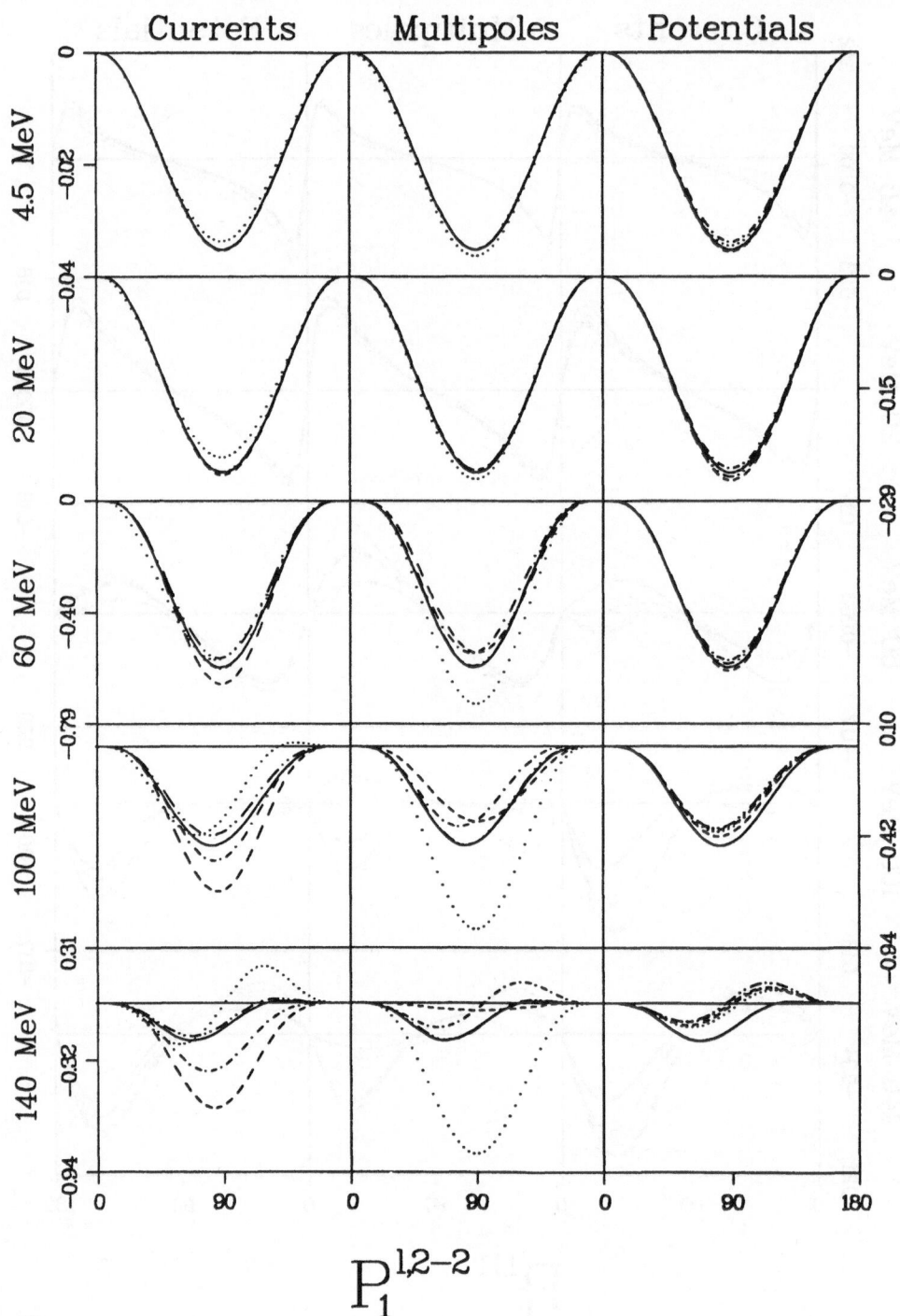

$$P_1^{1,2-2}$$

Target asymmetries $P_1^{l,IM}$

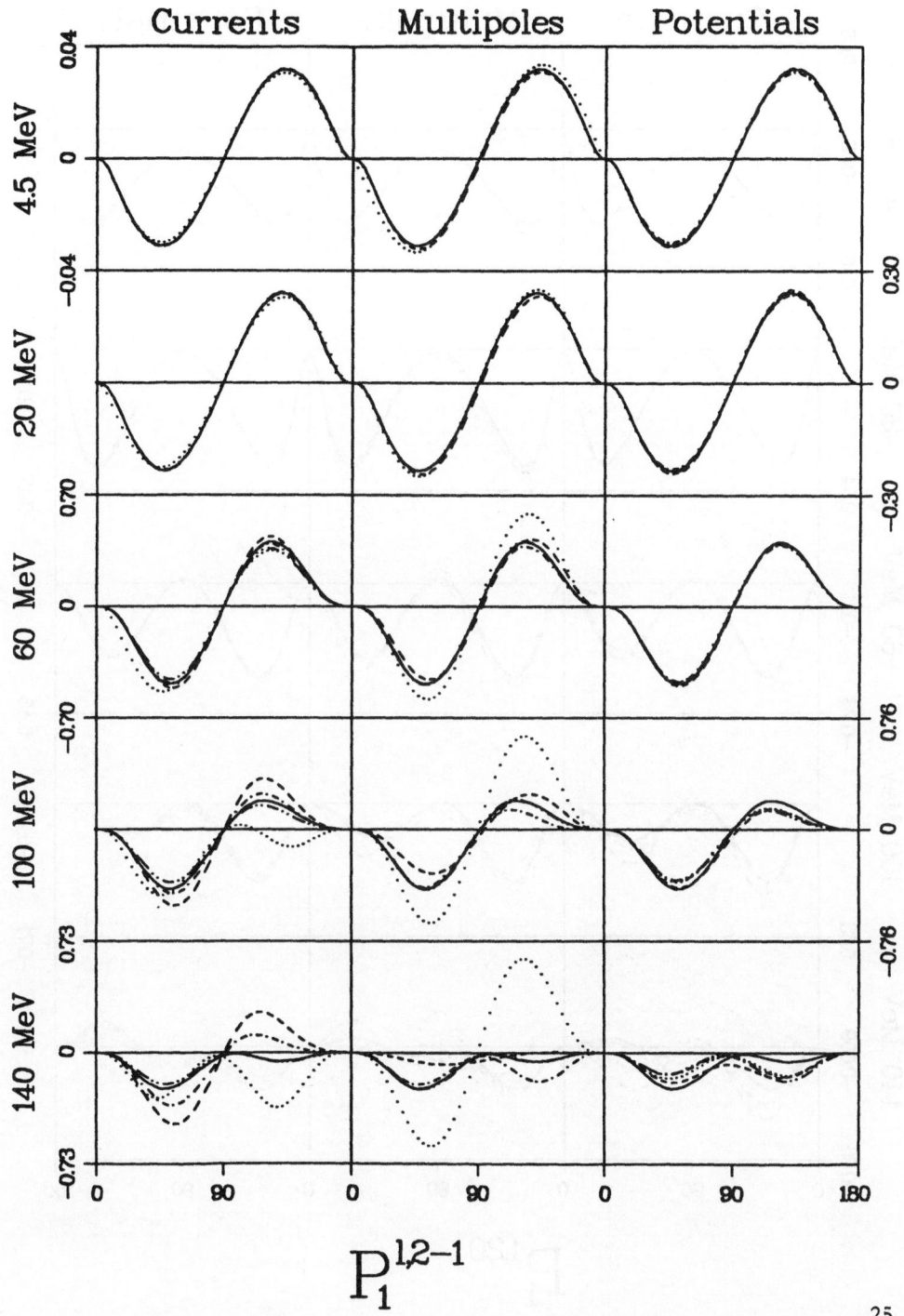

$$P_1^{1,2-1}$$

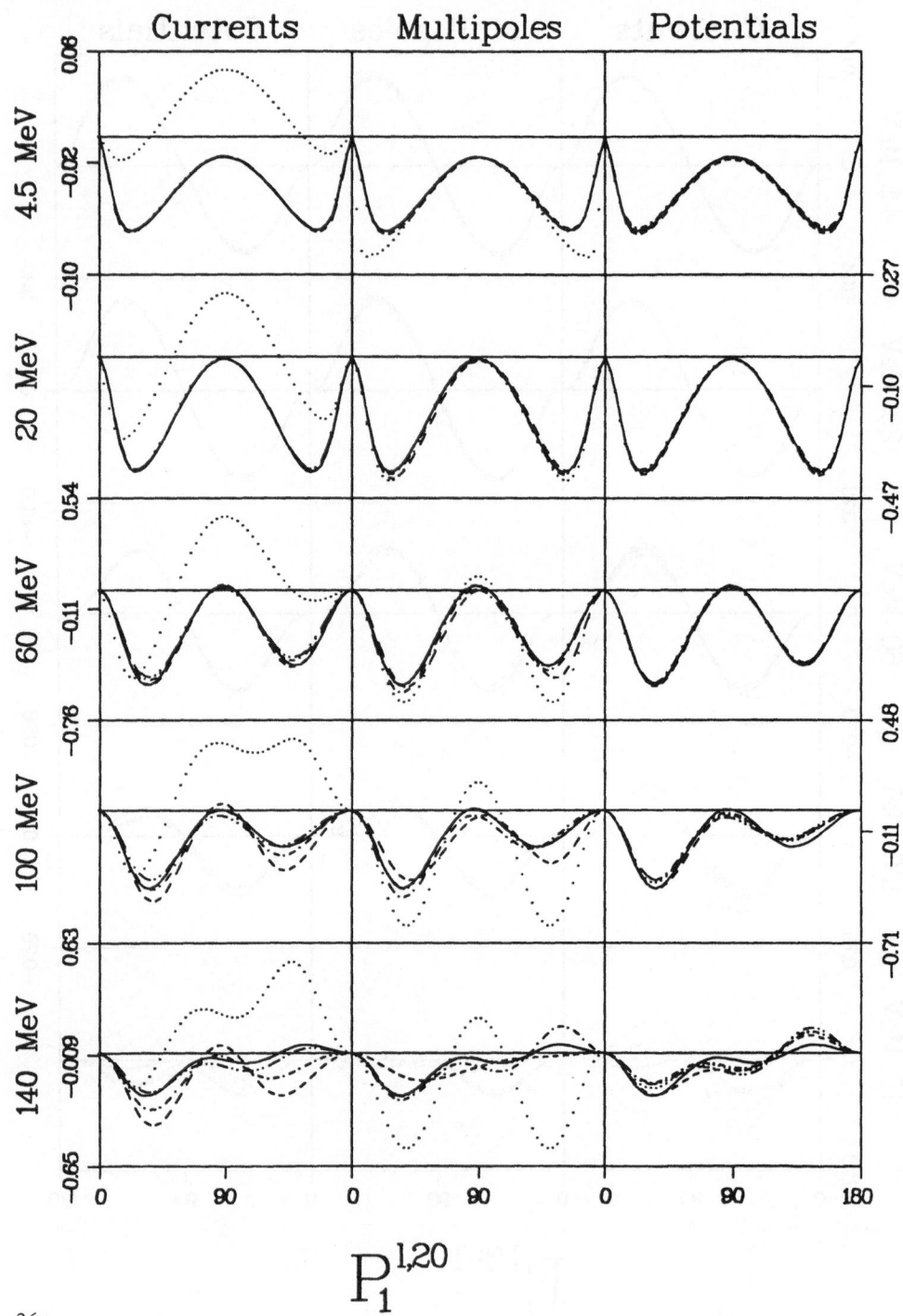

$$P_1^{1,20}$$

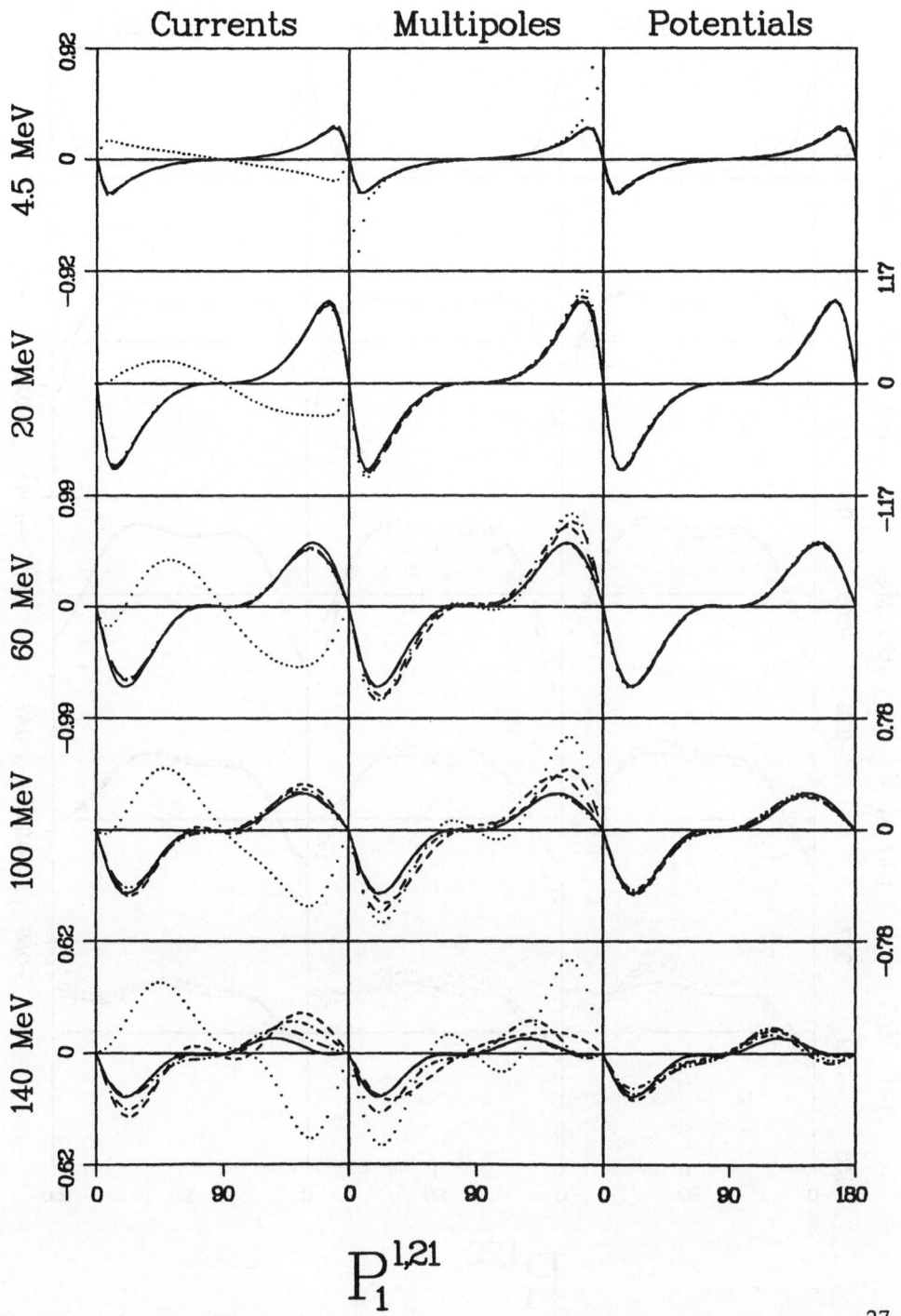

$$P_1^{1,21}$$

Target asymmetries $P_1^{l,IM}$

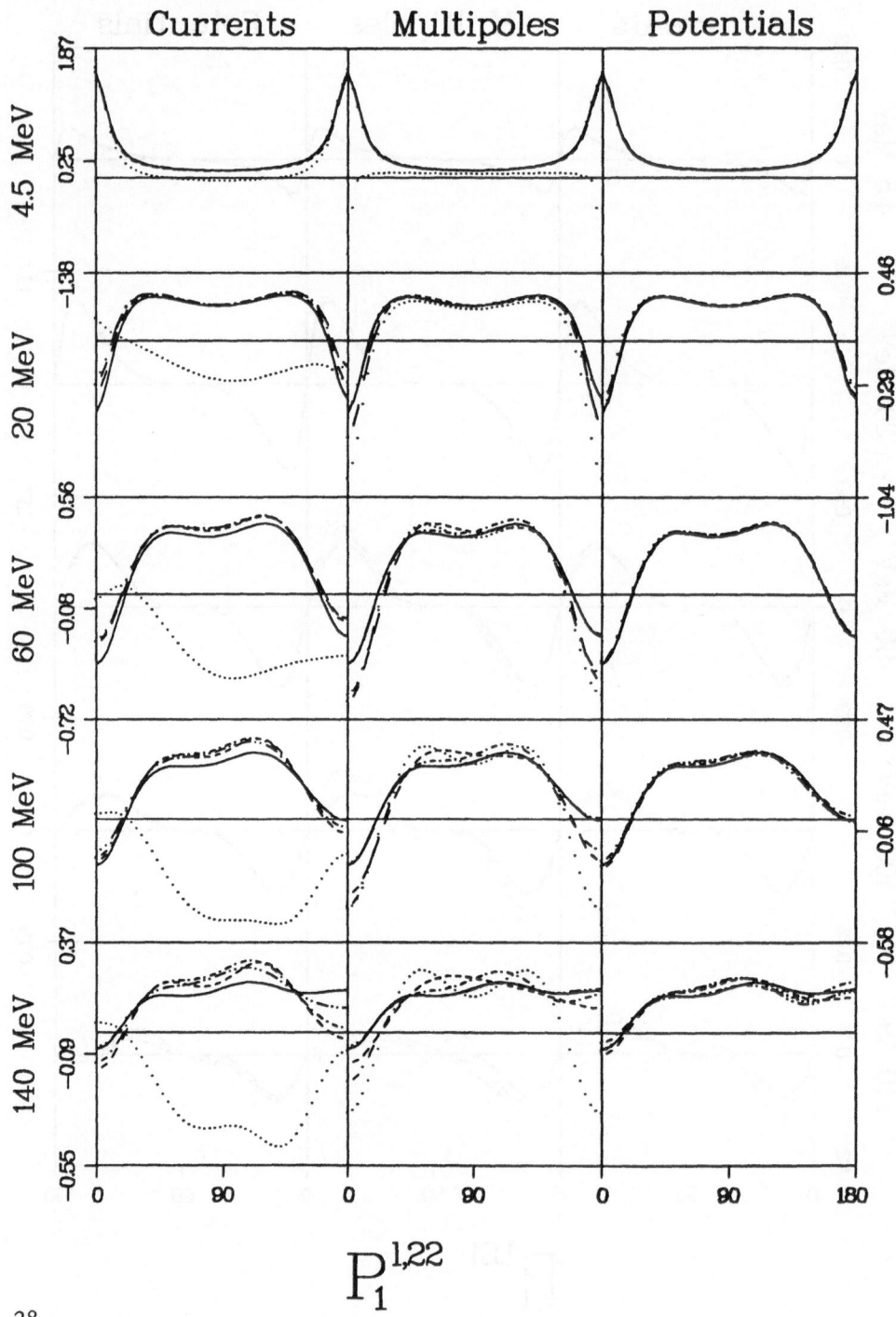

Currents   Multipoles   Potentials

4.5 MeV
20 MeV
60 MeV
100 MeV
140 MeV

$$P_1^{1,22}$$

5.5 Proton Polarization $P_x^{0,IM}(p)$

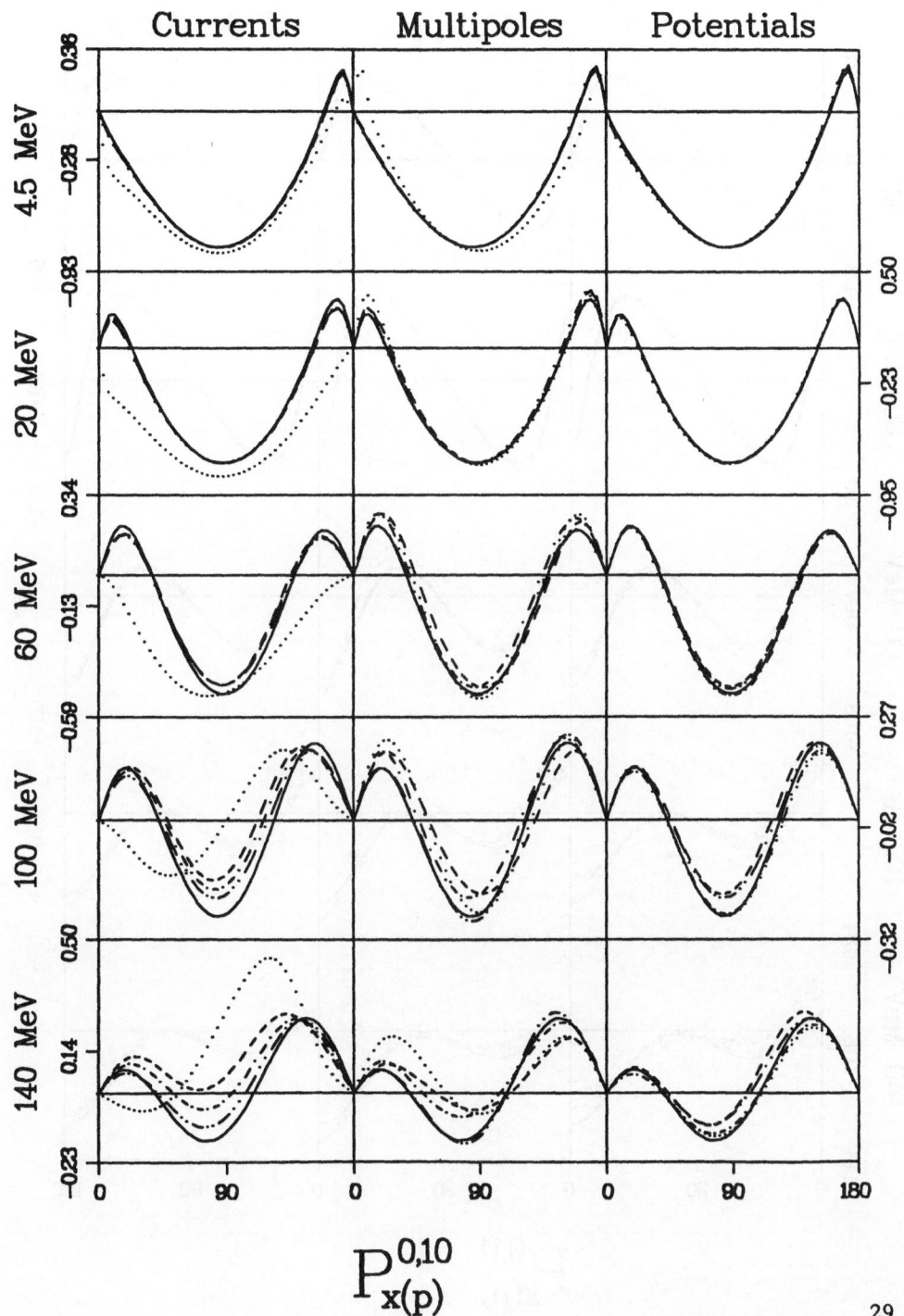

$$P_{x(p)}^{0,10}$$

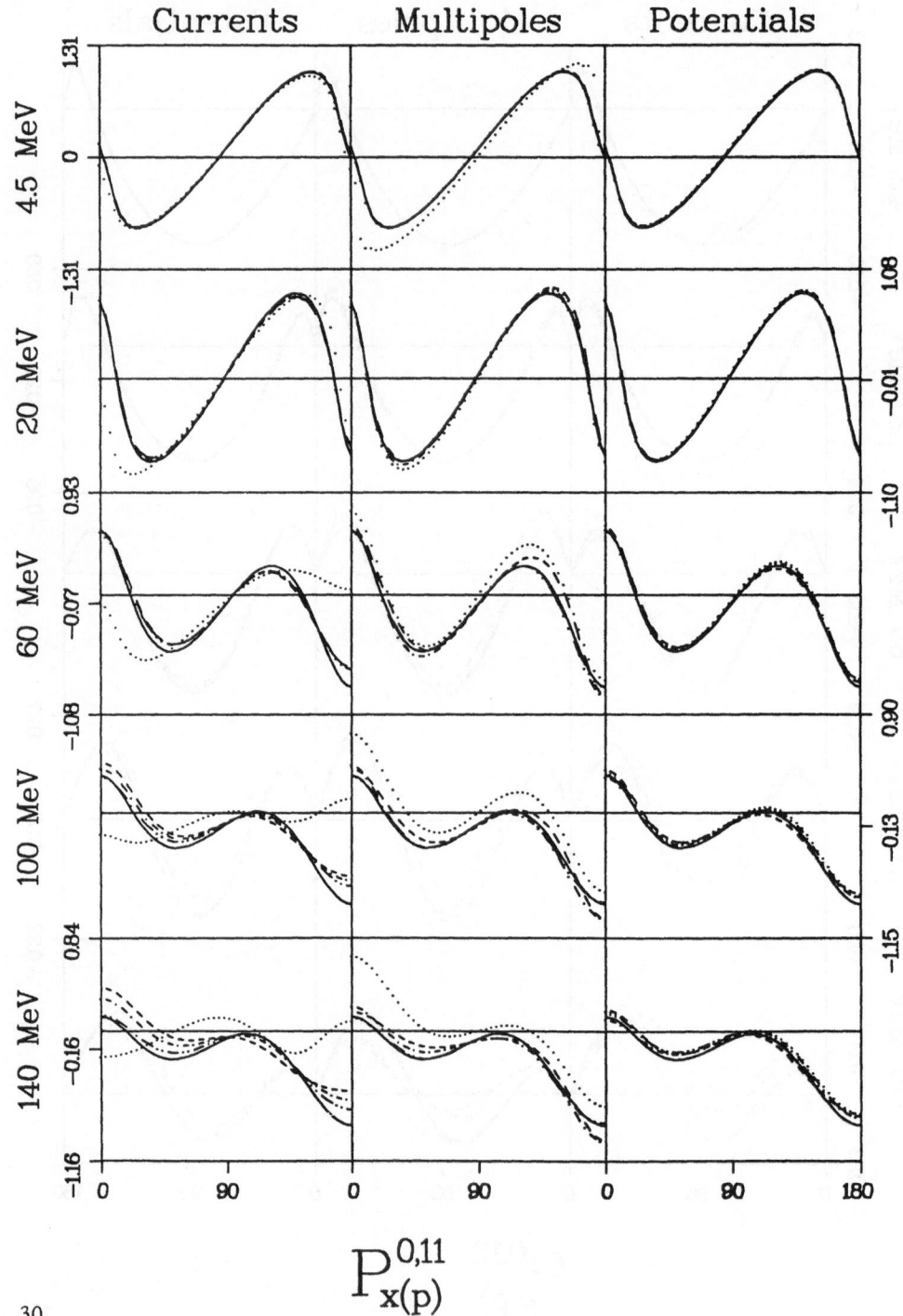

$$P_{x(p)}^{0,11}$$

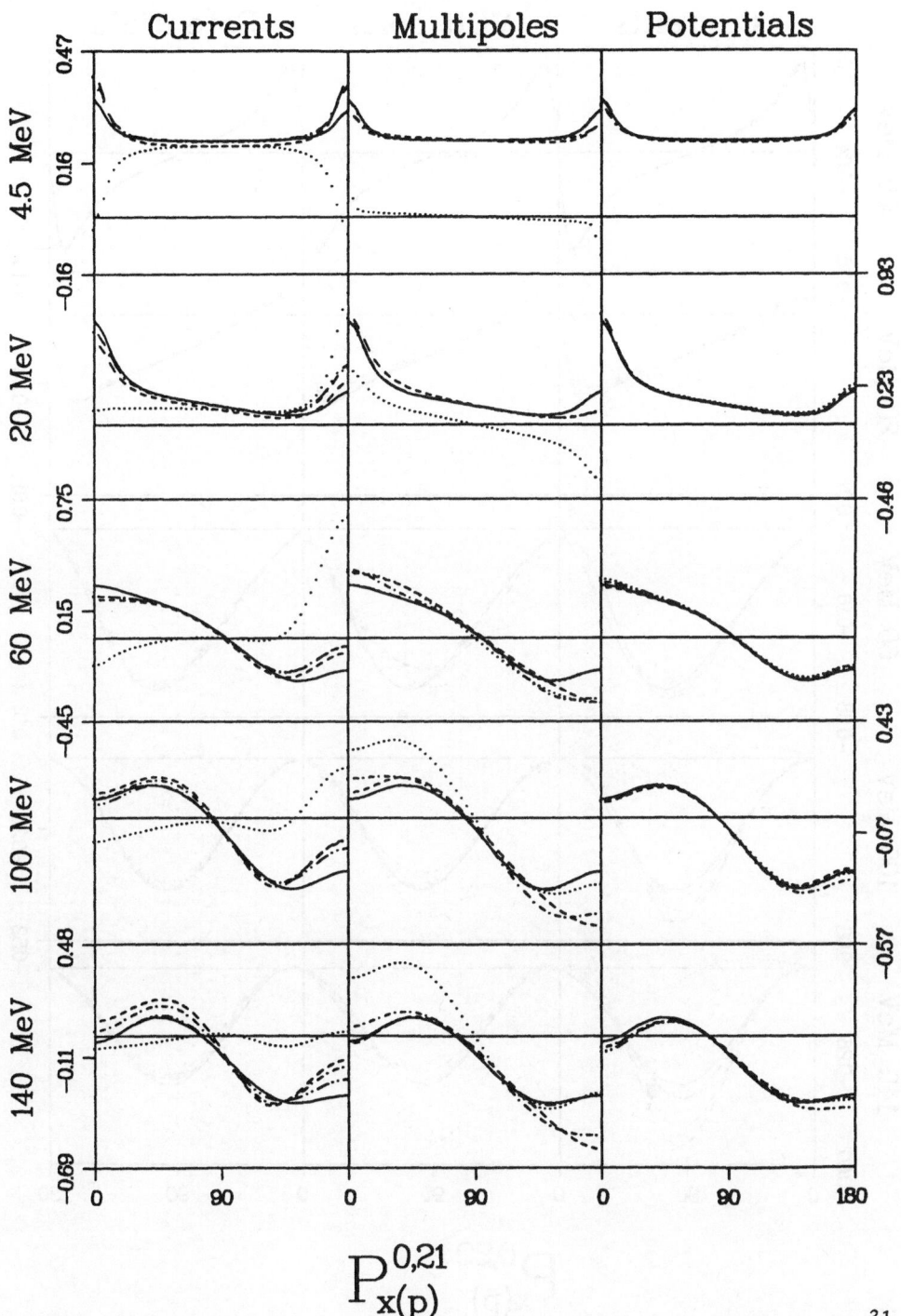

$$P_{x(p)}^{0,21}$$

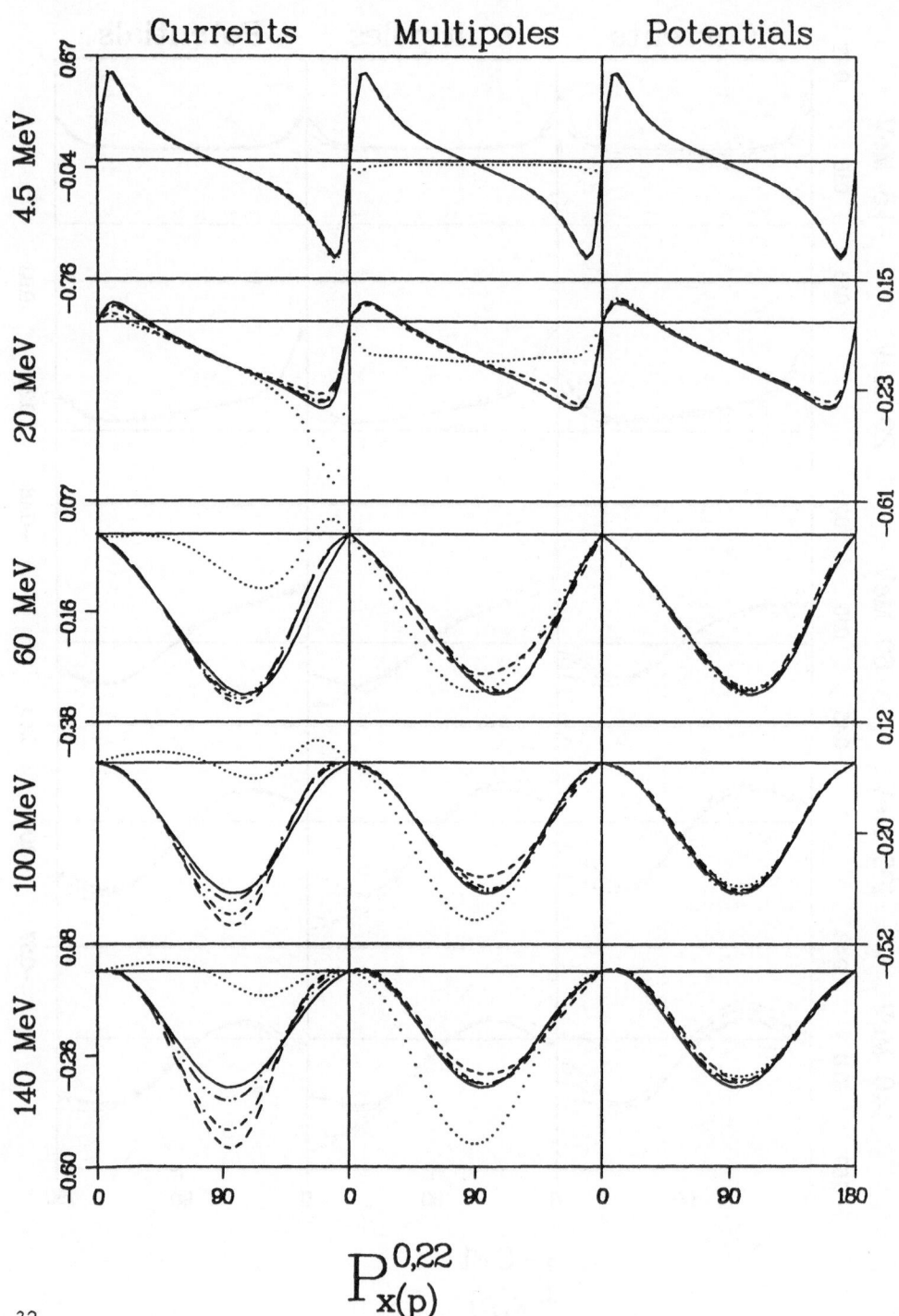

$$P_{x(p)}^{0,22}$$

5.6 Proton Polarization $P_x^{c,IM}(p)$

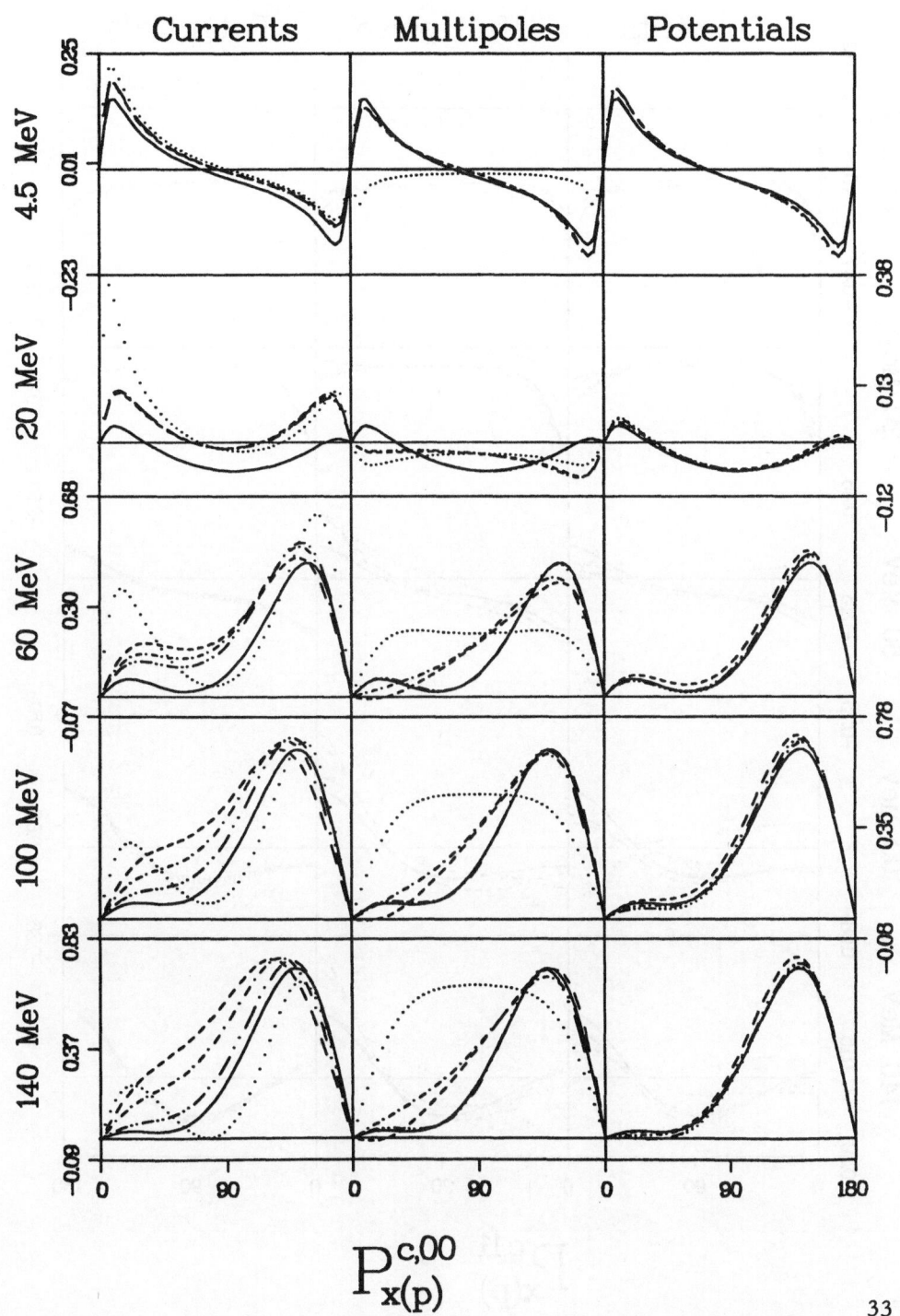

$$P_{x(p)}^{c,00}$$

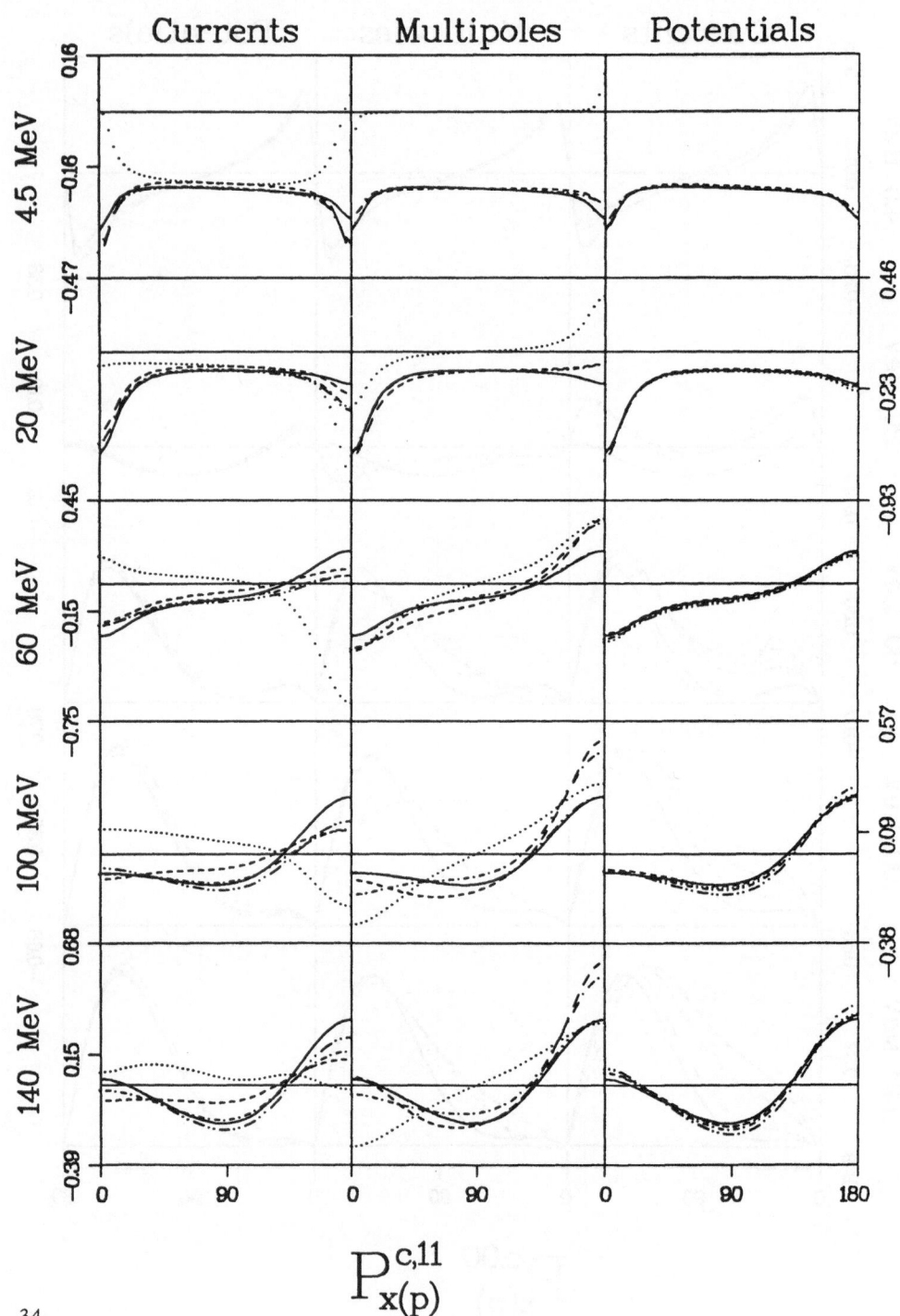

$$P_{x(p)}^{c,11}$$

Proton Polarization $P_x^{c,IM}(p)$

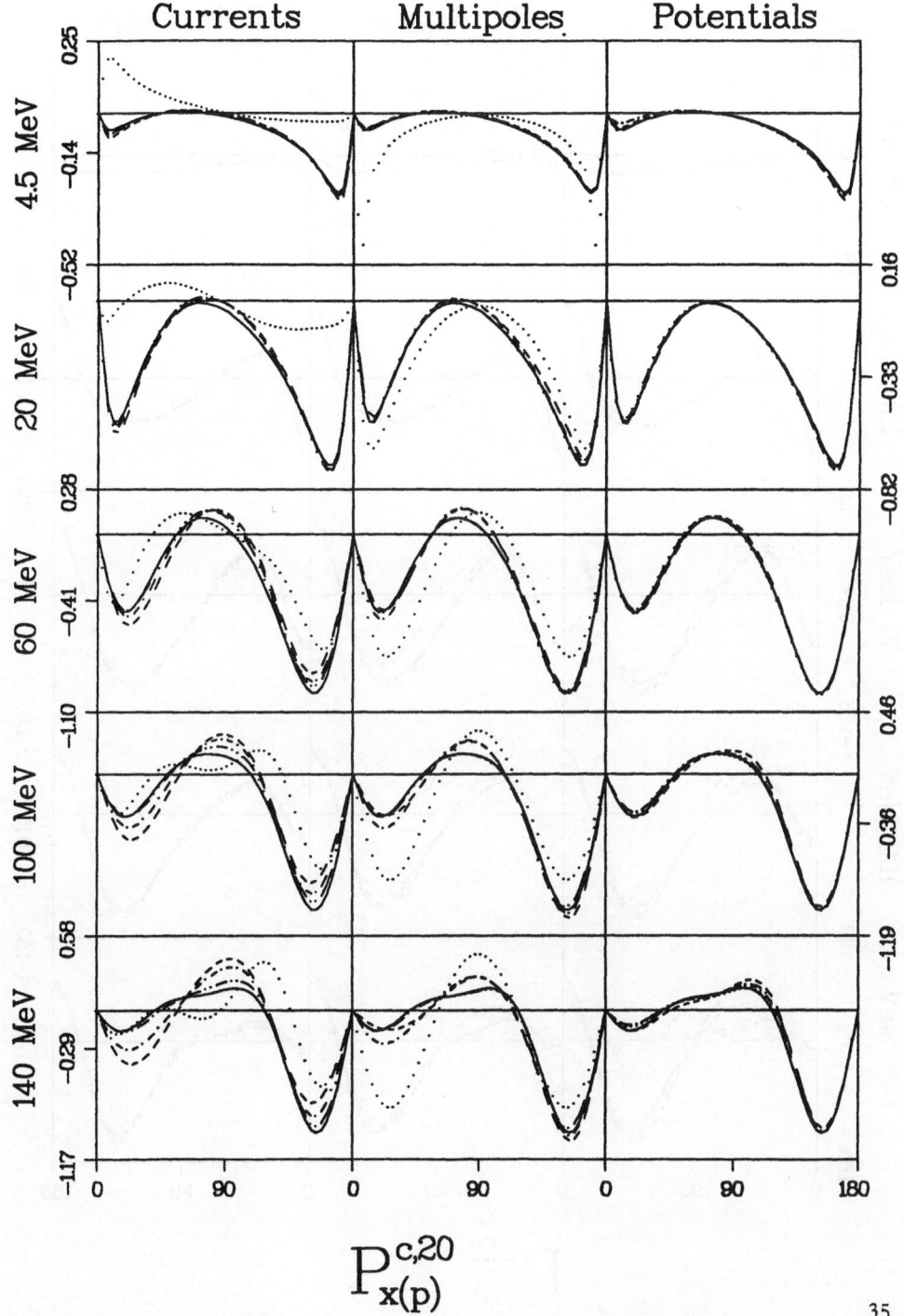

Currents   Multipoles   Potentials

4.5 MeV

20 MeV

60 MeV

100 MeV

140 MeV

$$P_{x(p)}^{c,20}$$

35

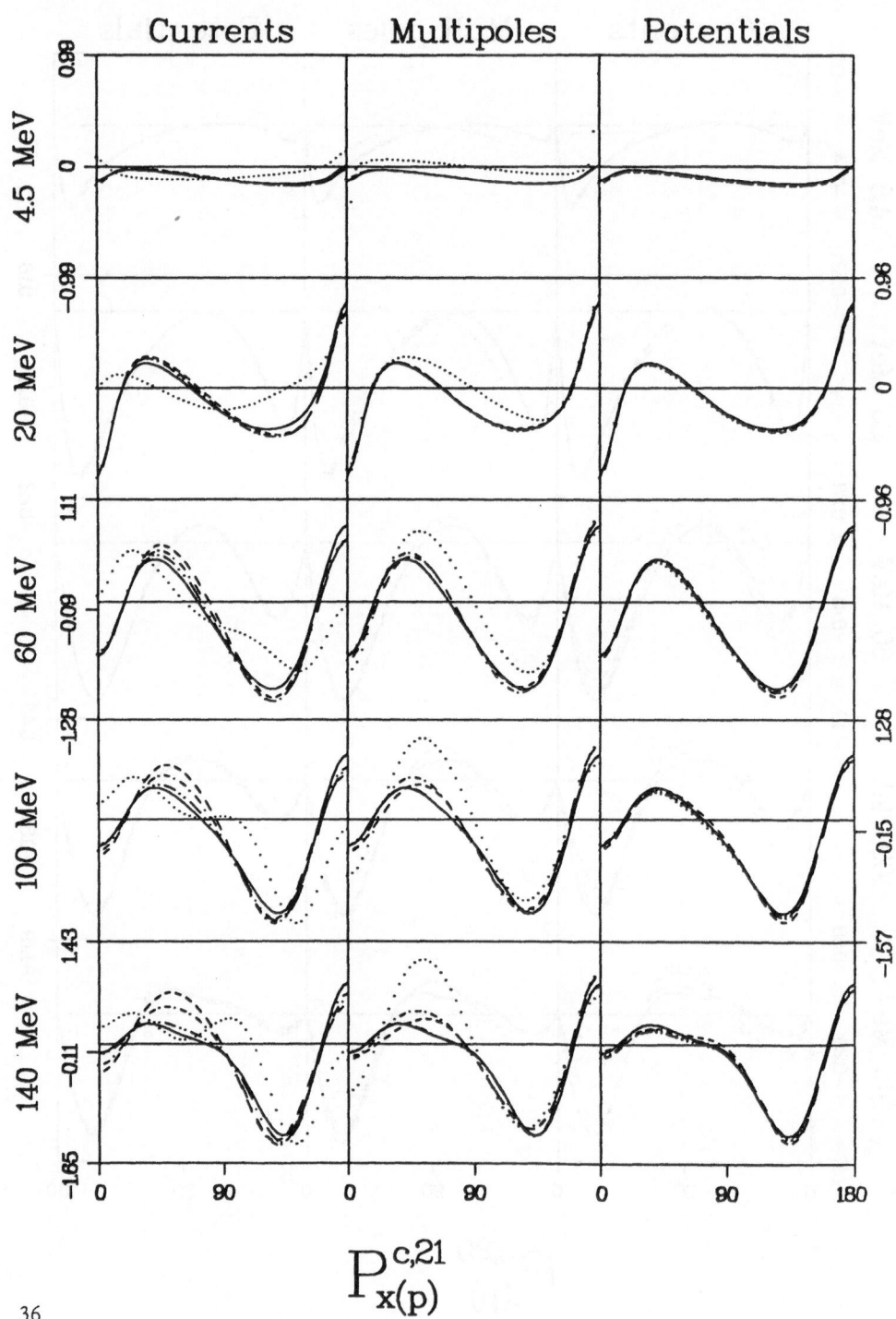

$$P_{x(p)}^{c,21}$$

Proton Polarization $P_x^{c,IM}(p)$

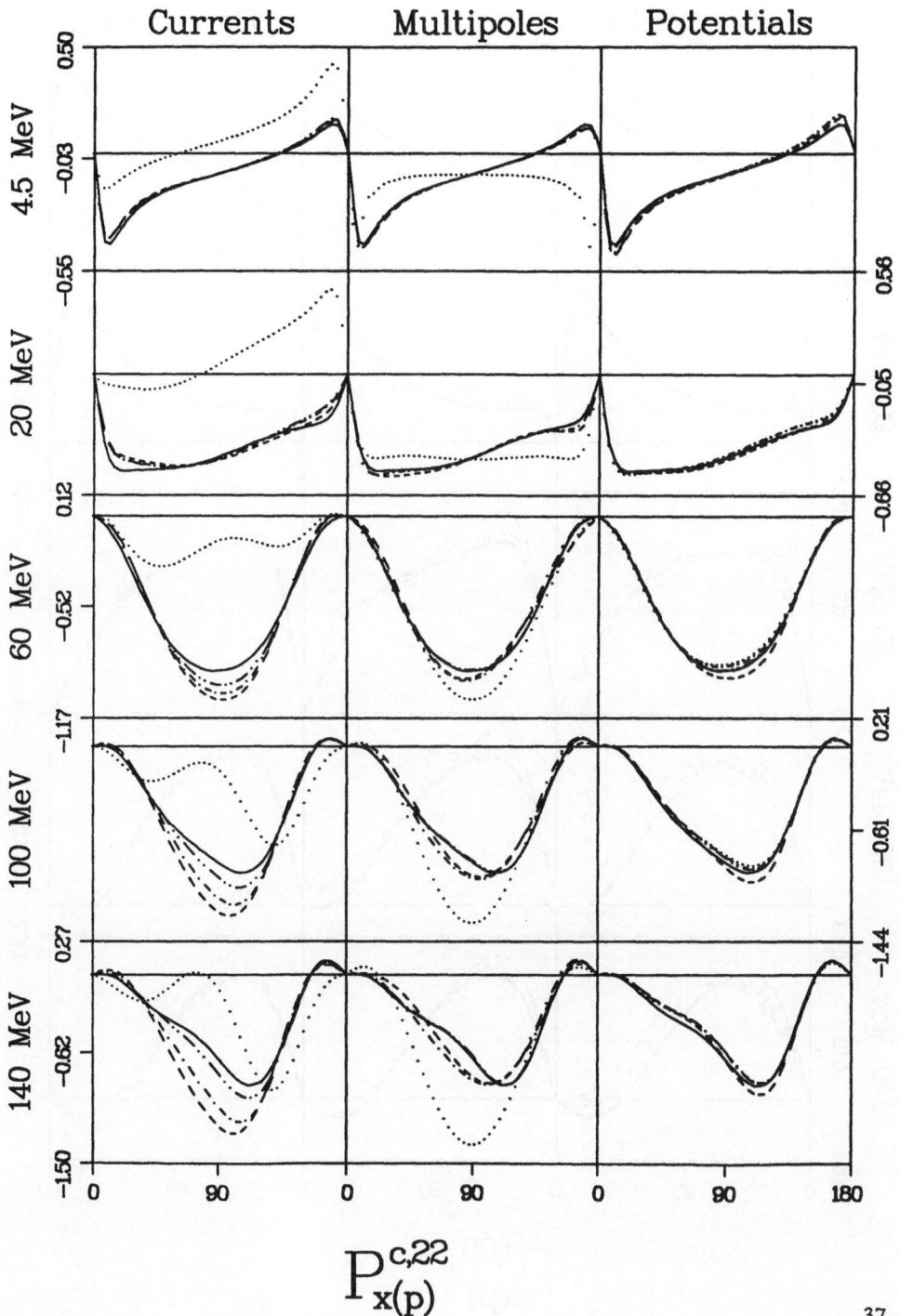

$$P_{x(p)}^{c,22}$$

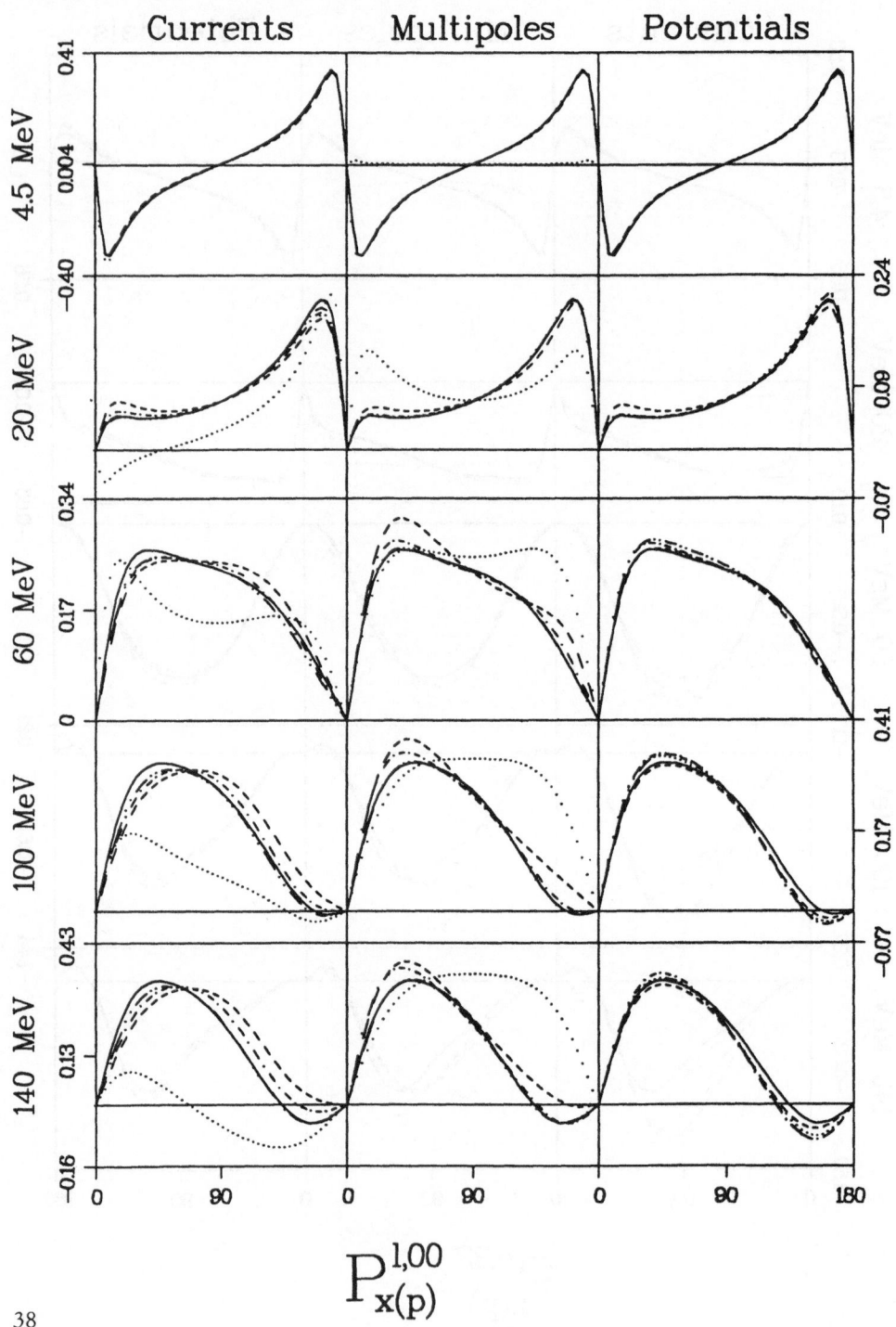

$$P_{x(p)}^{l,00}$$

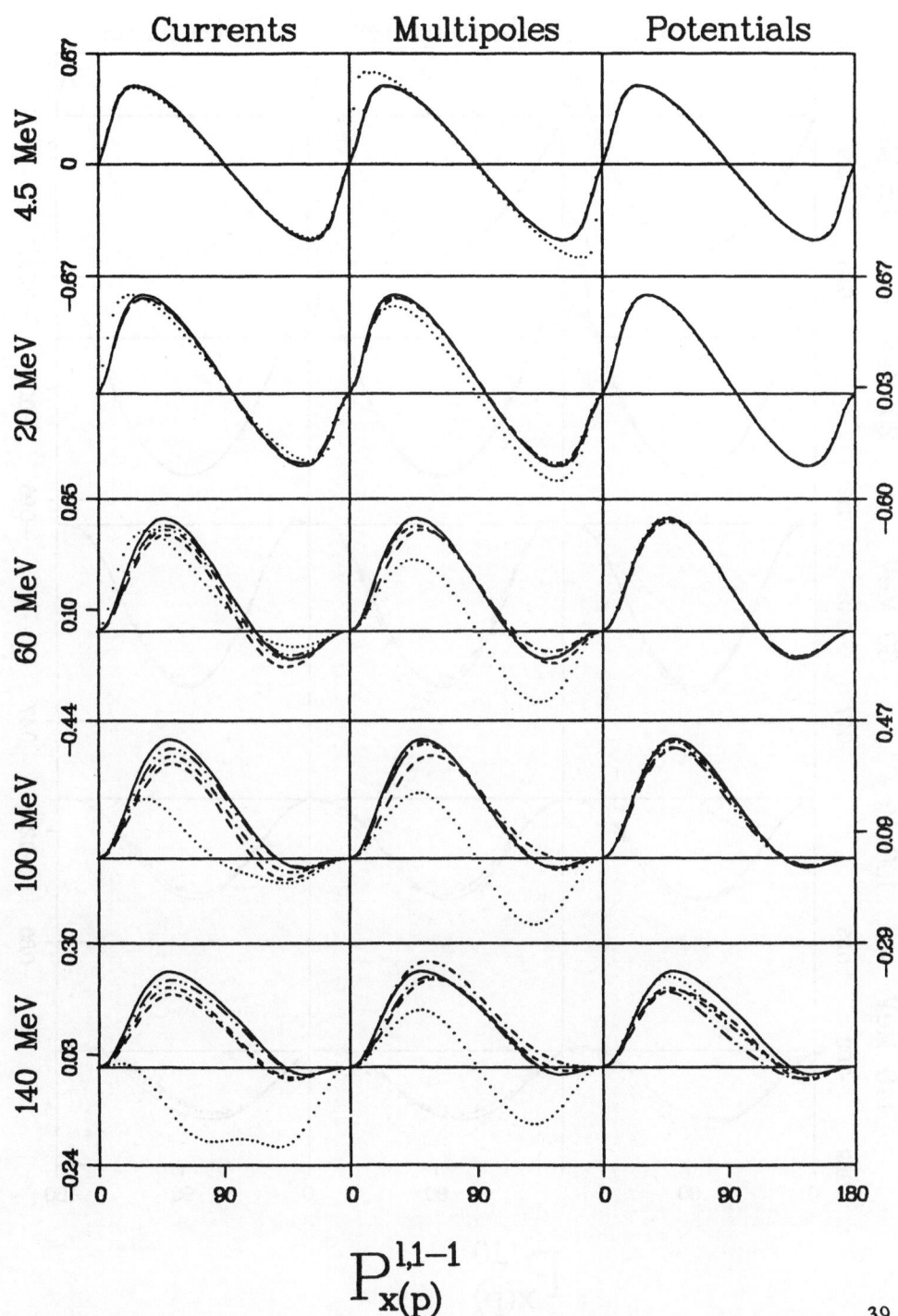

$$P_{x(p)}^{1,1-1}$$

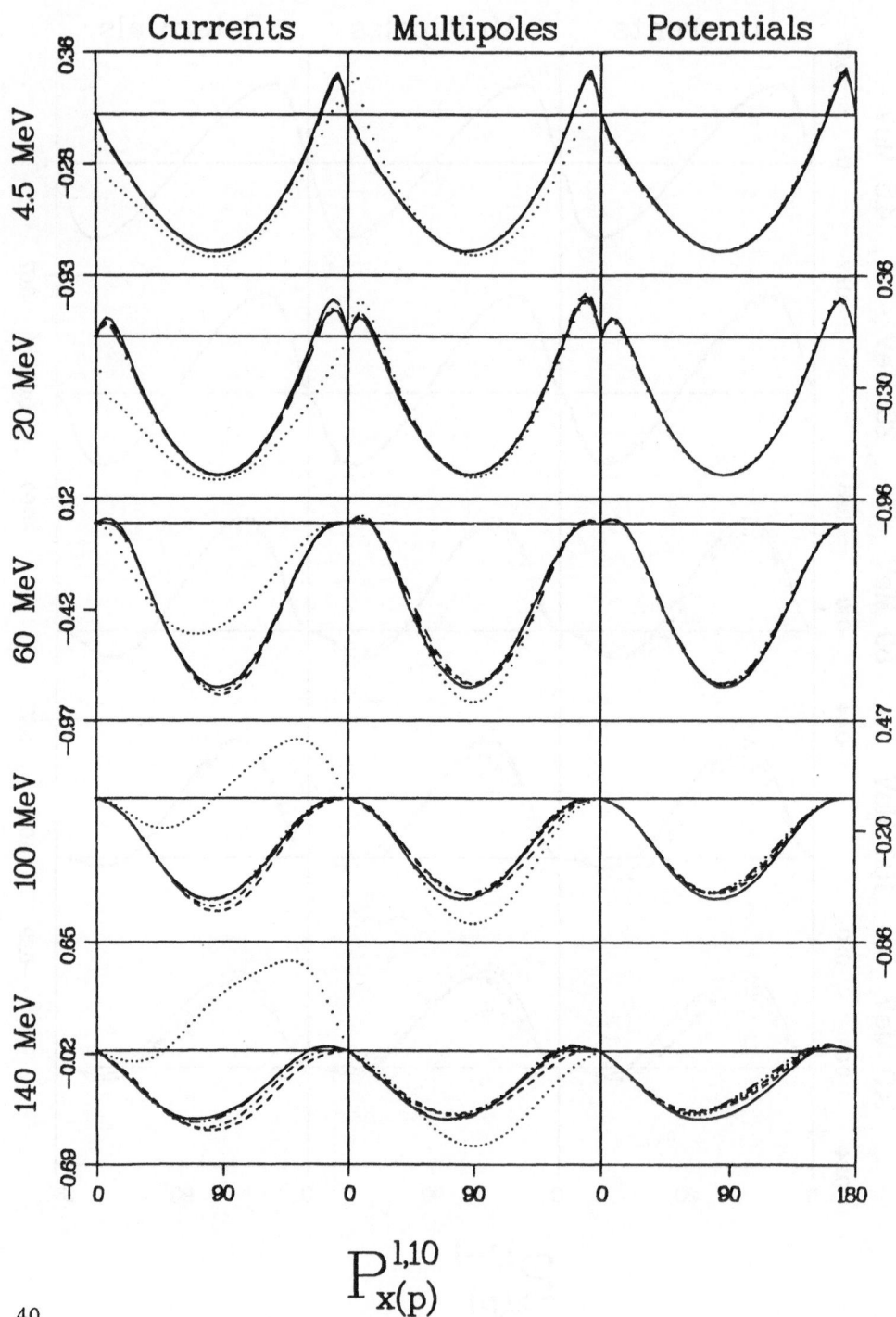

$$\mathrm{P}_{\mathbf{x(p)}}^{l,10}$$

Proton Polarization $P_x^{l,IM}(p)$

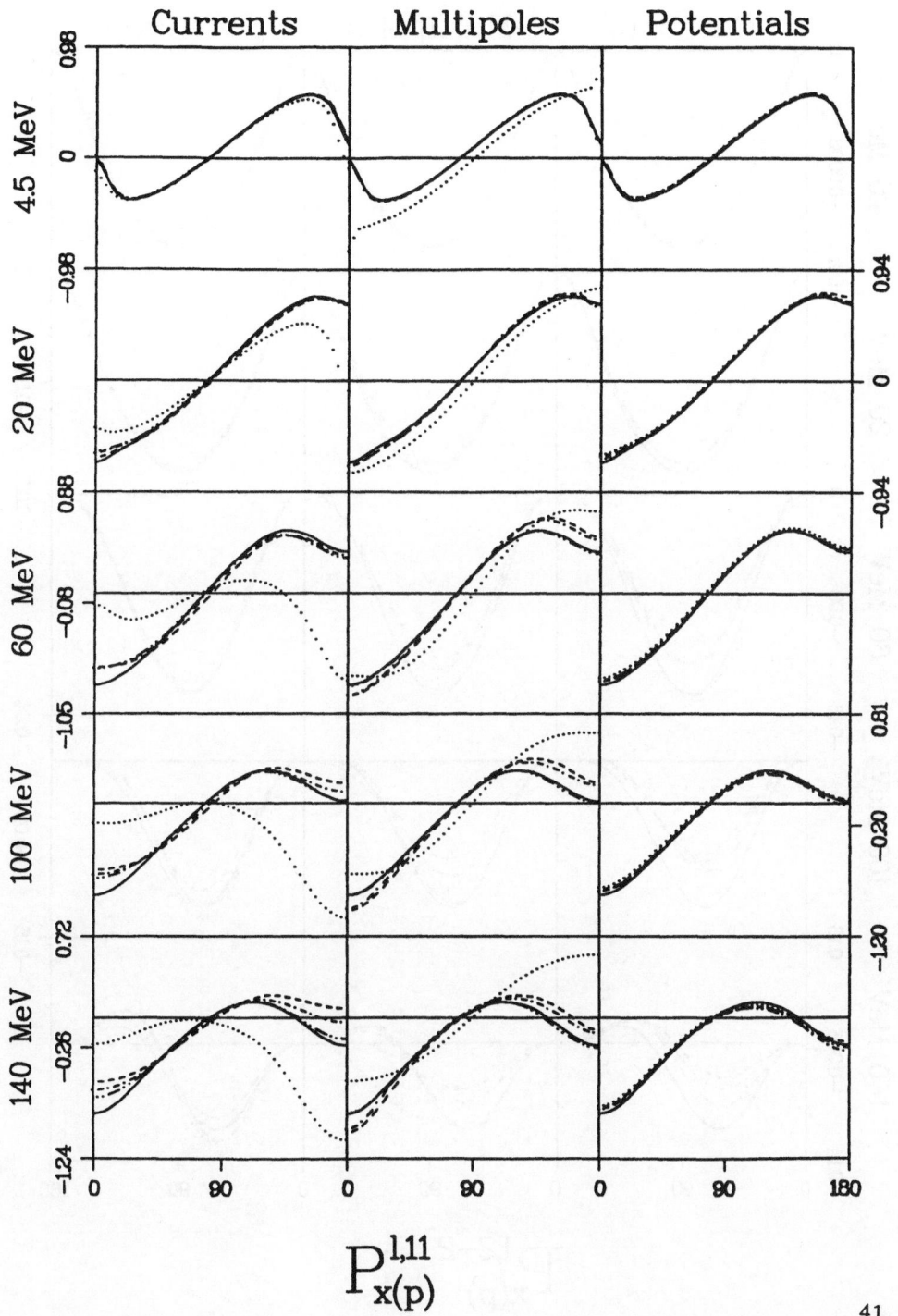

$$P_{x(p)}^{l,11}$$

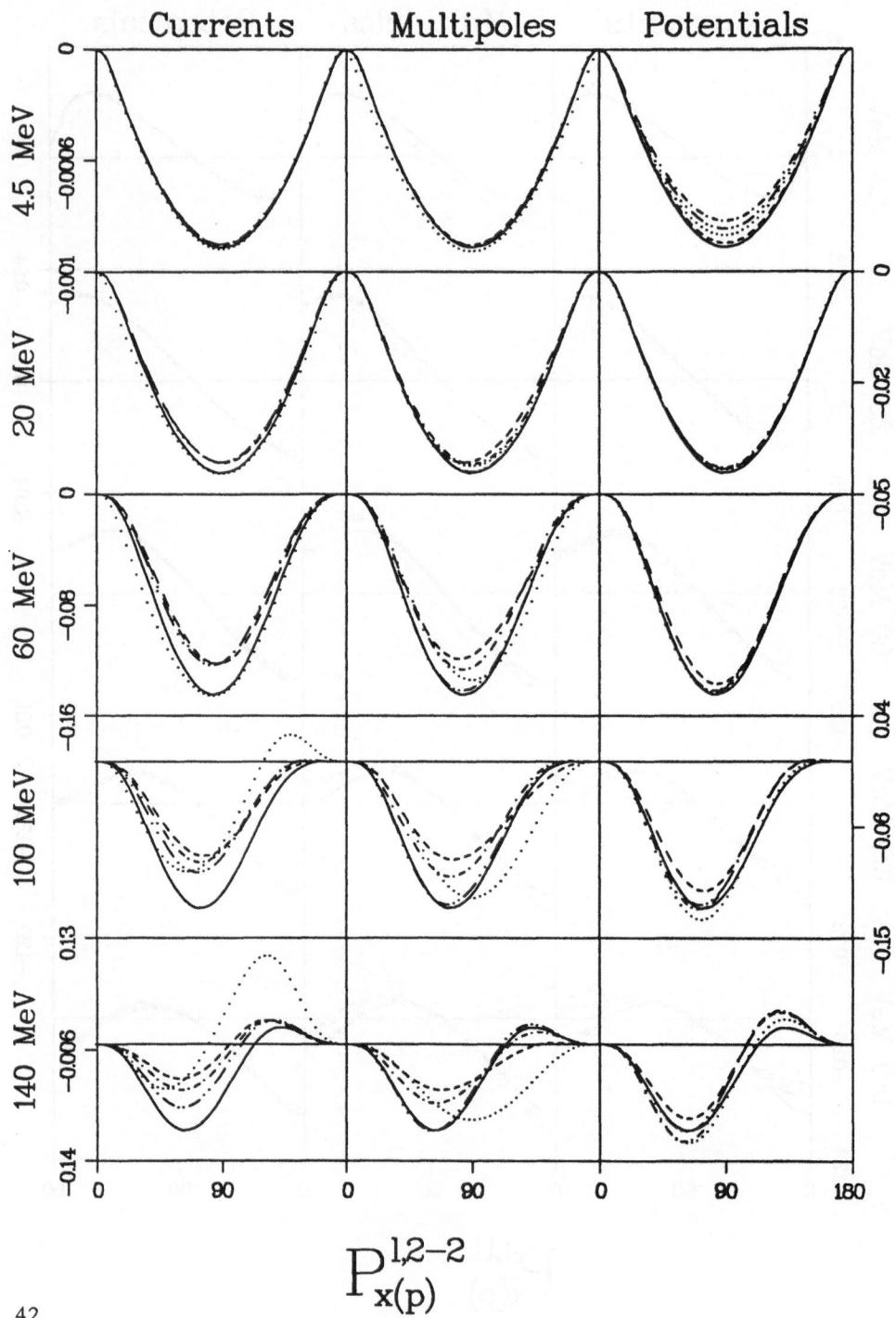

$$P_{x(p)}^{1,2-2}$$

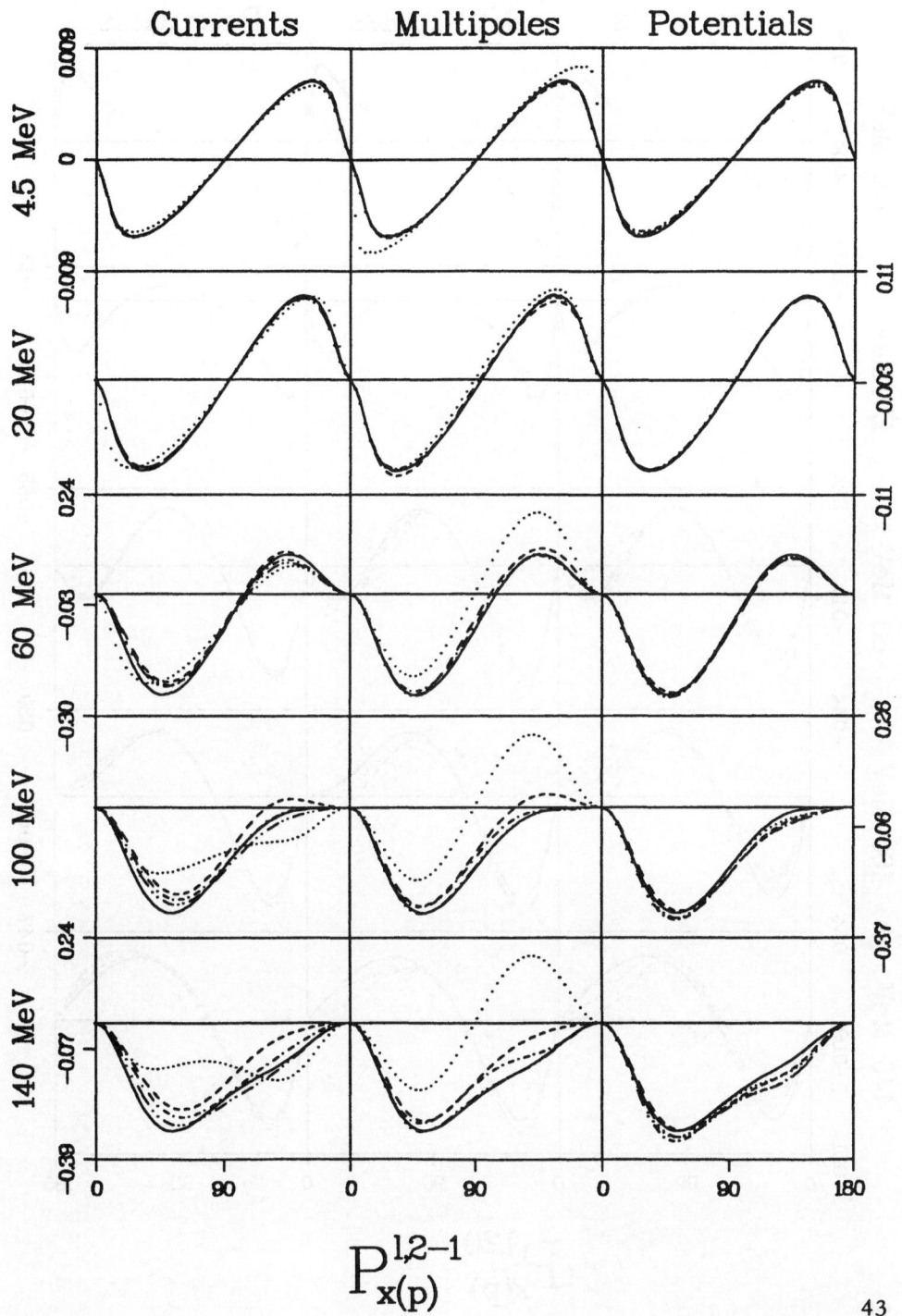

Currents   Multipoles   Potentials

4.5 MeV   20 MeV   60 MeV   100 MeV   140 MeV

$$P_{x(p)}^{1,2-1}$$

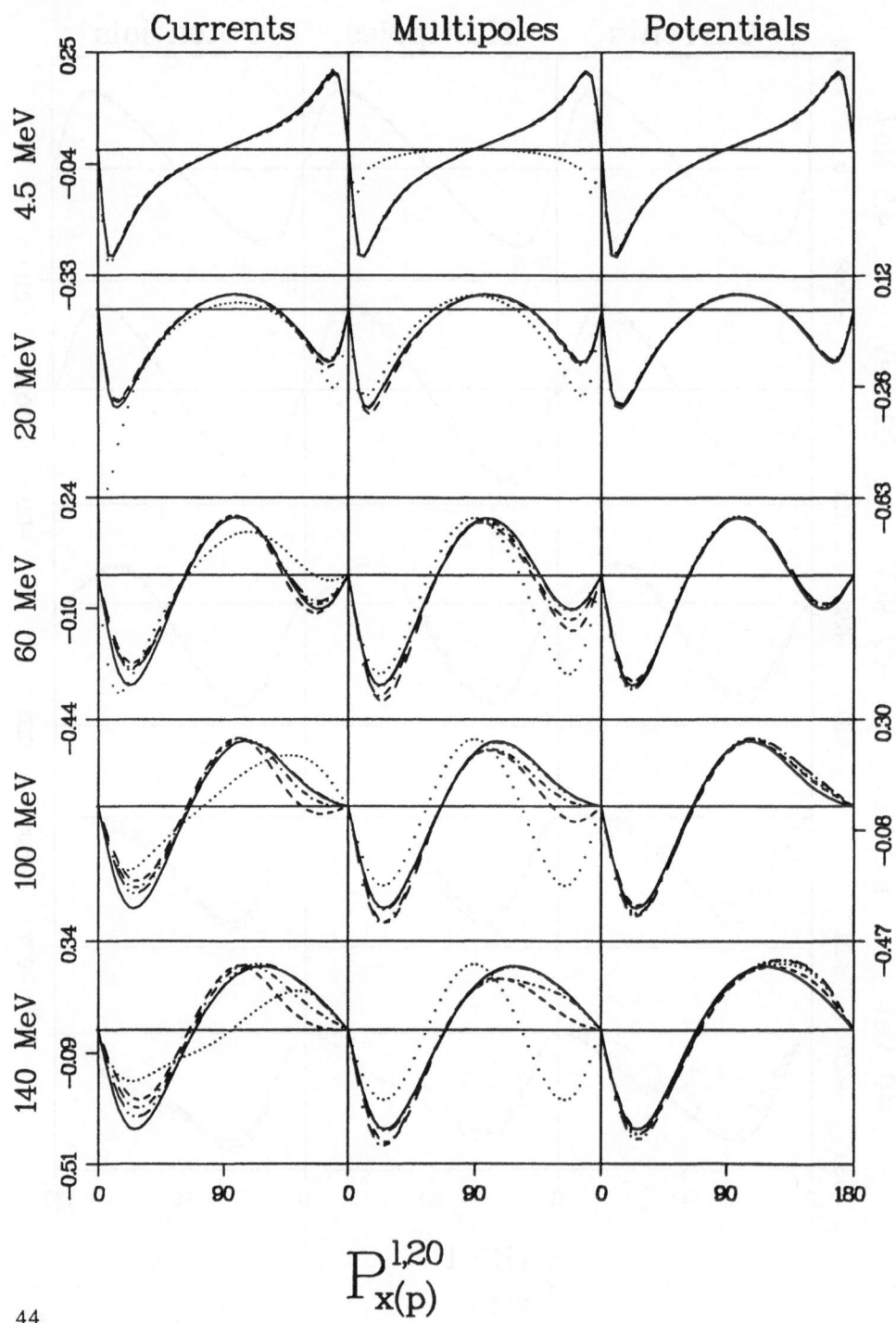

Currents Multipoles Potentials

$$P_{x(p)}^{1,20}$$

Proton Polarization $P_x^{l,IM}(p)$

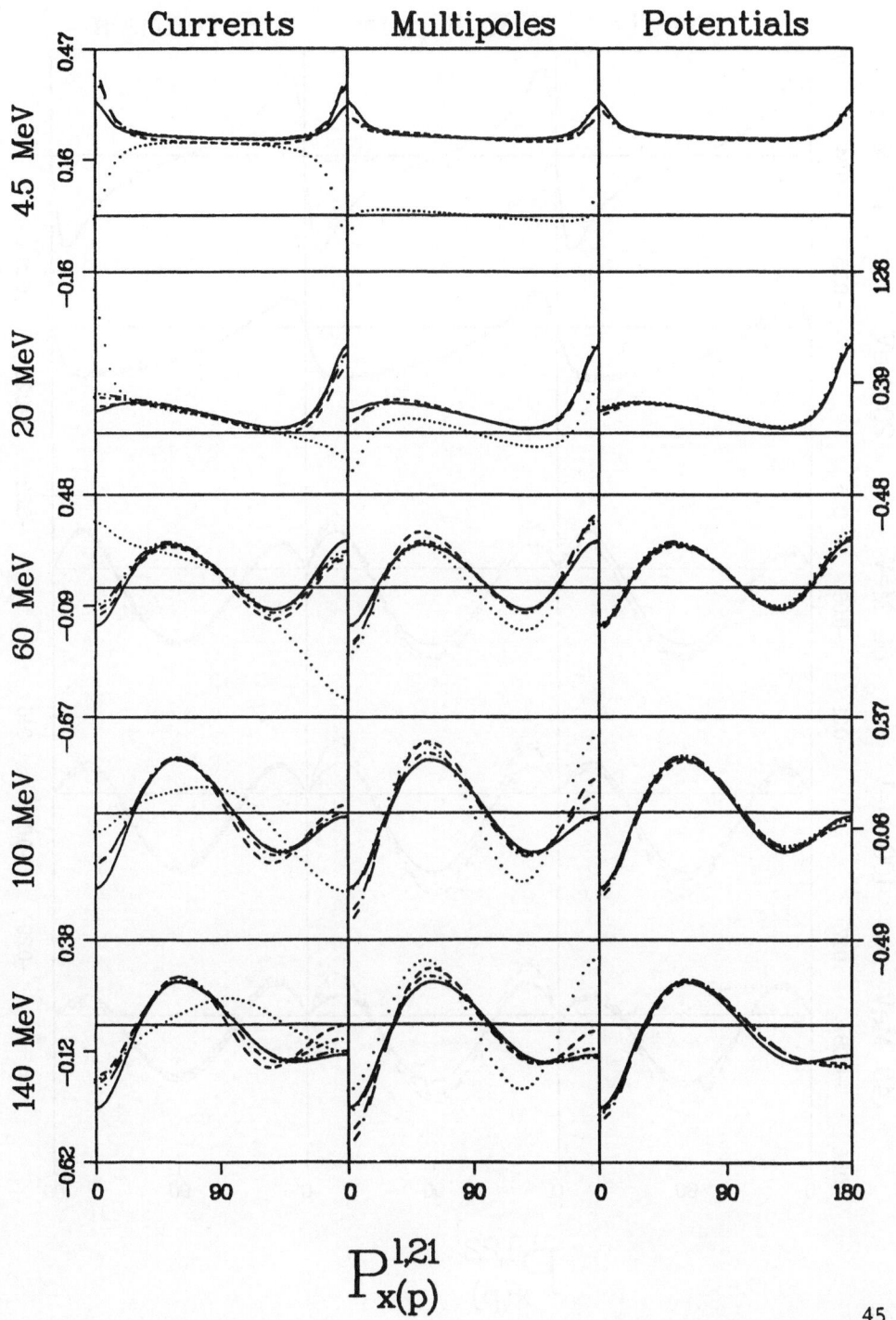

$$P_{x(p)}^{1,21}$$

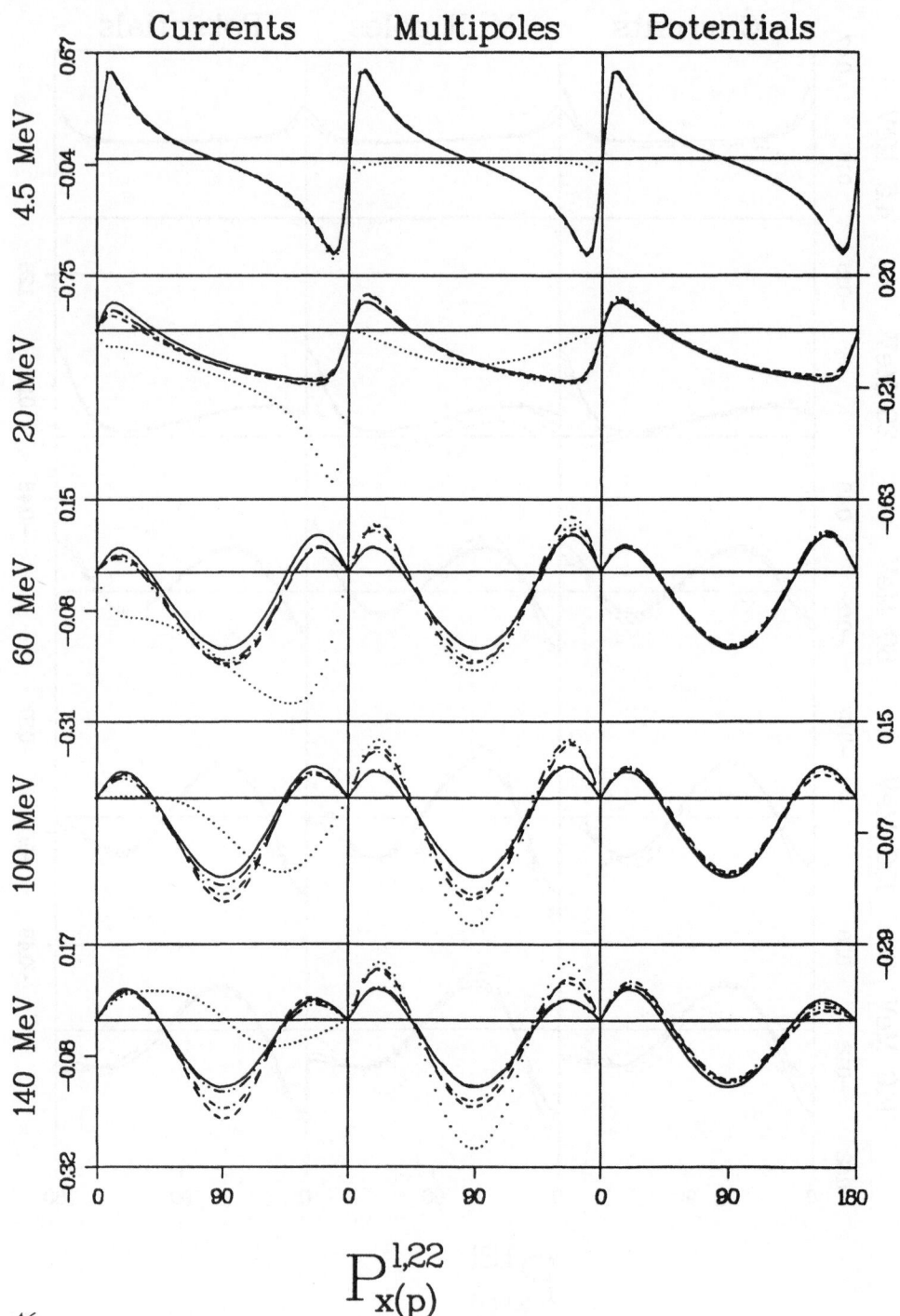

$$P_{x(p)}^{1,22}$$

5.8 Neutron Polarization $P_z^{0,IM}(n)$

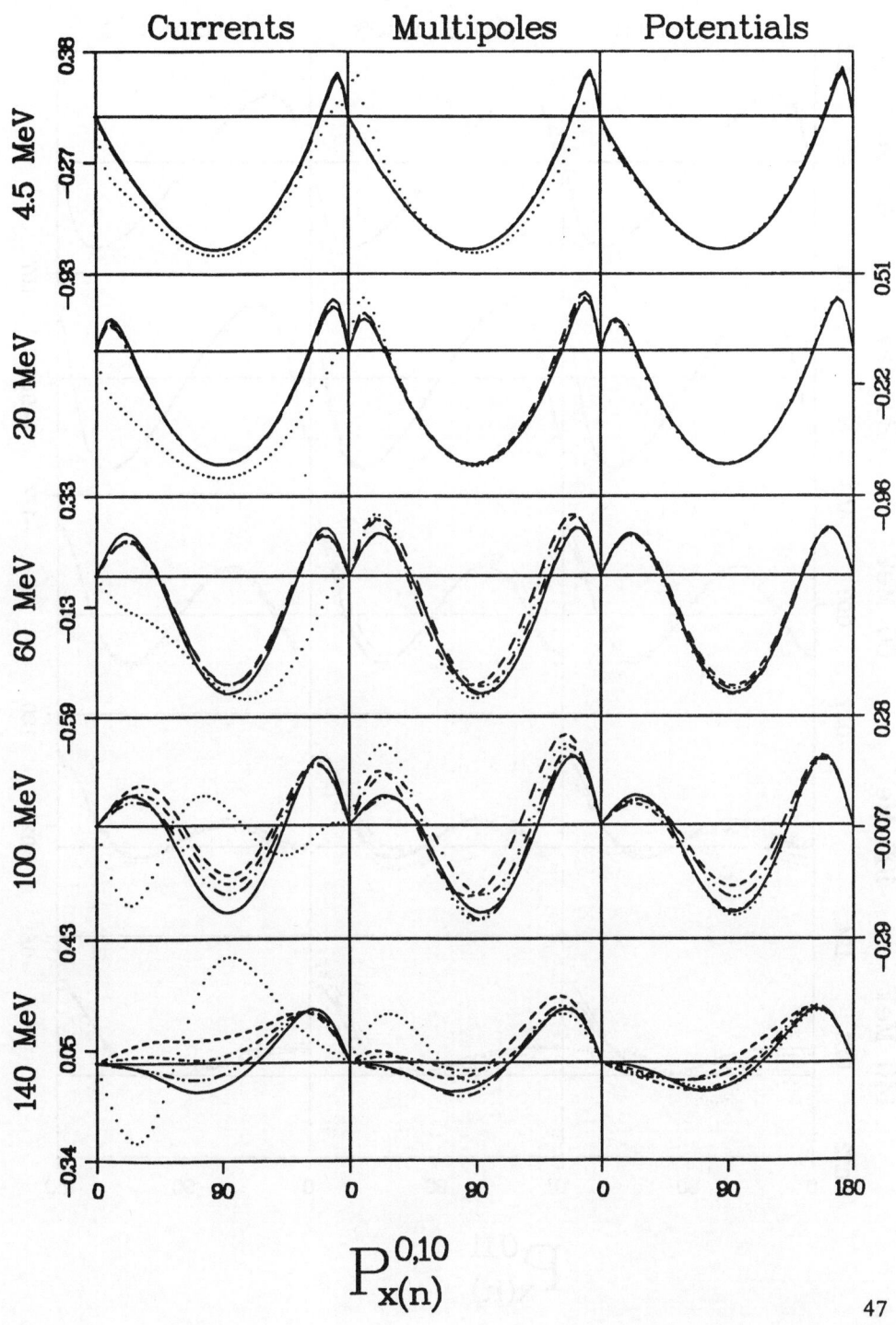

$$P_{x(n)}^{0,10}$$

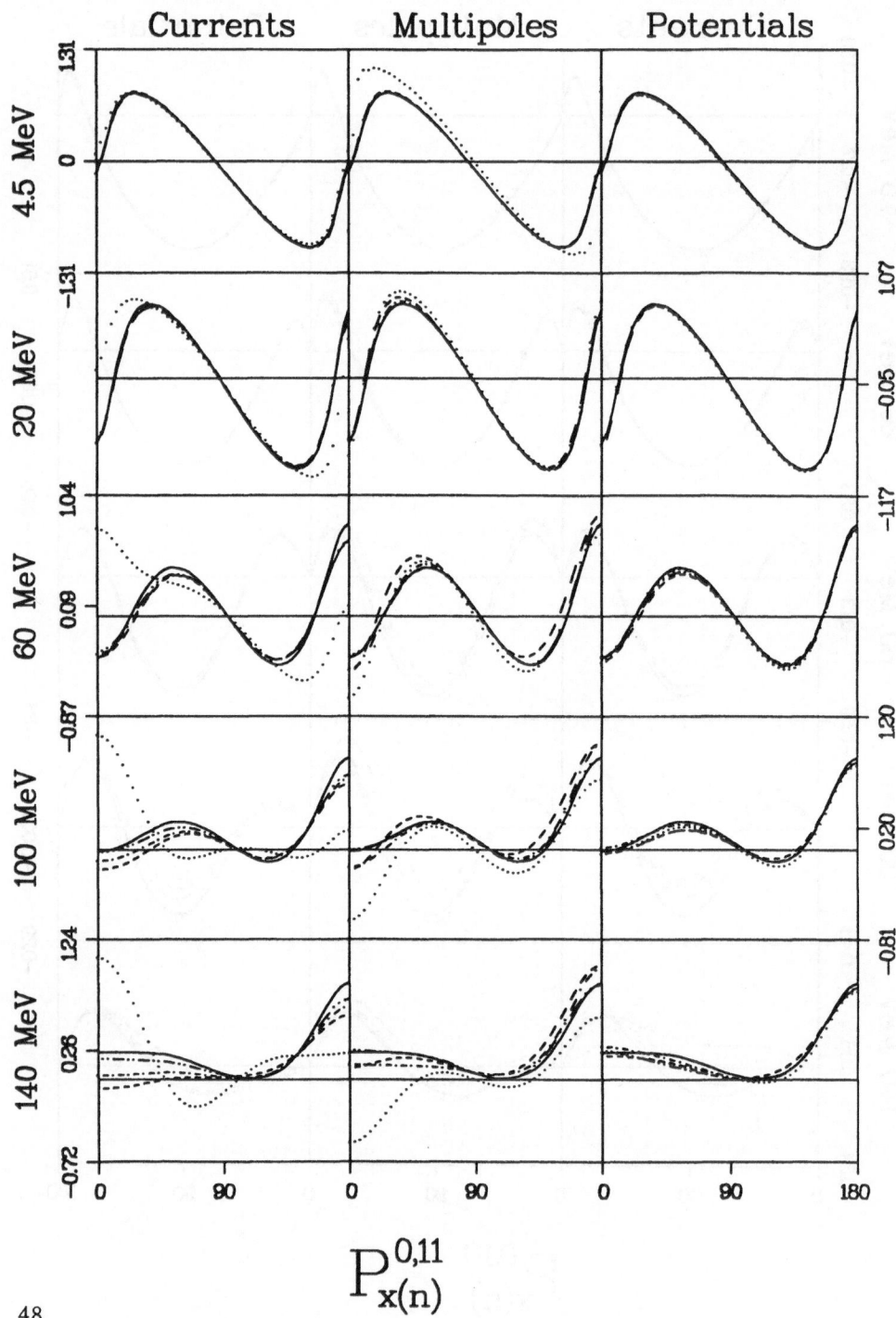

$$P_{x(n)}^{0,11}$$

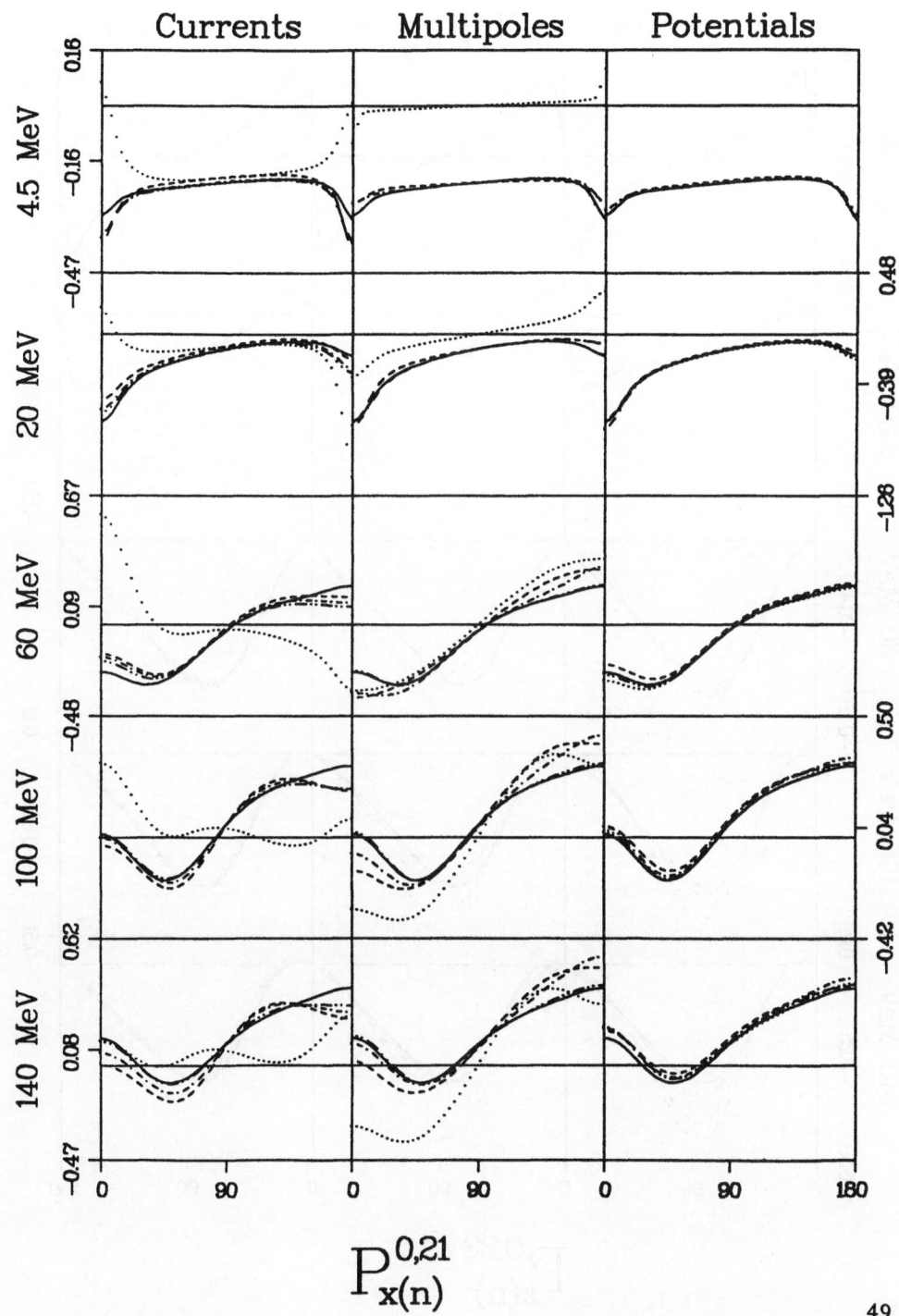

$$P_{x(n)}^{0,21}$$

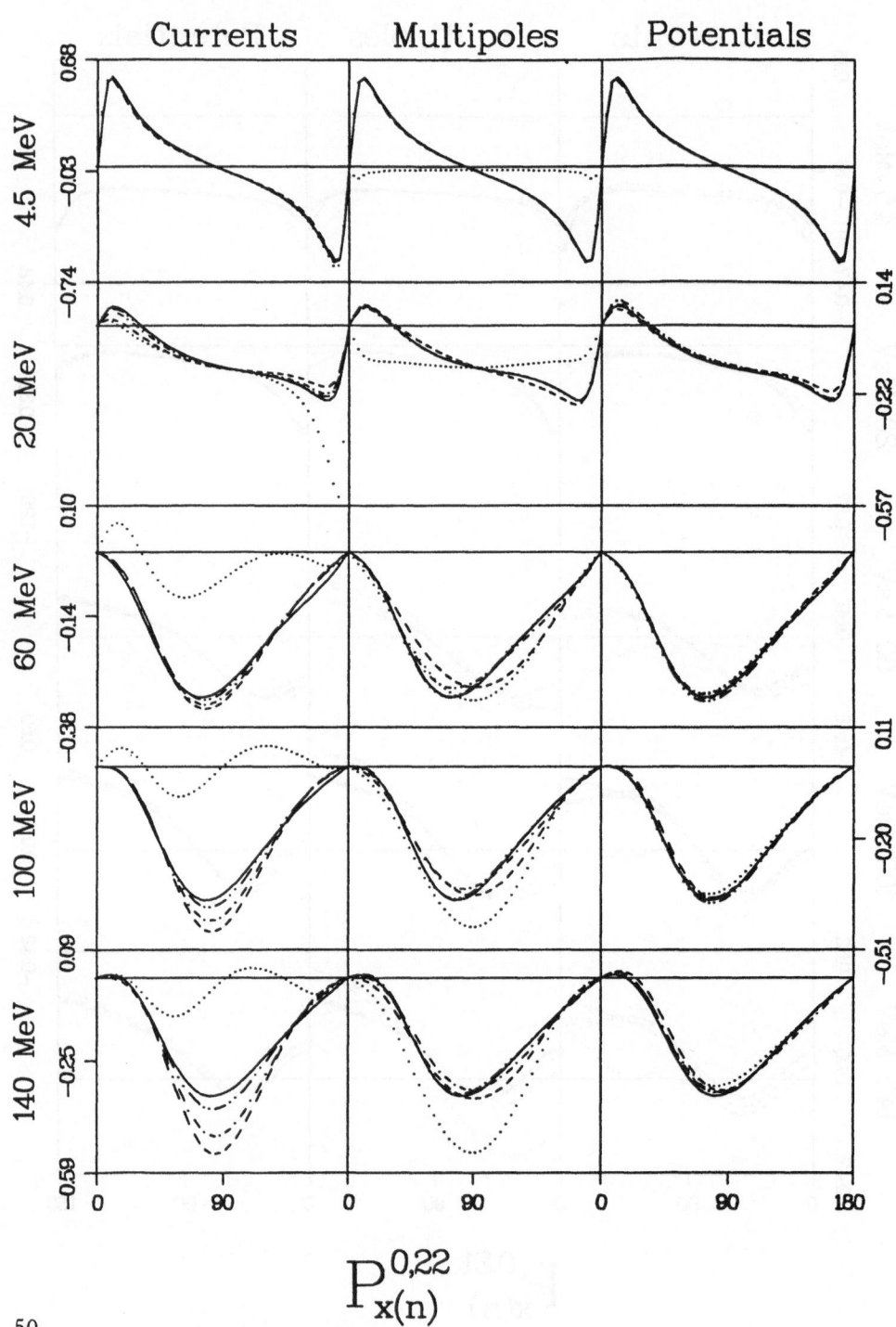

$$P_{x(n)}^{0,22}$$

5.9 Neutron Polarization $P_z^{c,IM}(n)$

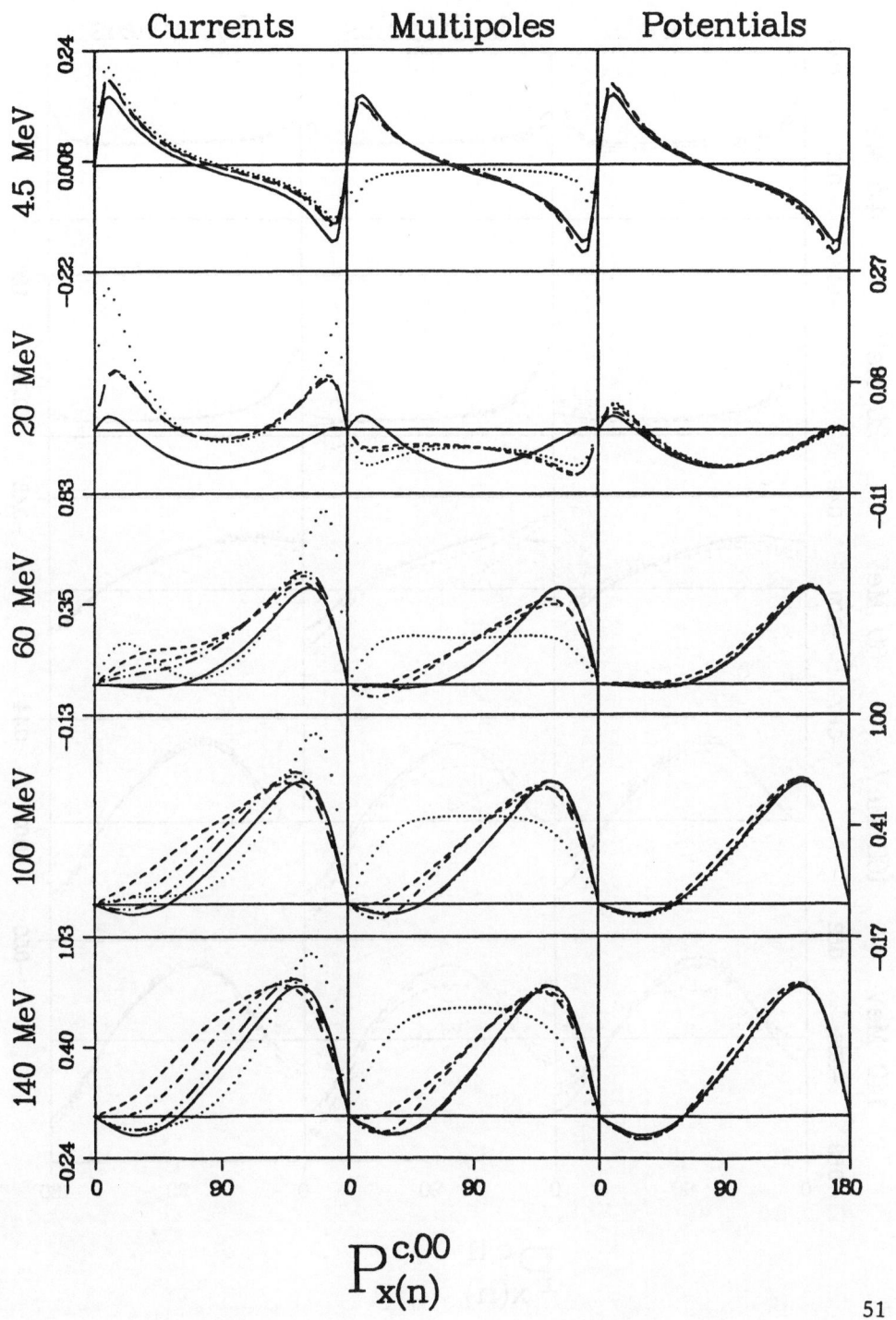

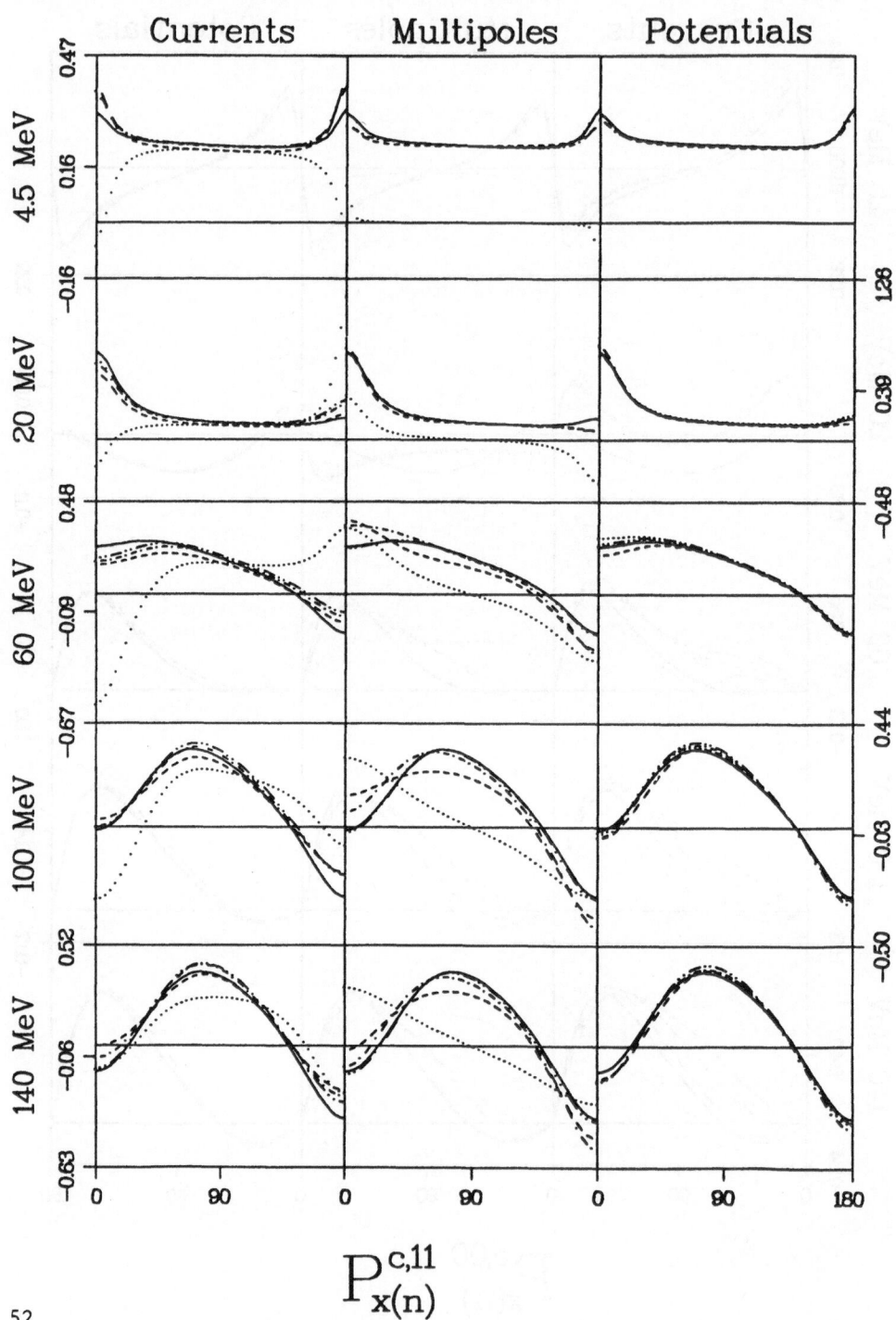

$$\mathrm{P}_{x(n)}^{c,11}$$

Neutron Polarization $P_x^{c,IM}(n)$

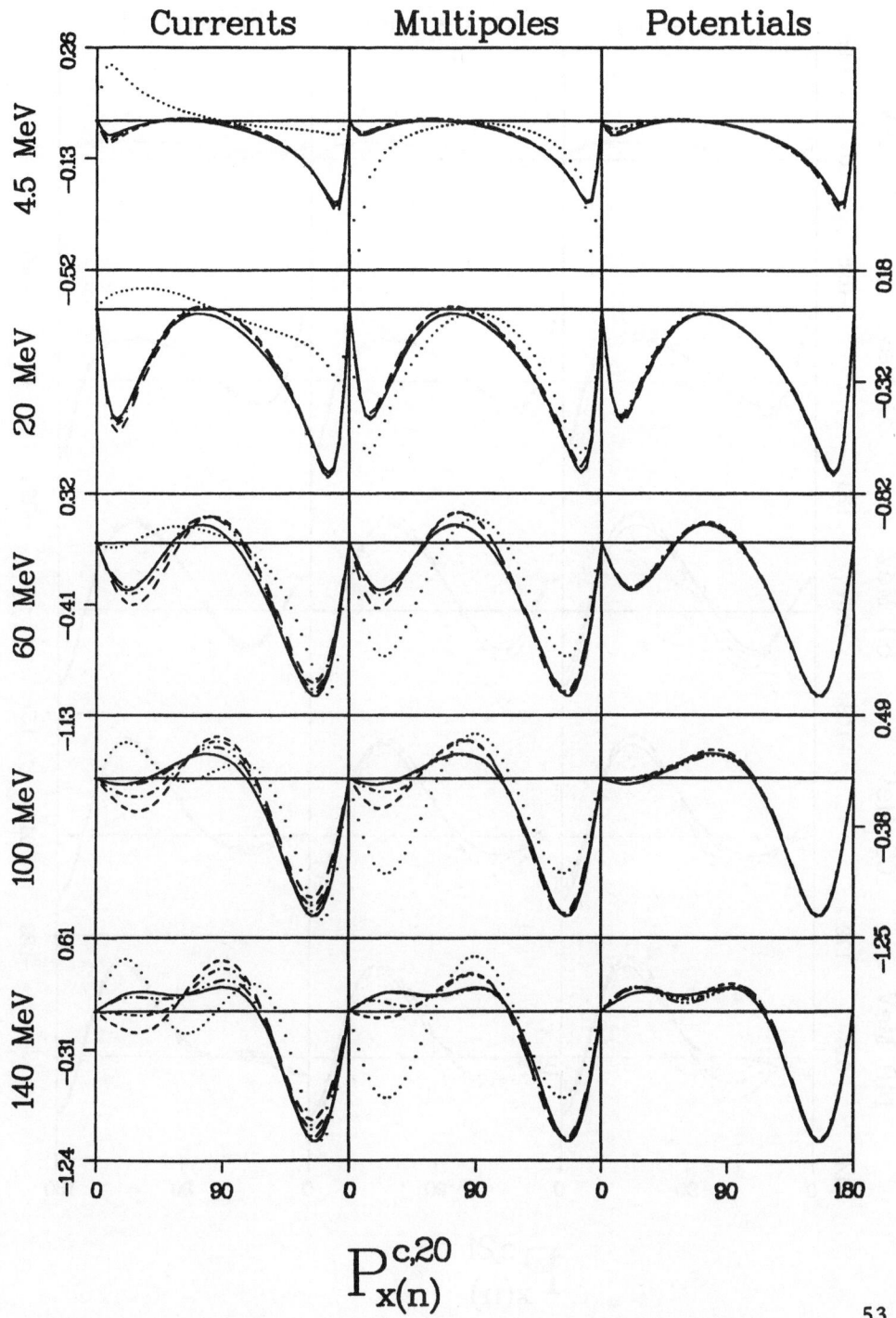

$$P_{x(n)}^{c,20}$$

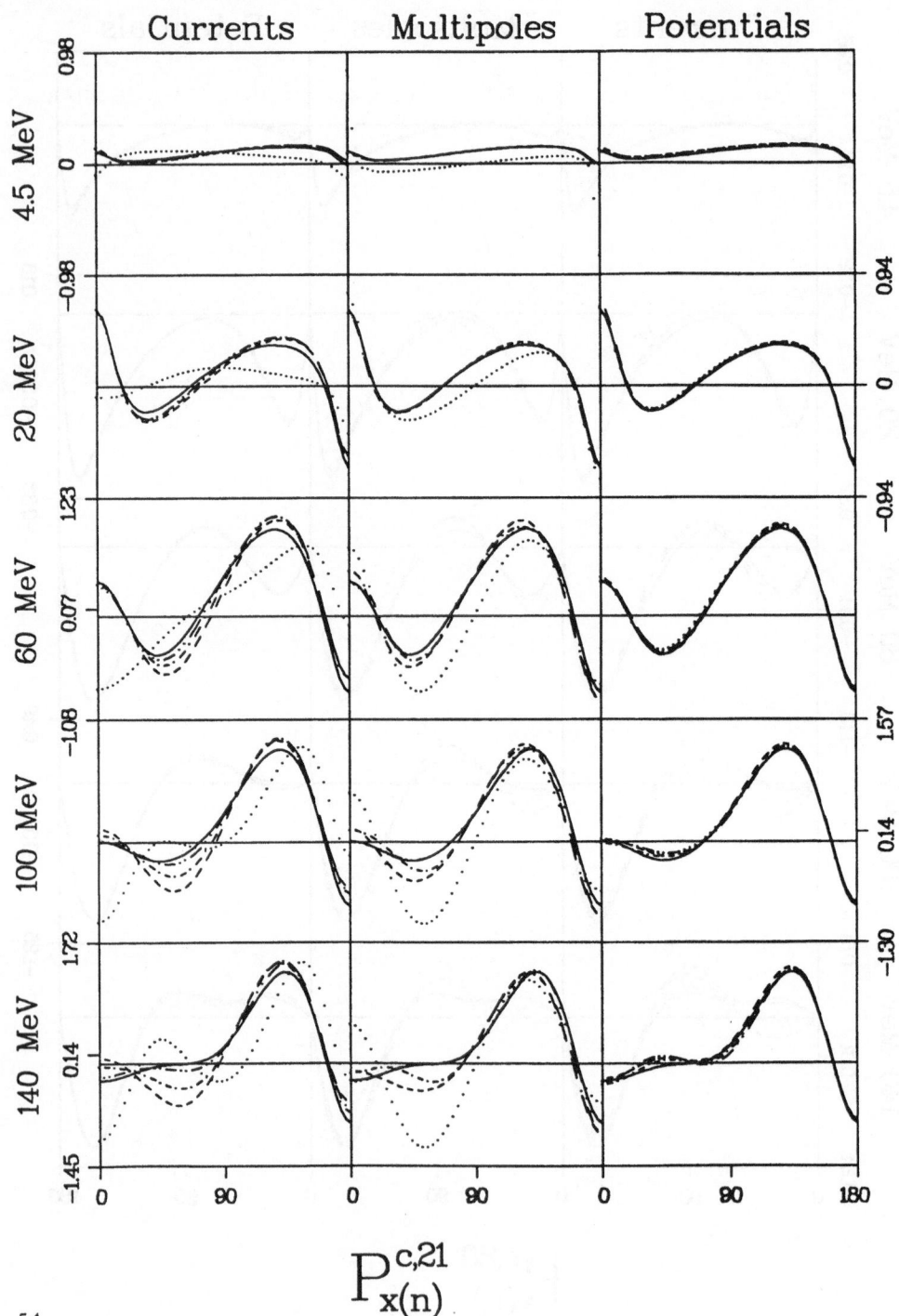

$$P_{x(n)}^{c,21}$$

54

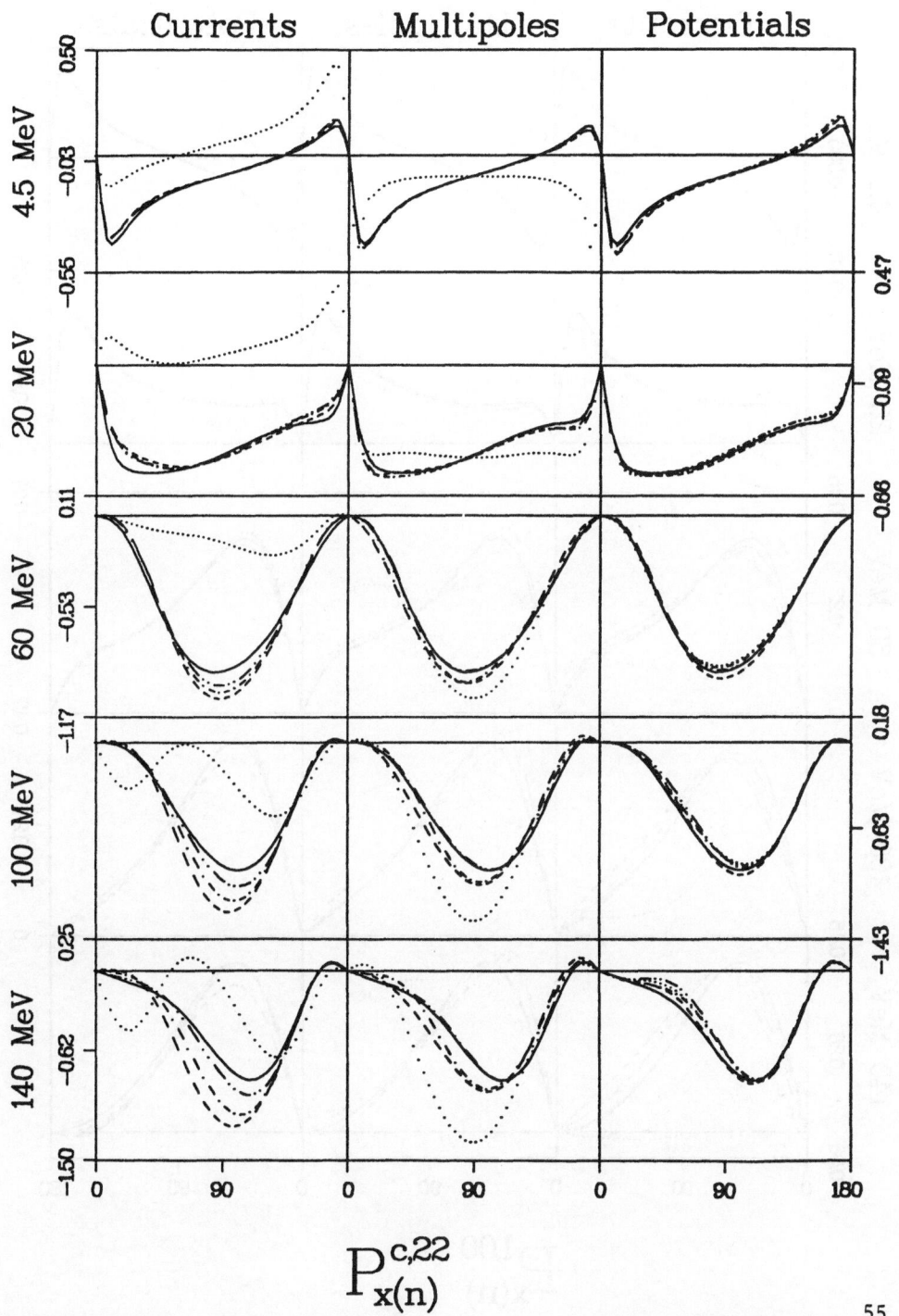

$$P_{x(n)}^{c,22}$$

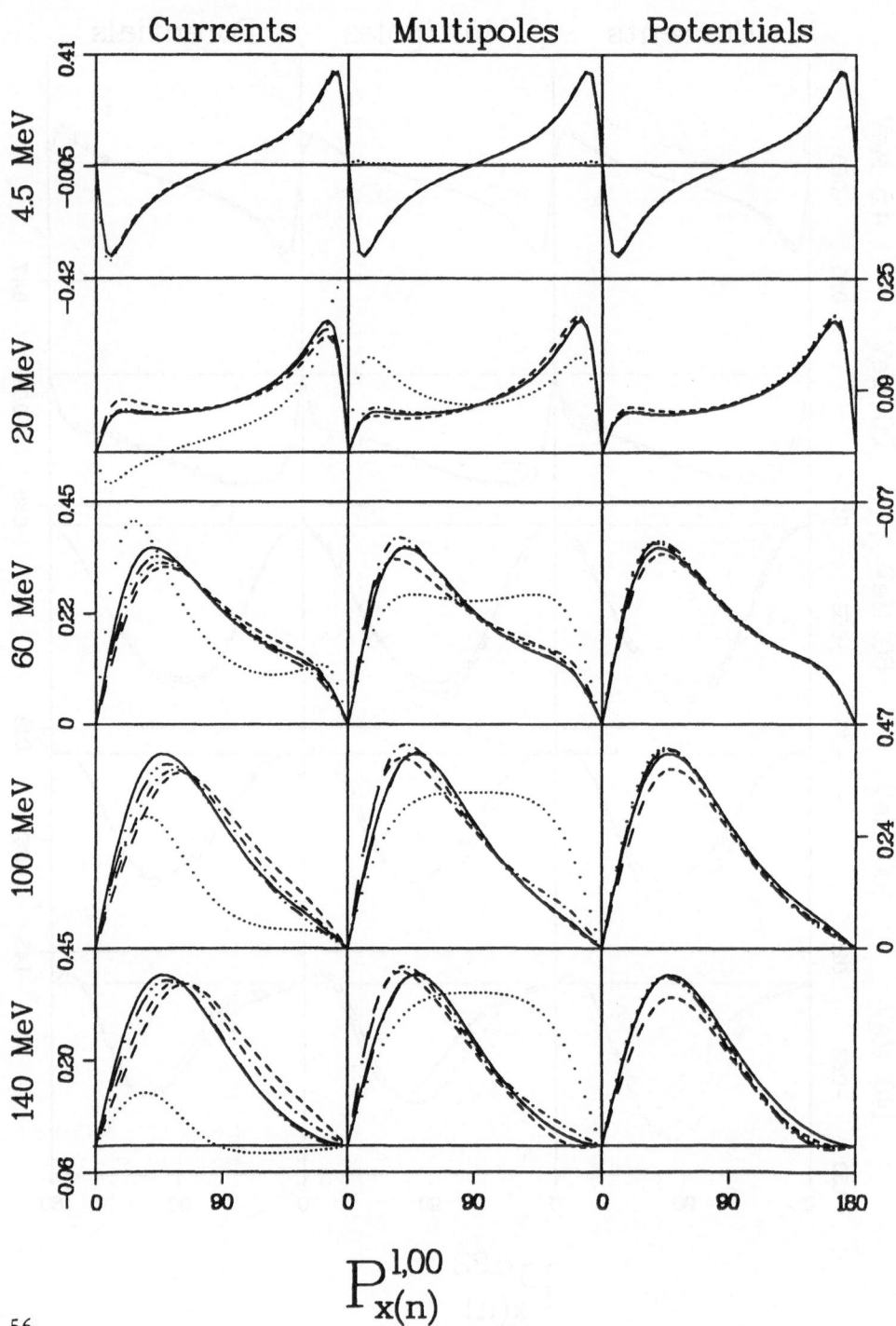

$$\mathrm{P}_{\mathrm{x(n)}}^{1,00}$$

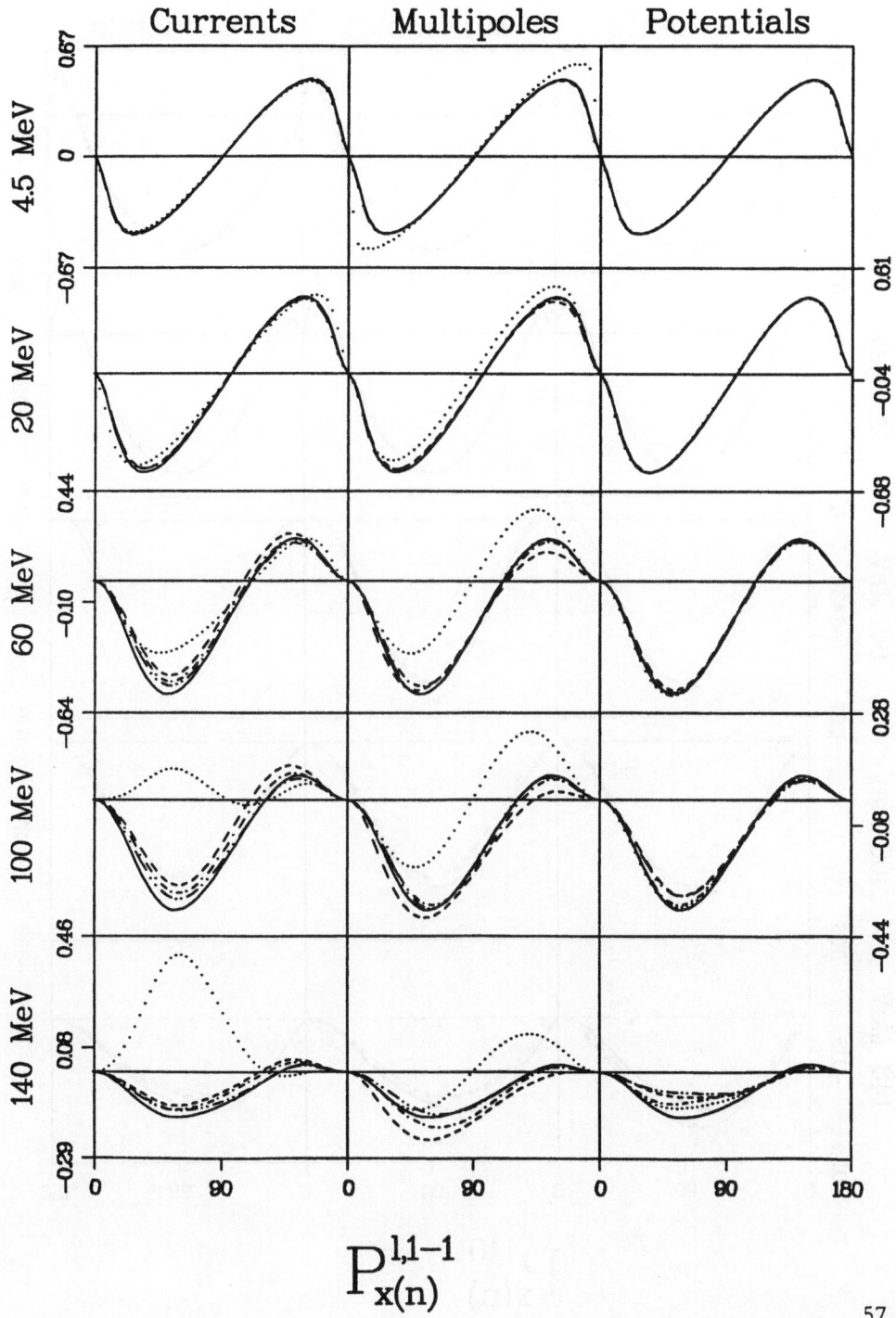

$$P_{x(n)}^{l,1-1}$$

Neutron Polarization $P_x^{l,IM}(n)$

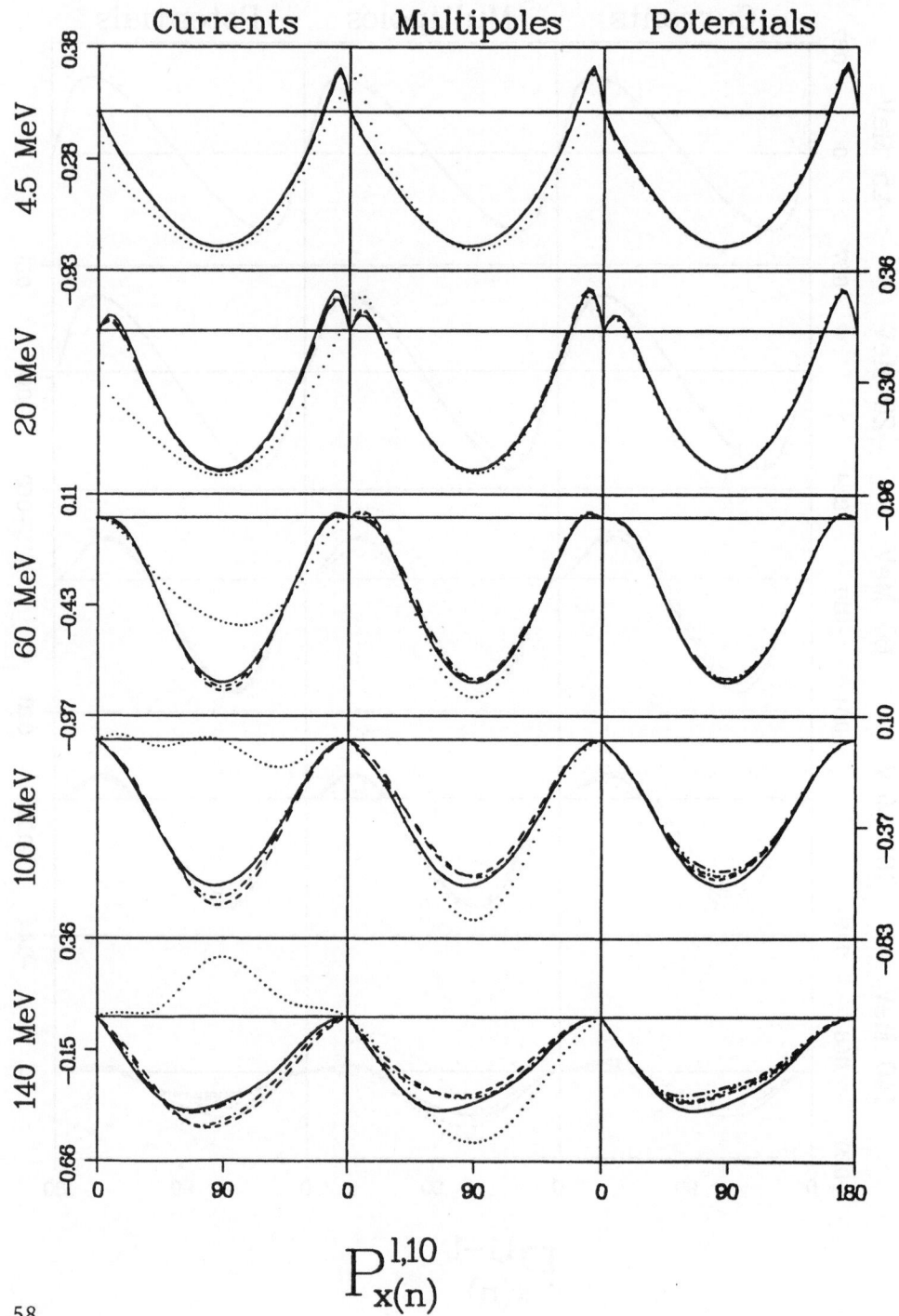

$$P_{x(n)}^{1,10}$$

58

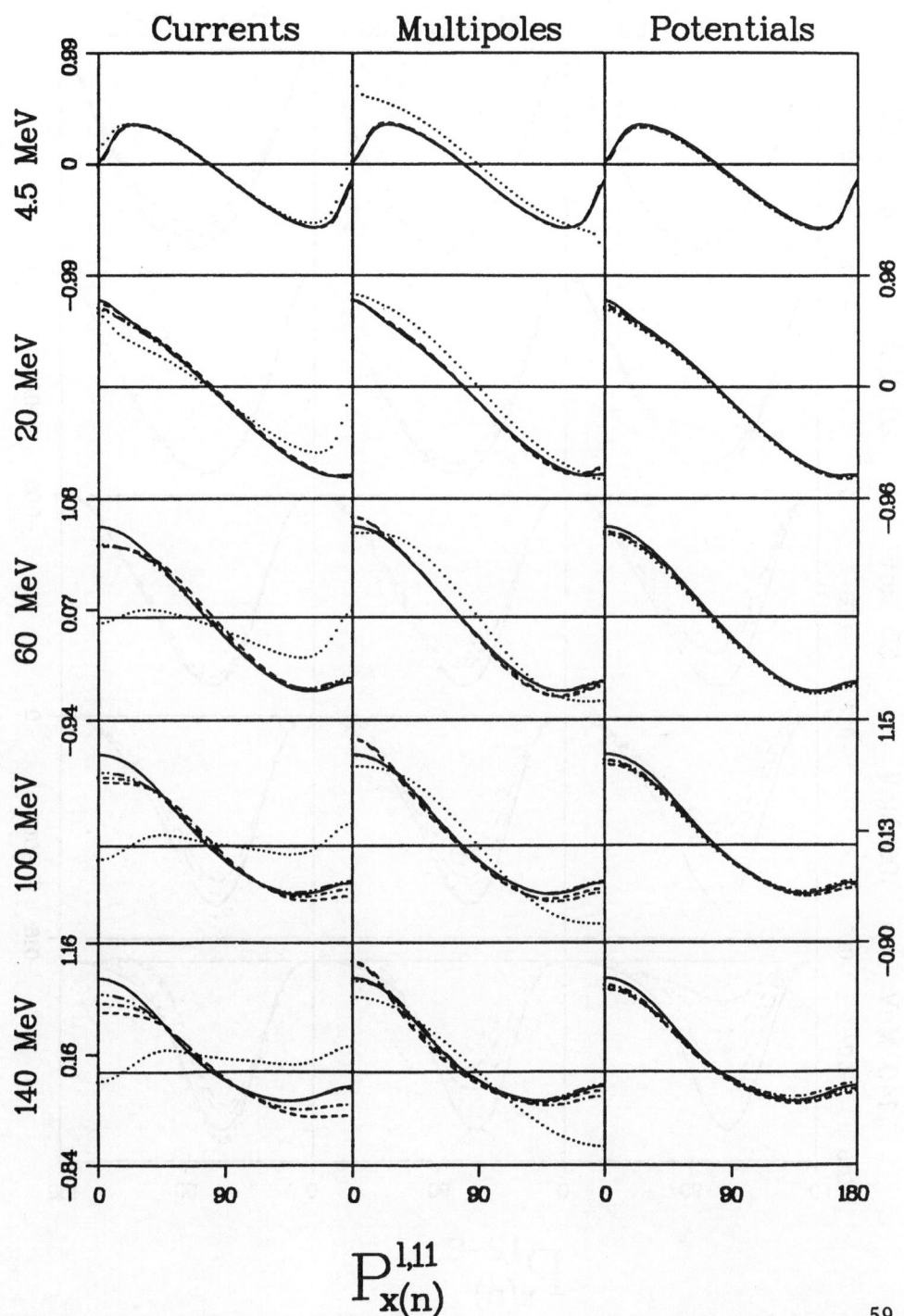

$$P_{x(n)}^{l,11}$$

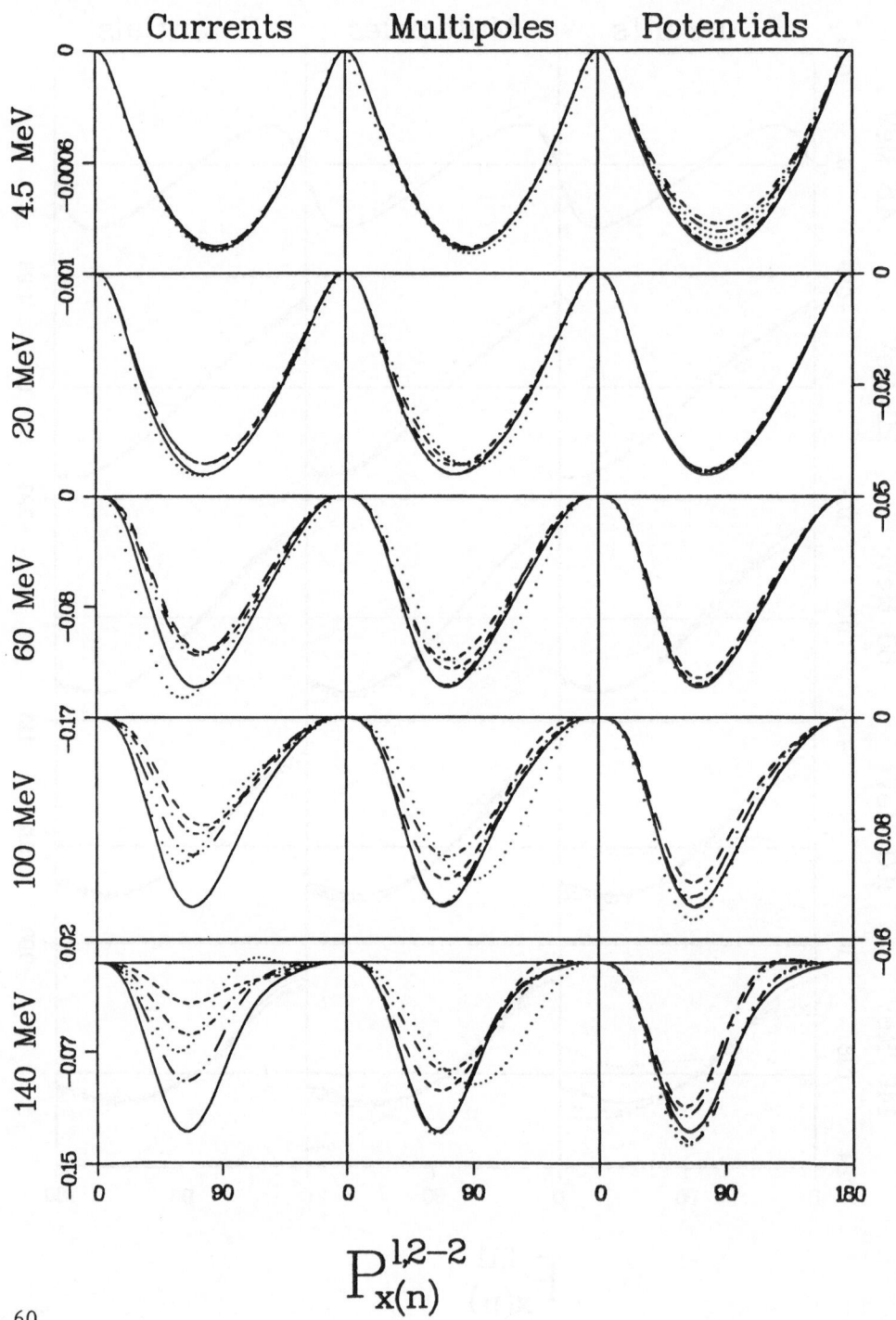

$$P^{1,2-2}_{x(n)}$$

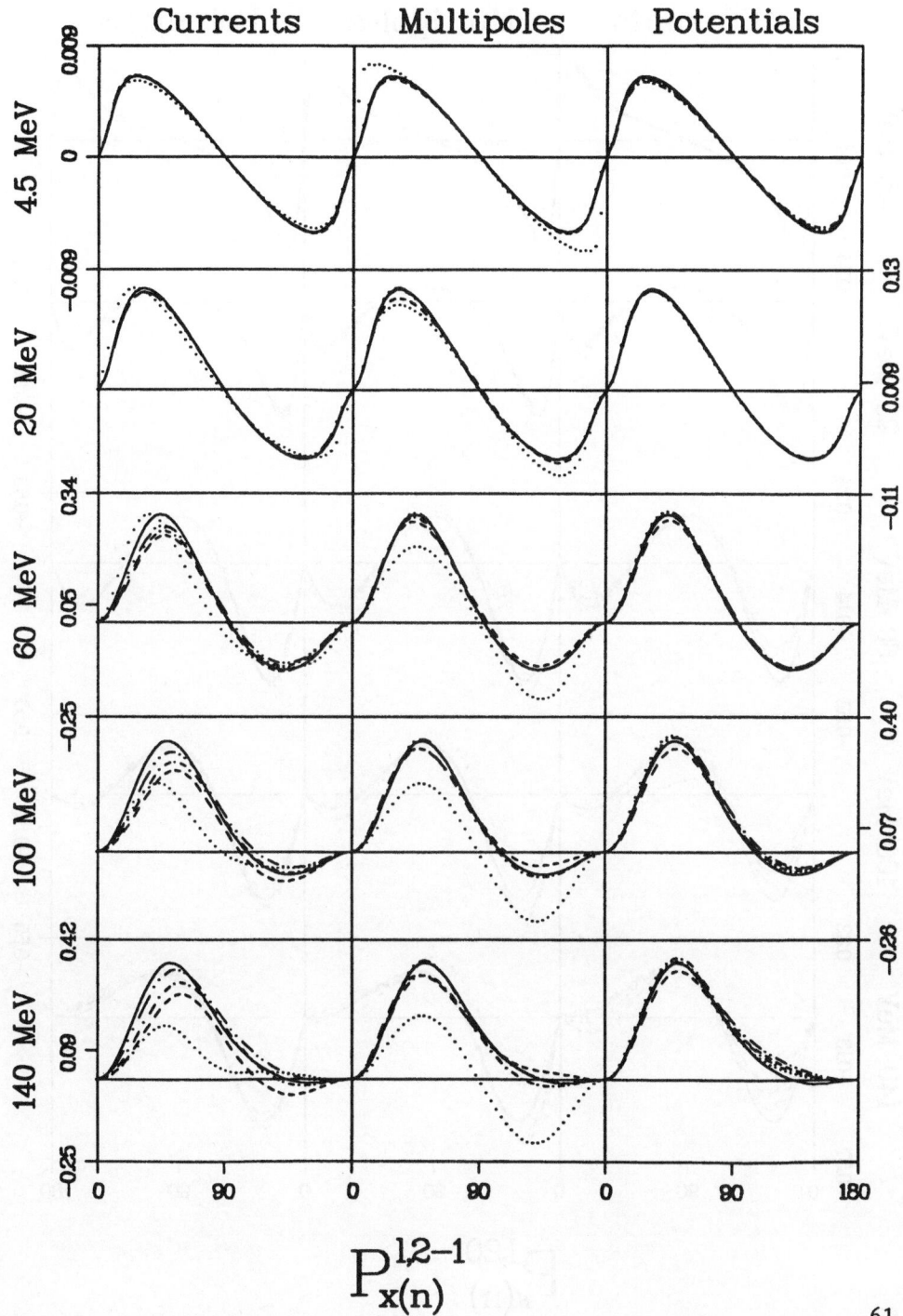

$$P_{x(n)}^{1,2-1}$$

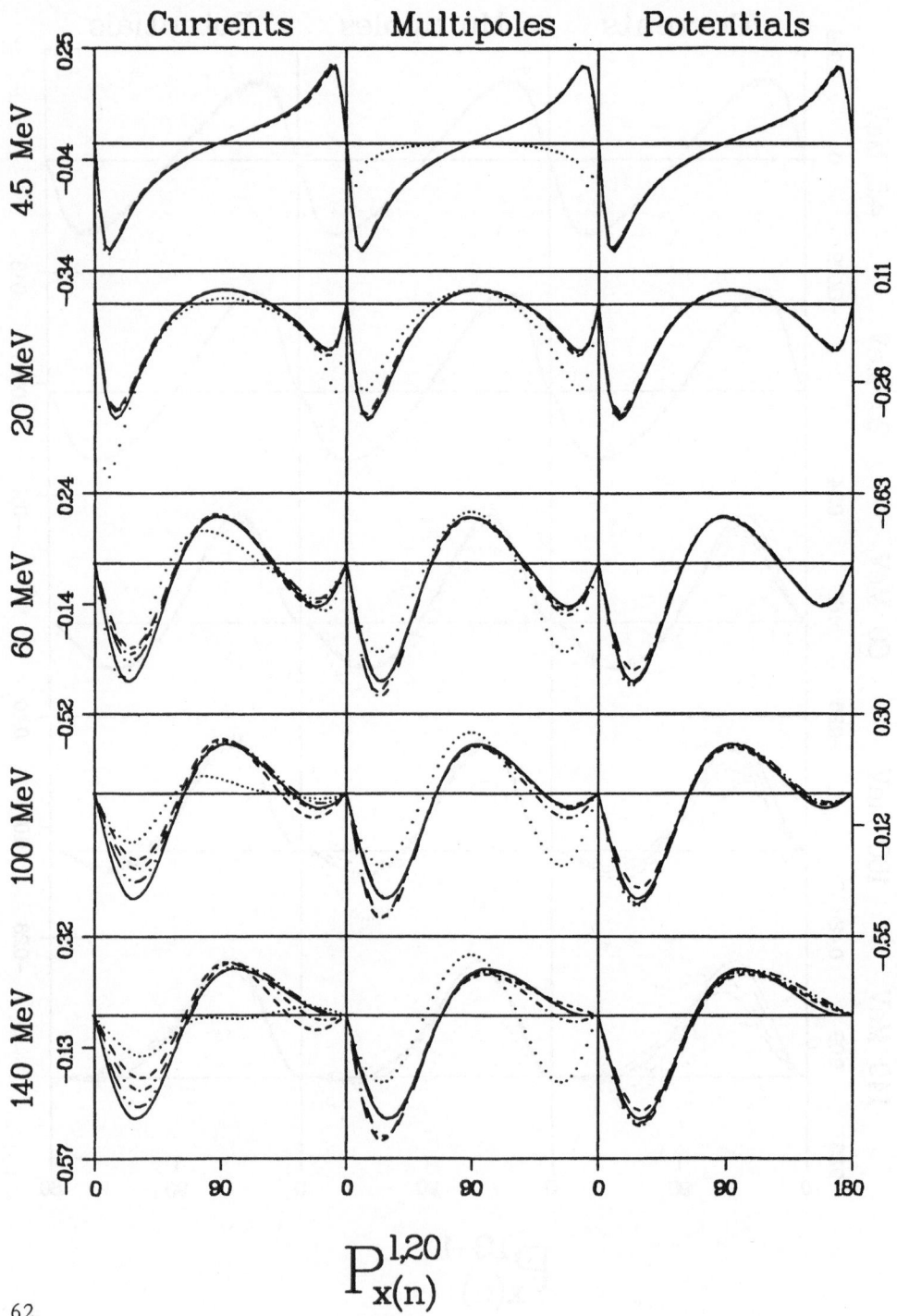

$$P_{x(n)}^{1,20}$$

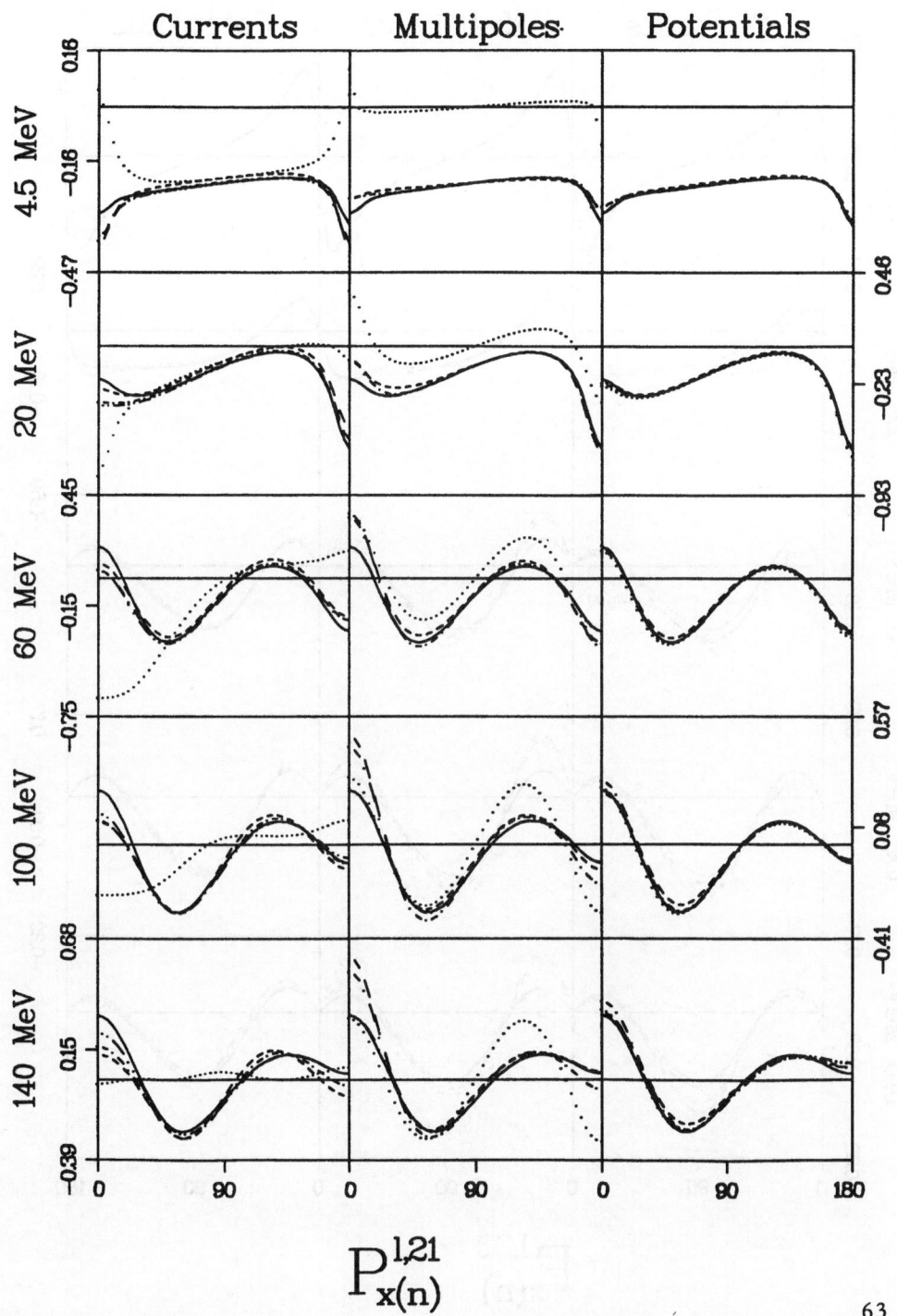

$$P_{x(n)}^{1,21}$$

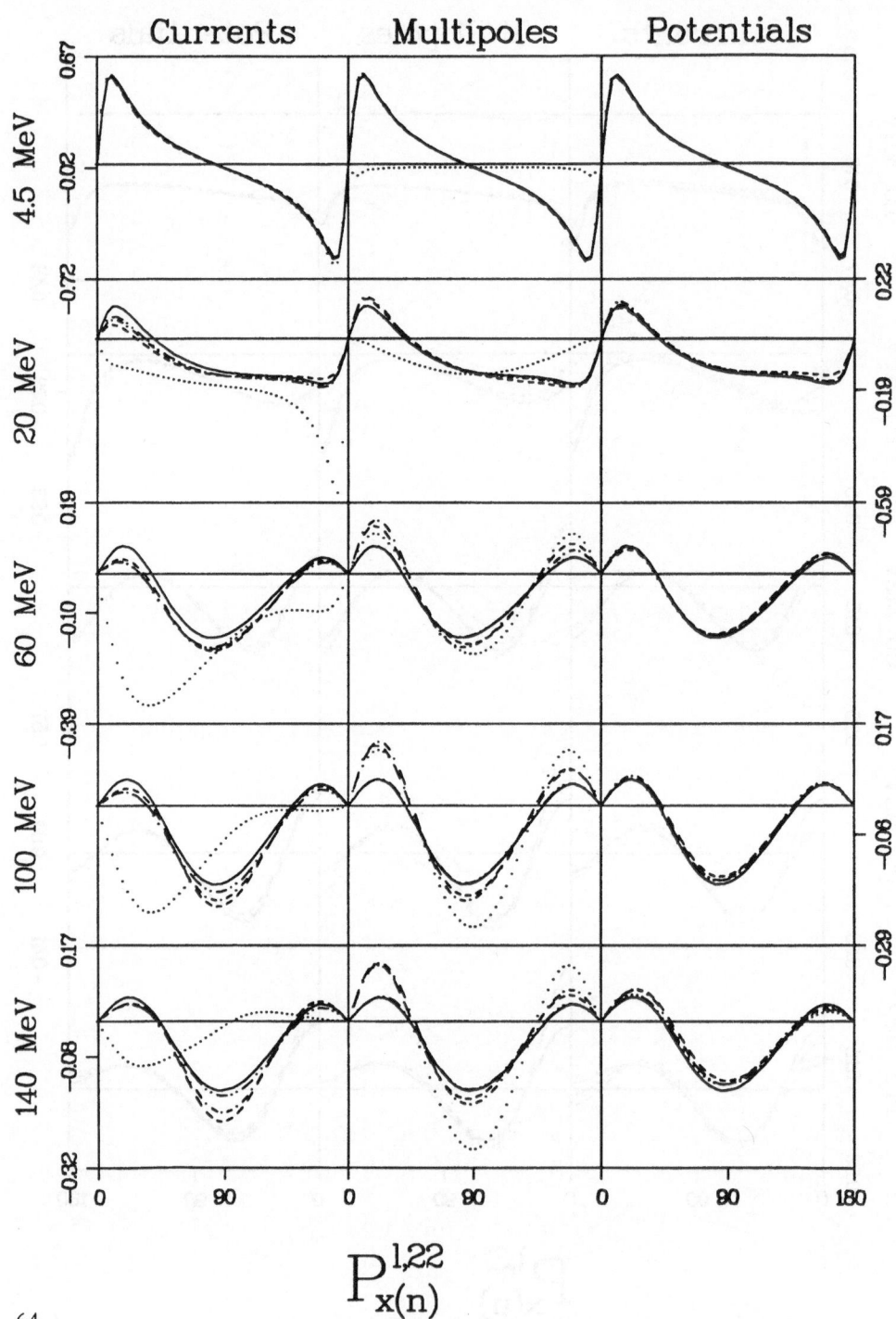

5.11 Proton Polarization $P_y^{0,IM}(p)$

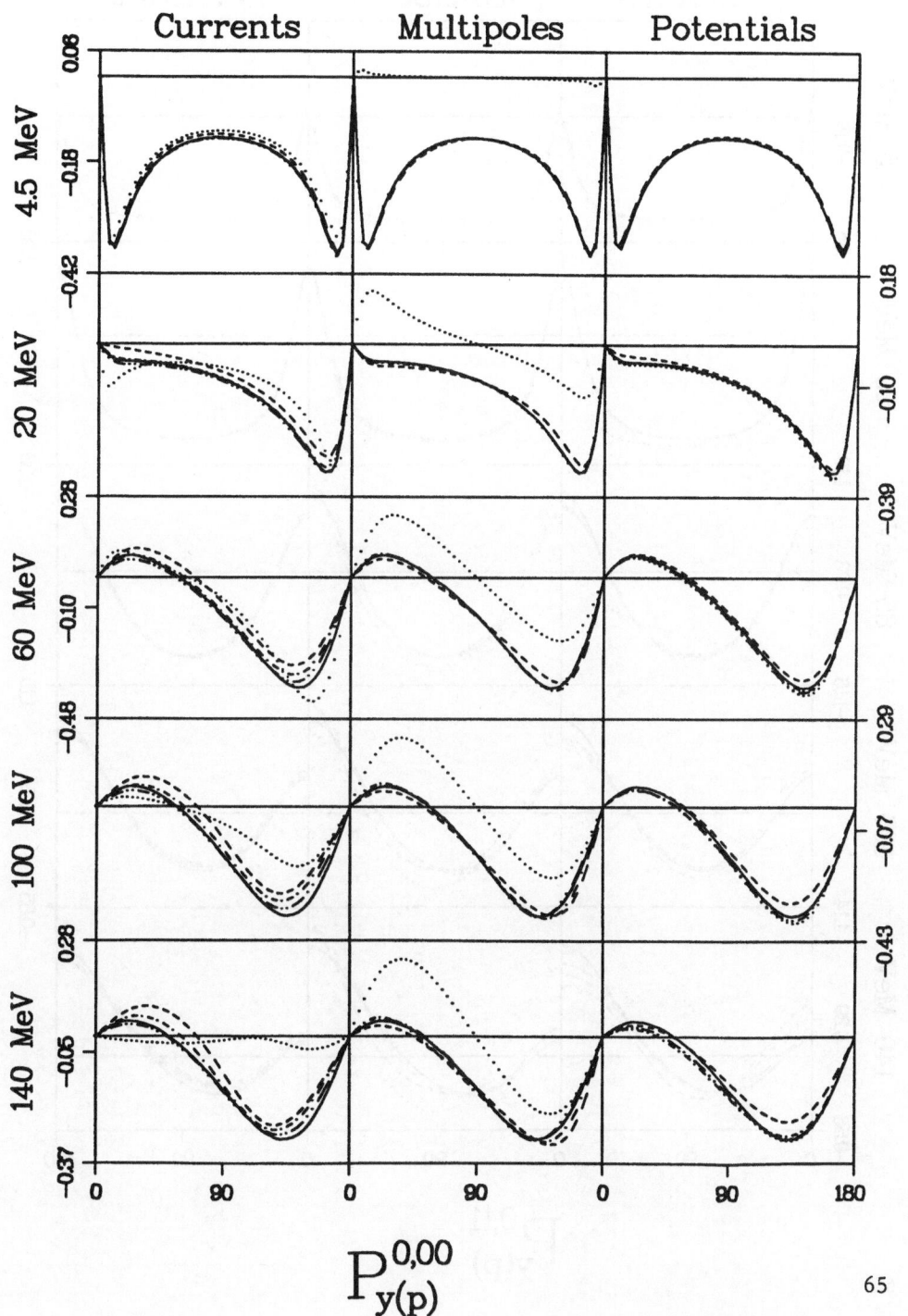

$$P_{y(p)}^{0,00}$$

65

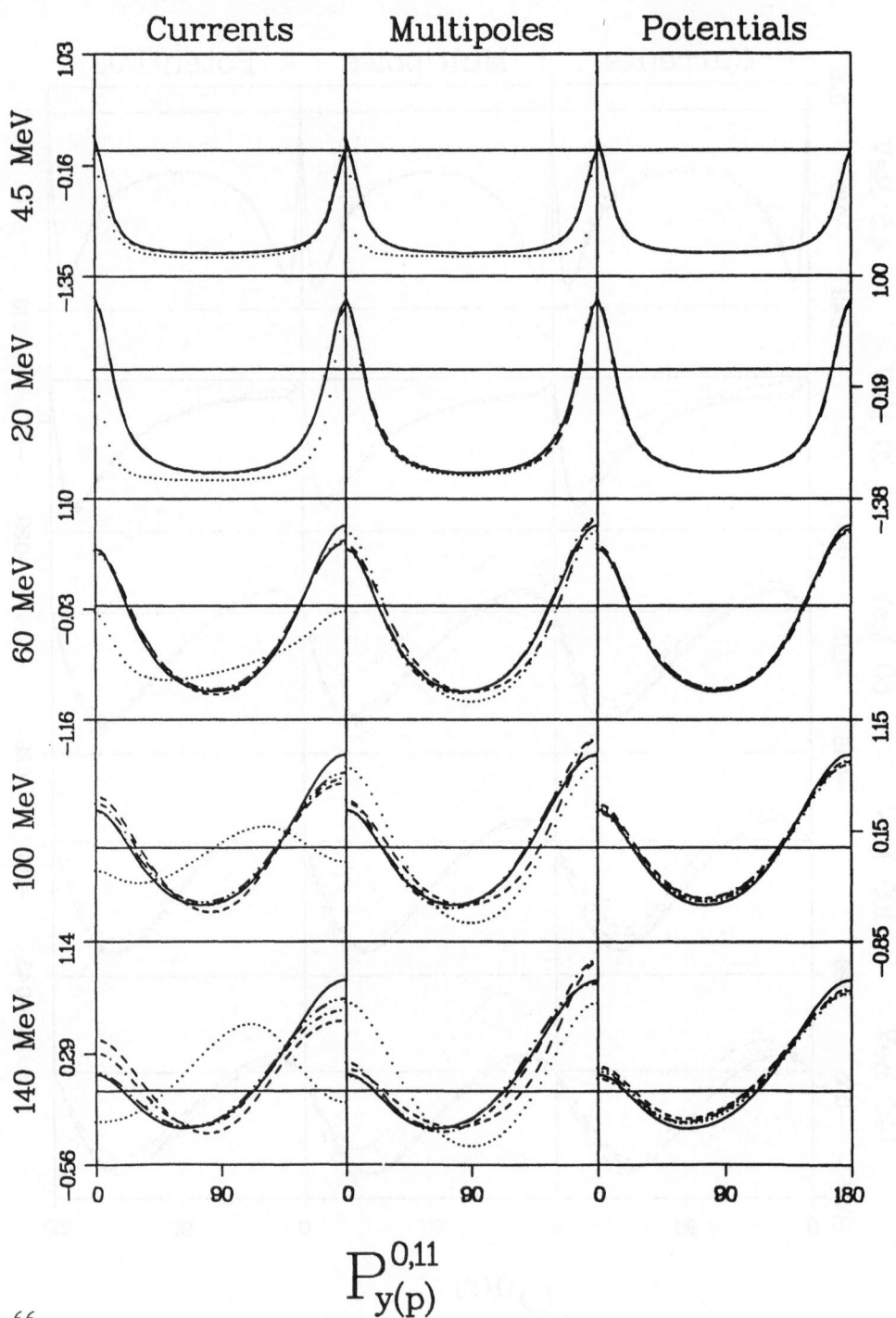

$$P_{y(p)}^{0,11}$$

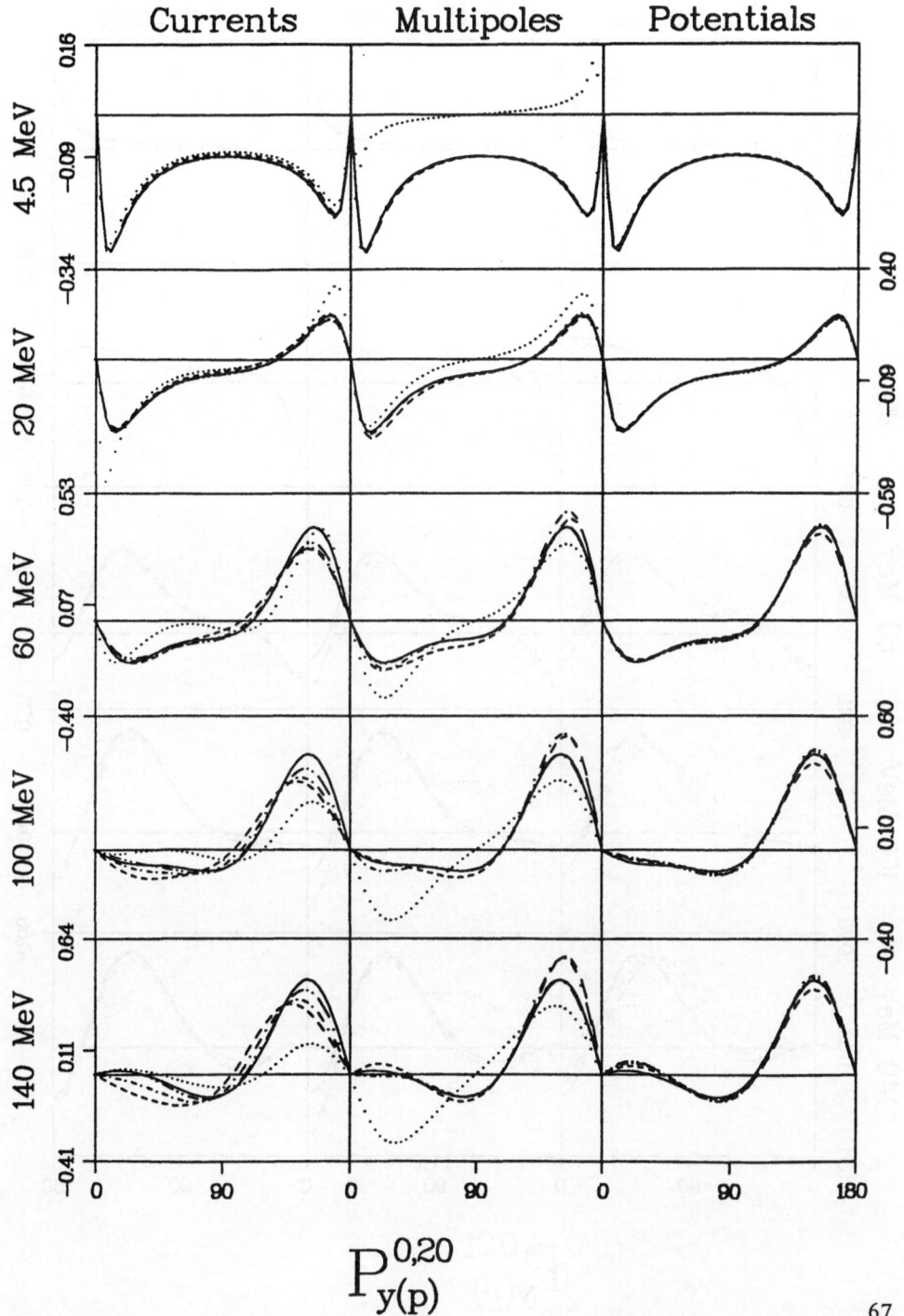

$$P_{y(p)}^{0,20}$$

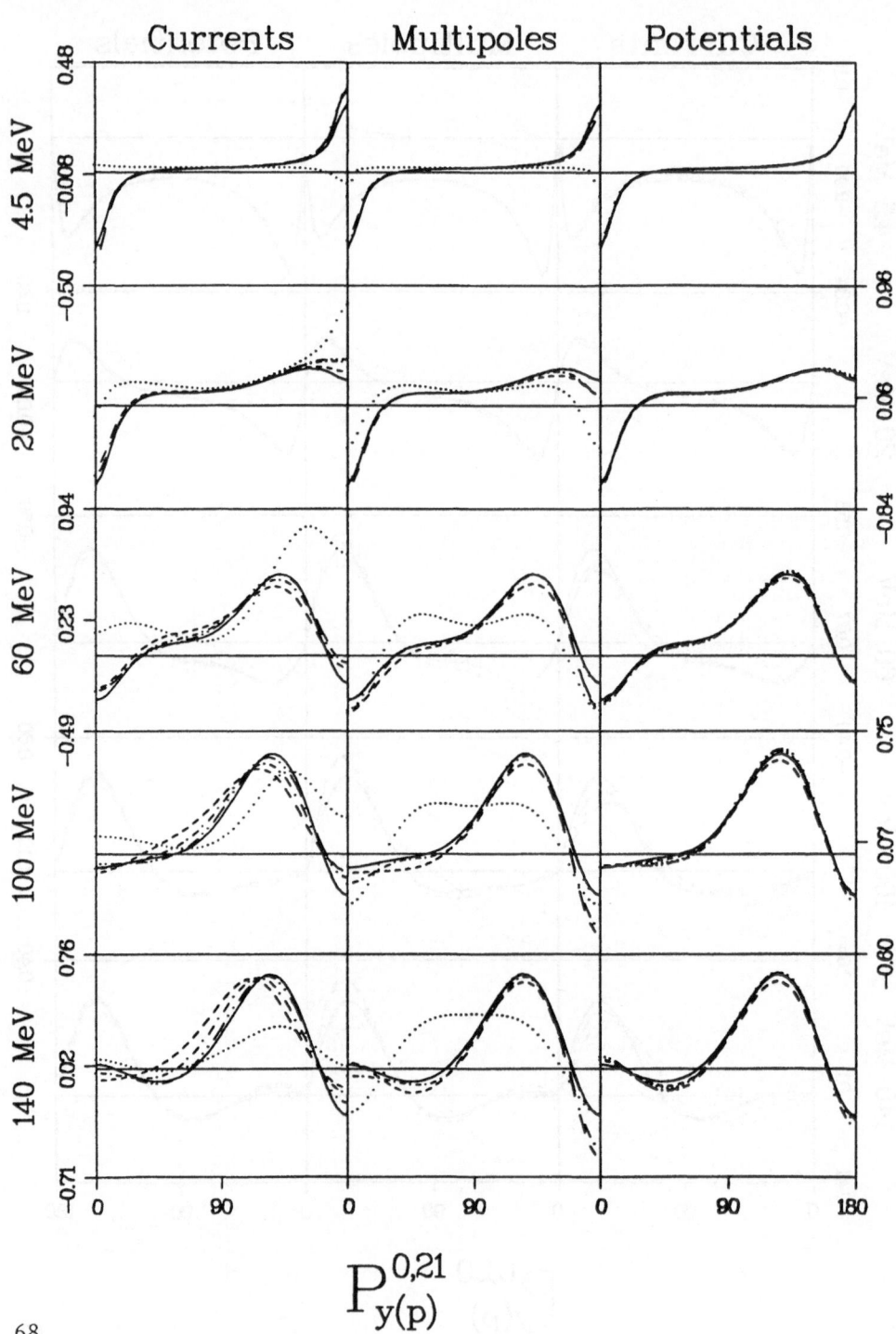

$$P_{y(p)}^{0,21}$$

Proton Polarization $P_y^{0,IM}(p)$

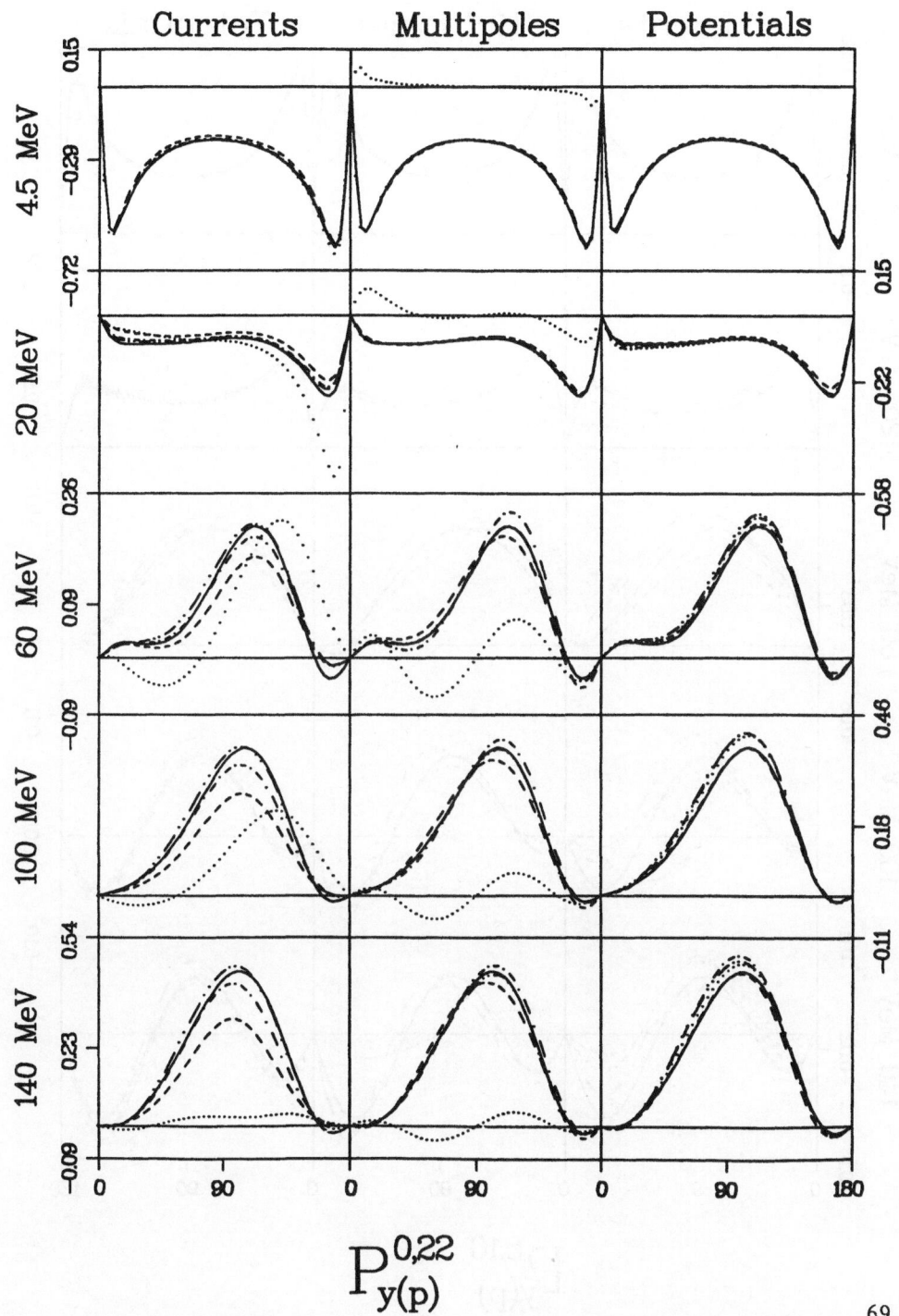

$$P_{y(p)}^{0,22}$$

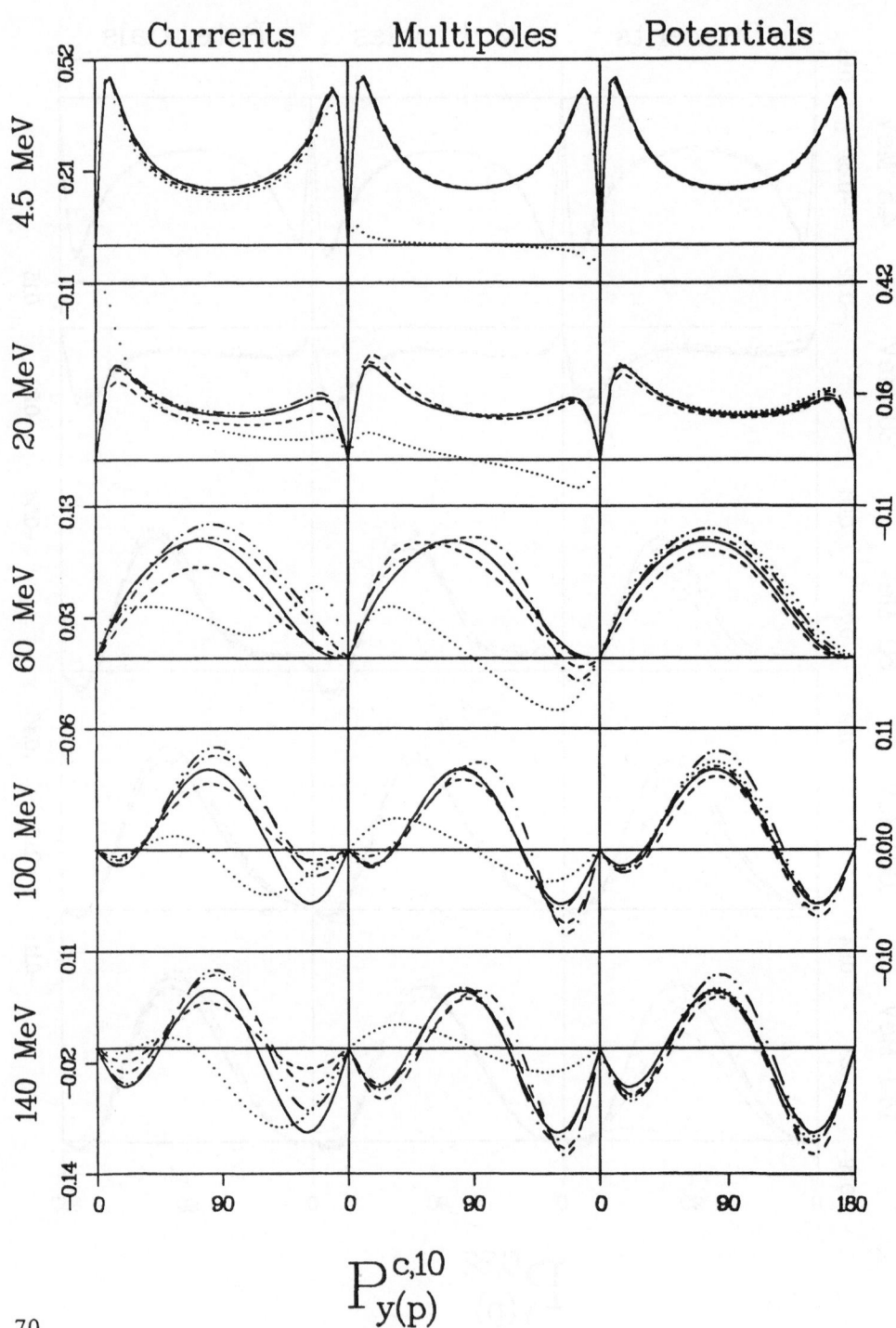

$$P_{y(p)}^{c,10}$$

Proton Polarization $P_y^{c,IM}(p)$

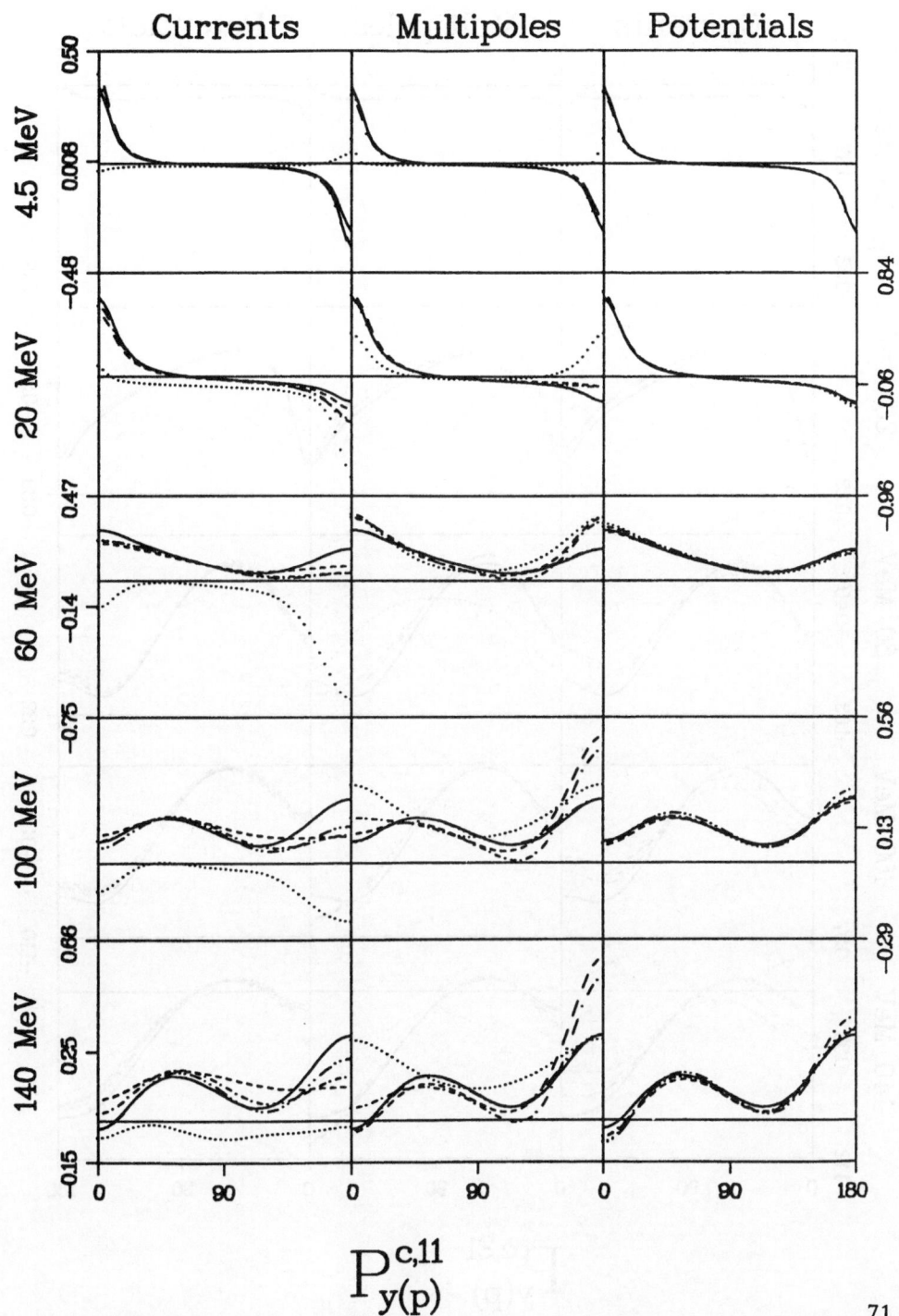

$$P_{y(p)}^{c,11}$$

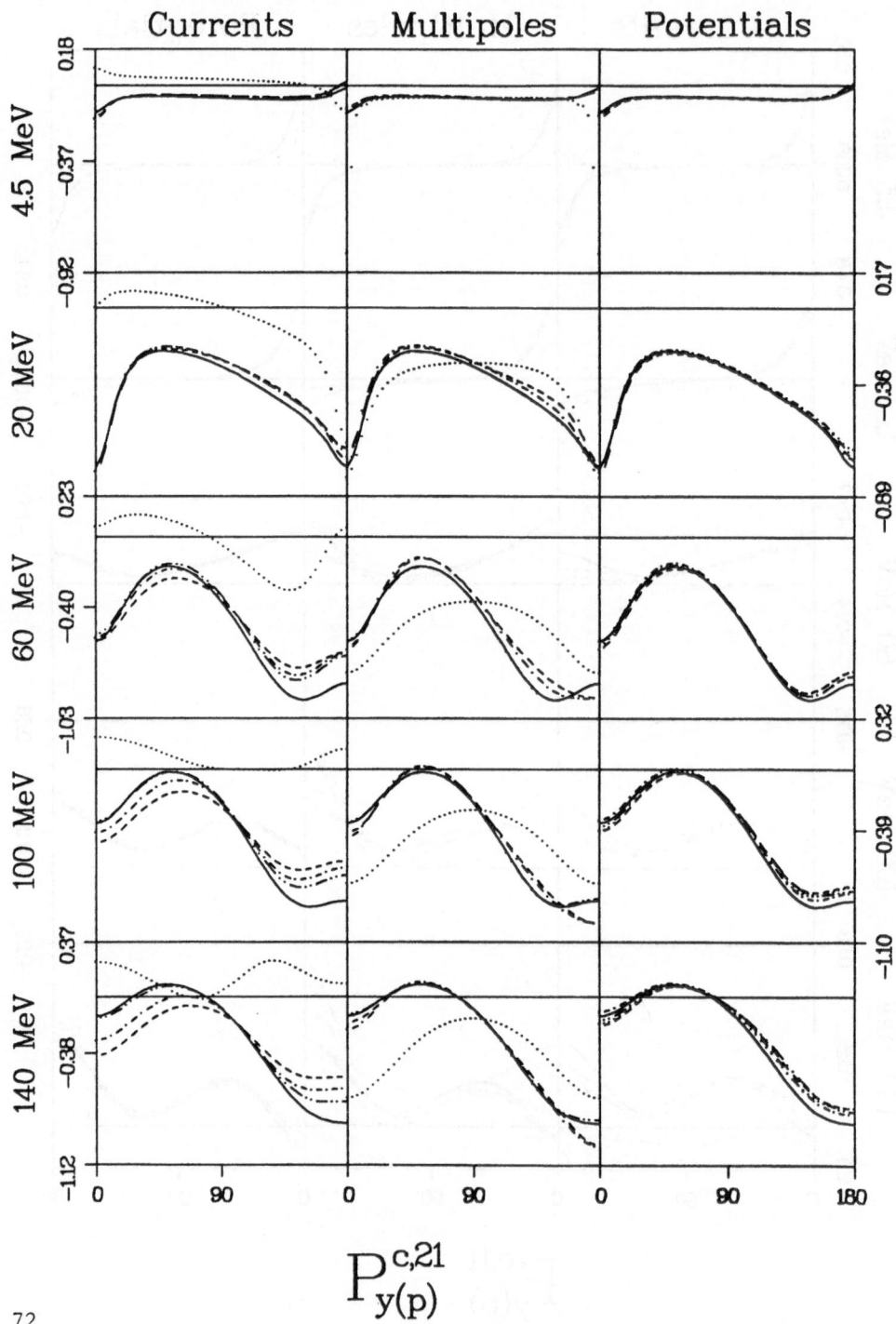

$$P_{y(p)}^{c,21}$$

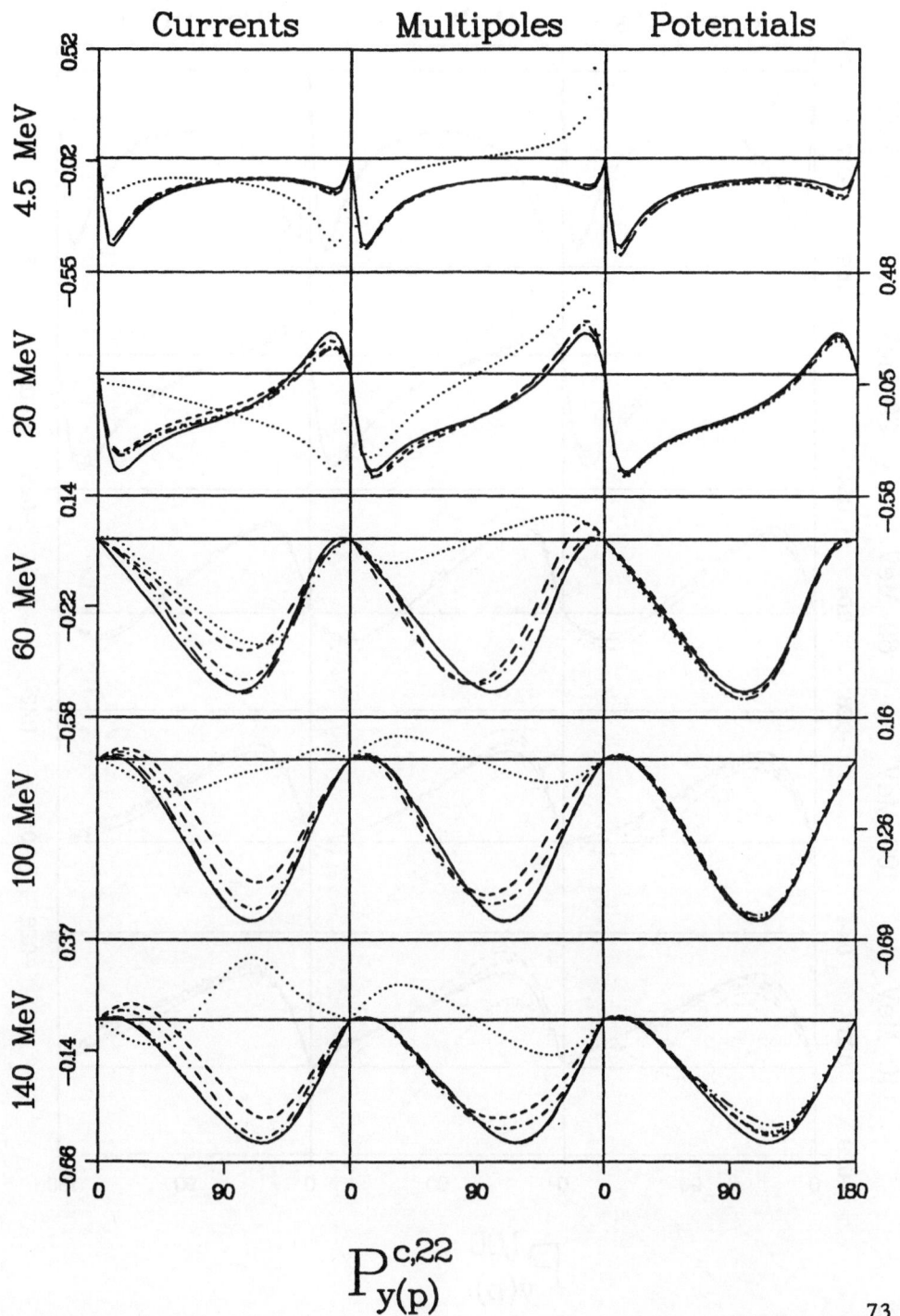

$$P_{y(p)}^{c,22}$$

5.13 Proton Polarization $P_y^{l,IM}(p)$

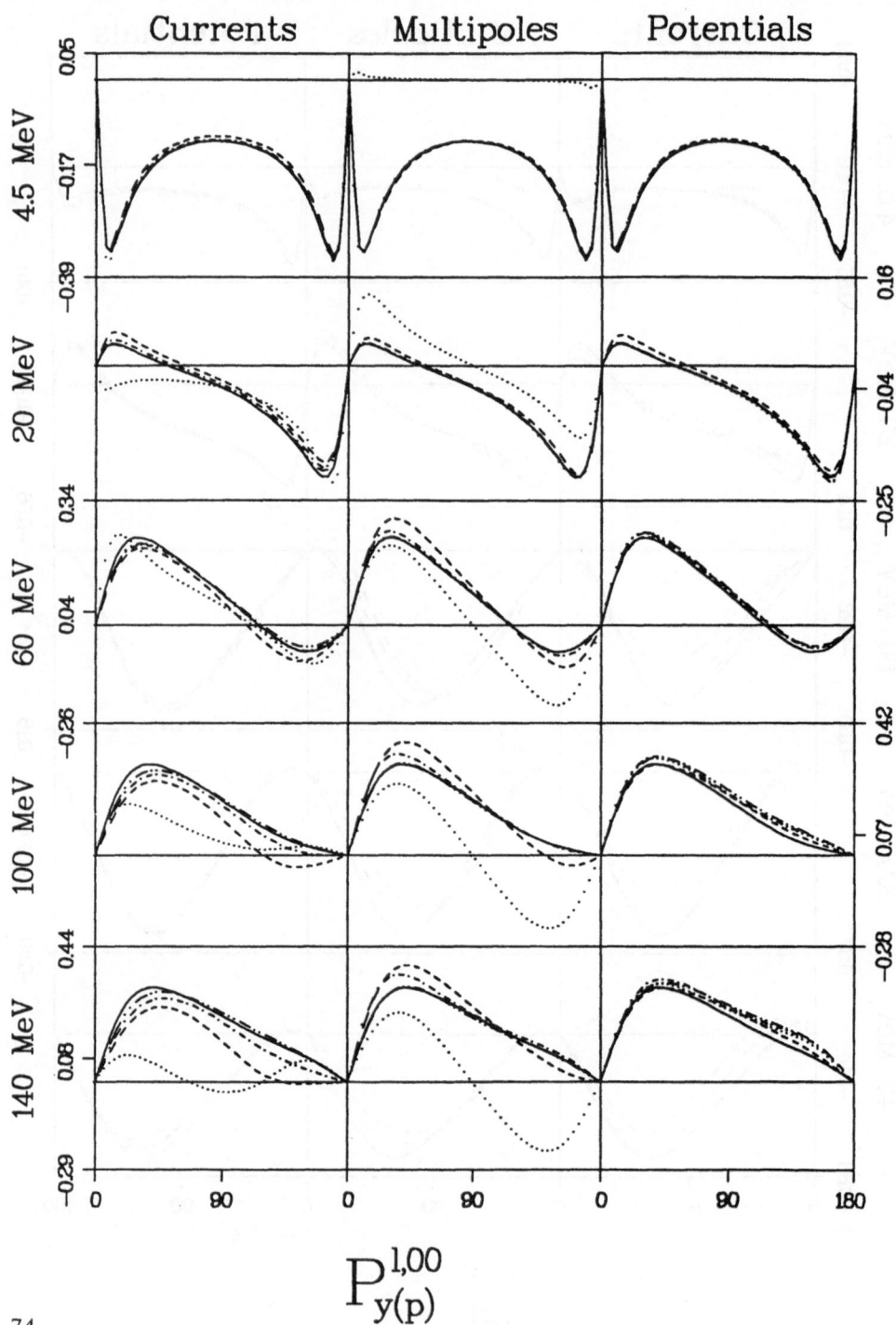

$$P_{y(p)}^{1,00}$$

Proton Polarization $P_y^{l,IM}(p)$

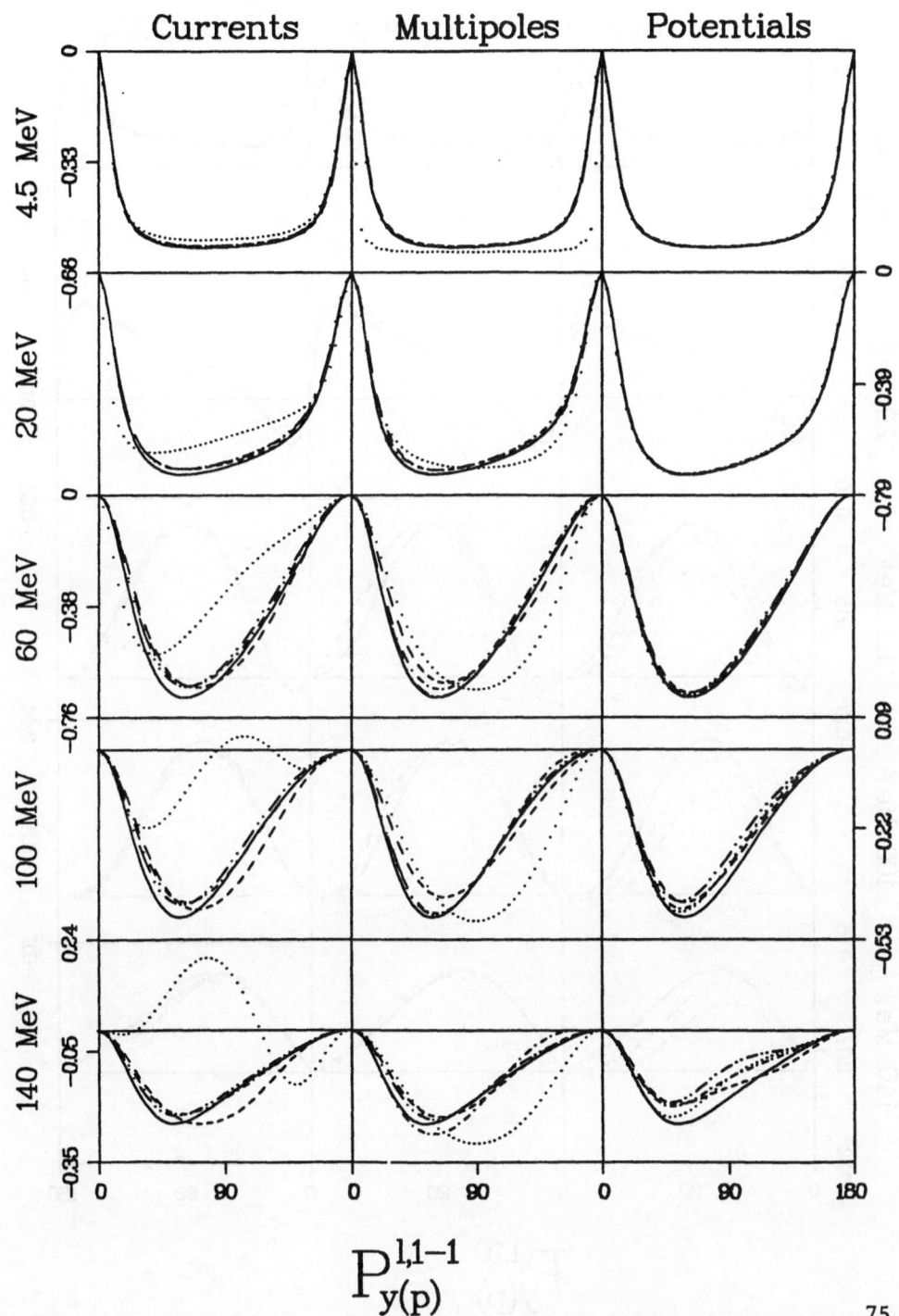

$$P_{y(p)}^{l,1-1}$$

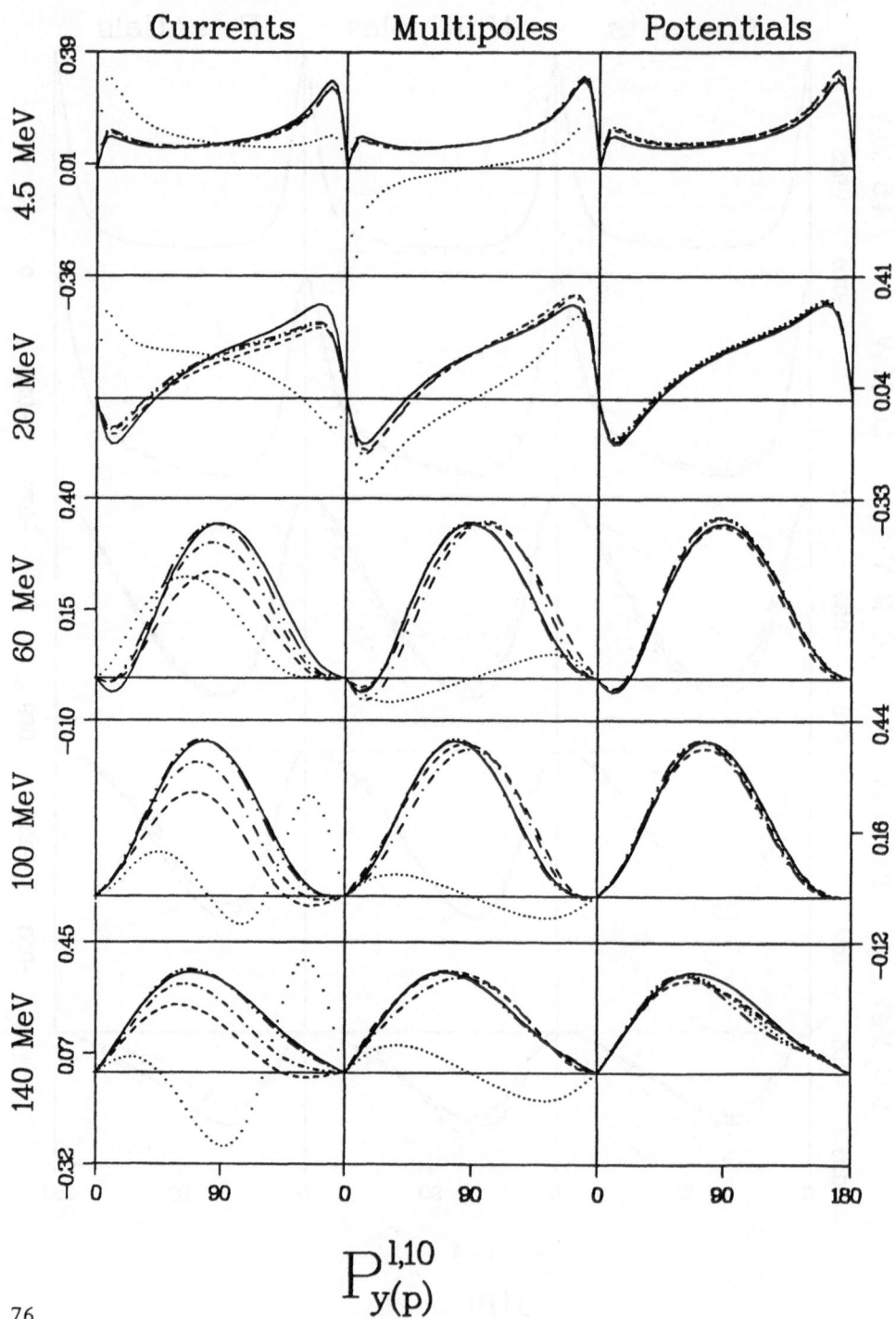

Currents   Multipoles   Potentials

4.5 MeV   20 MeV   60 MeV   100 MeV   140 MeV

$$P_{y(p)}^{l,10}$$

Proton Polarization $P_y^{l,IM}(p)$

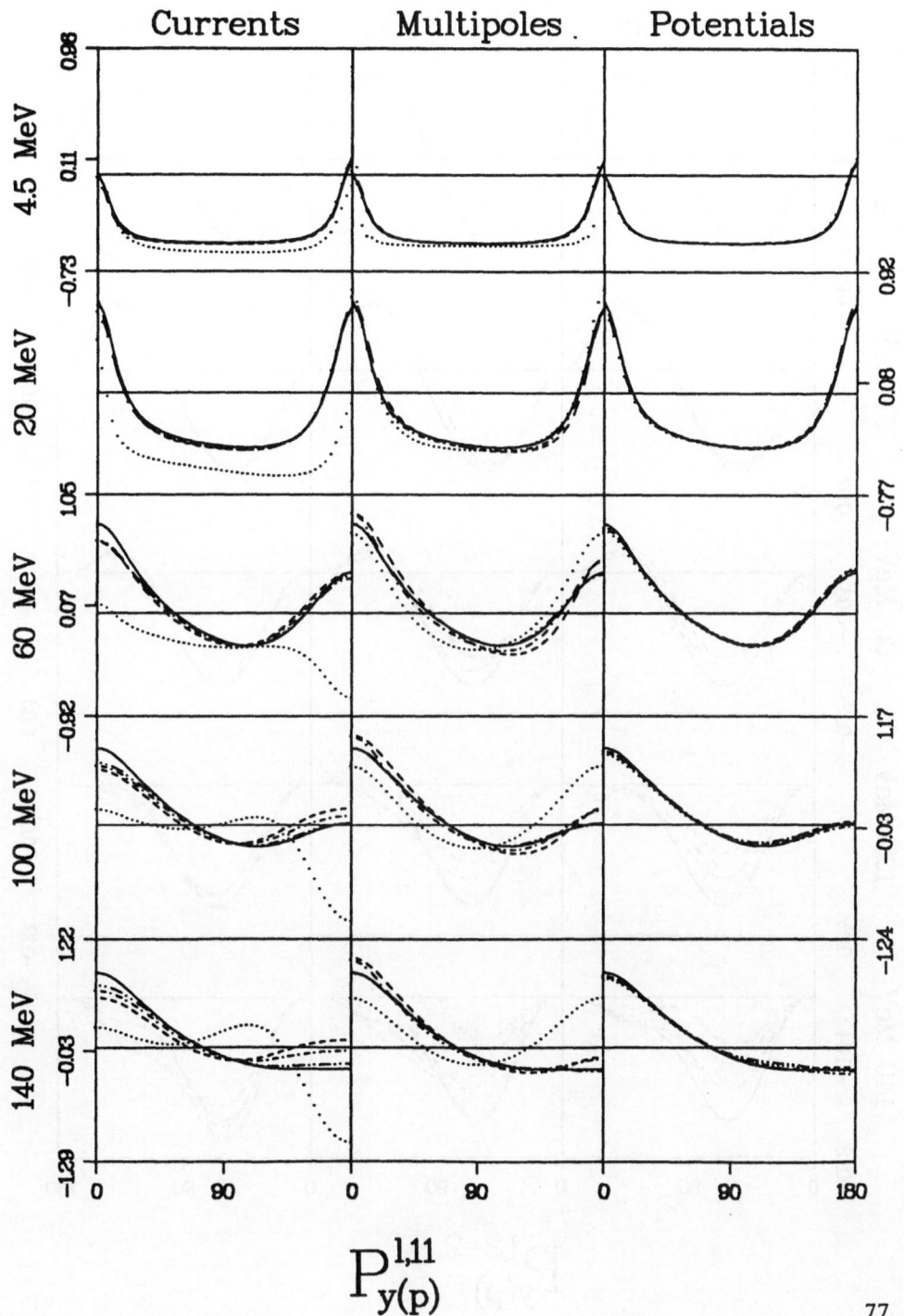

$$P_{y(p)}^{1,11}$$

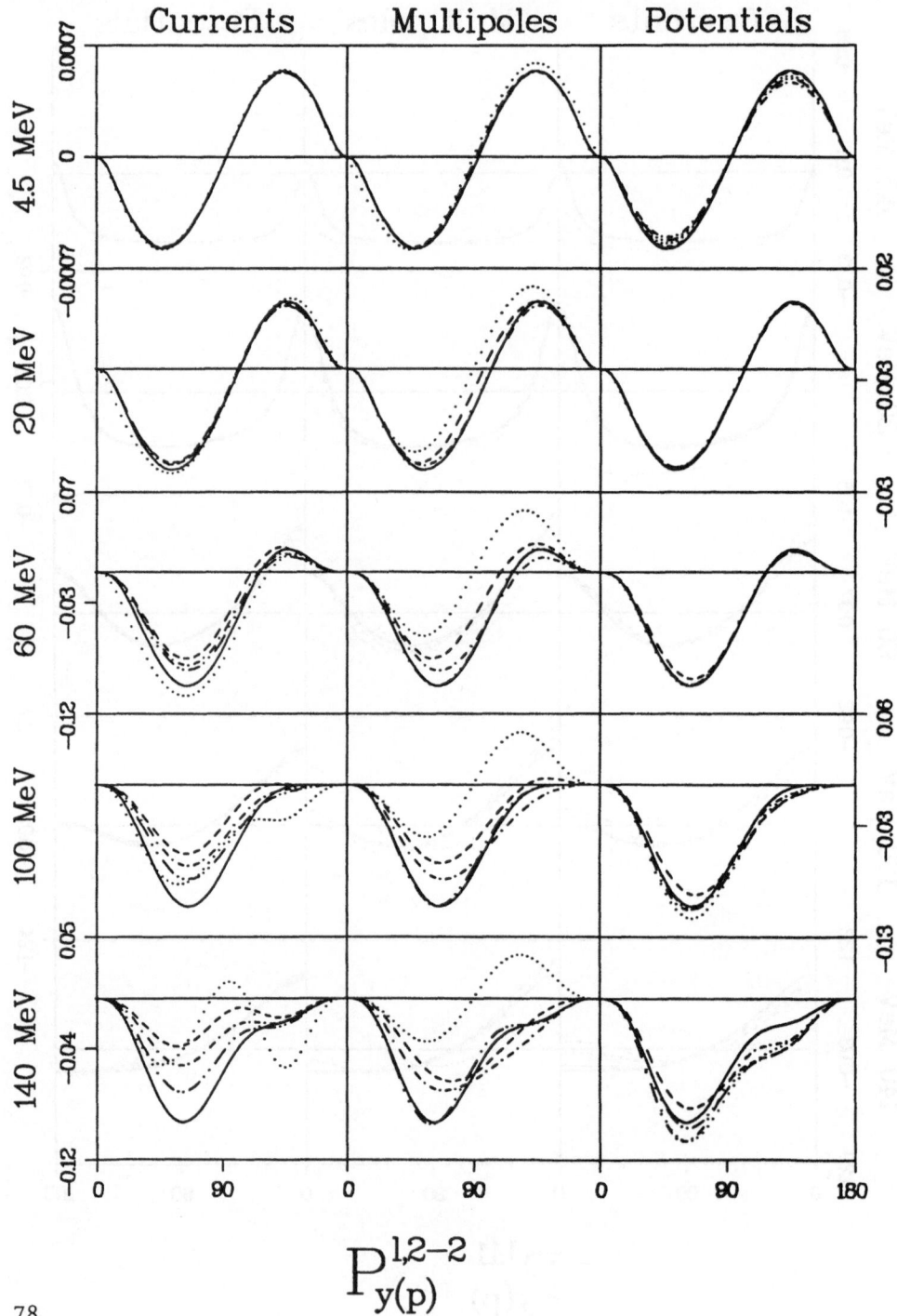

$$P_{y(p)}^{1,2-2}$$

Proton Polarization $P_y^{l,IM}(p)$

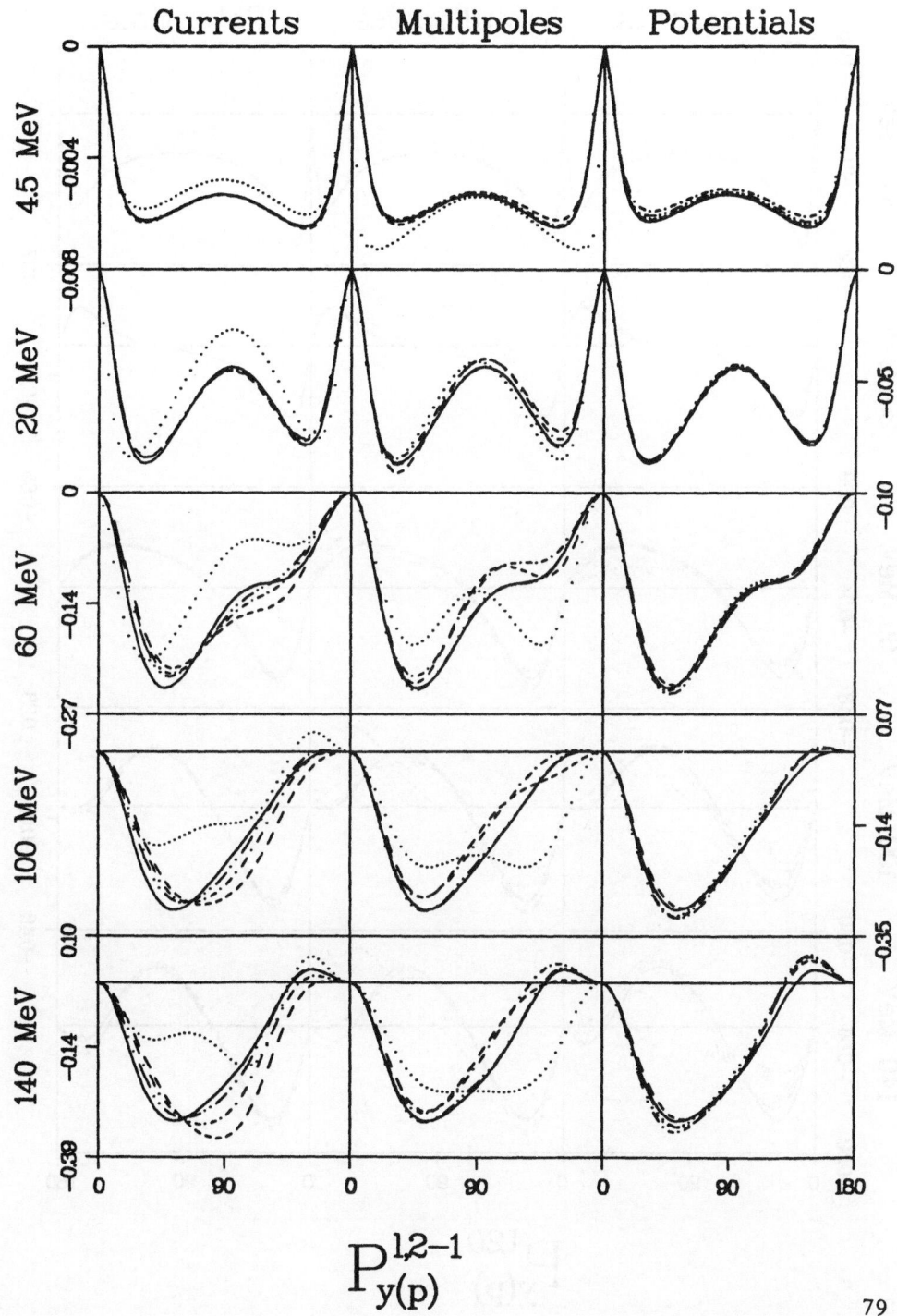

$$P_{y(p)}^{12-1}$$

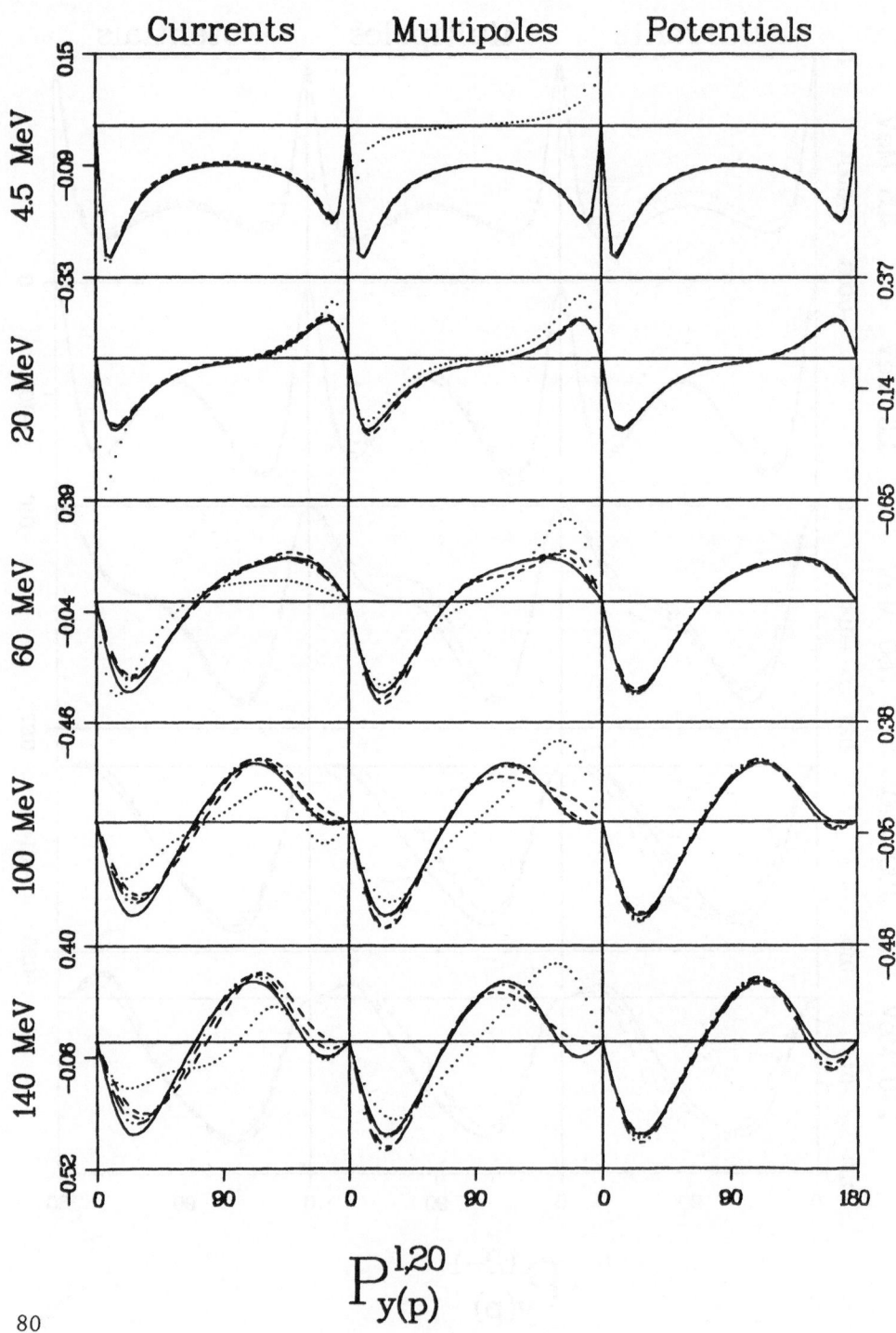

$$P_{y(p)}^{l,20}$$

Proton Polarization $P_y^{l,IM}(p)$

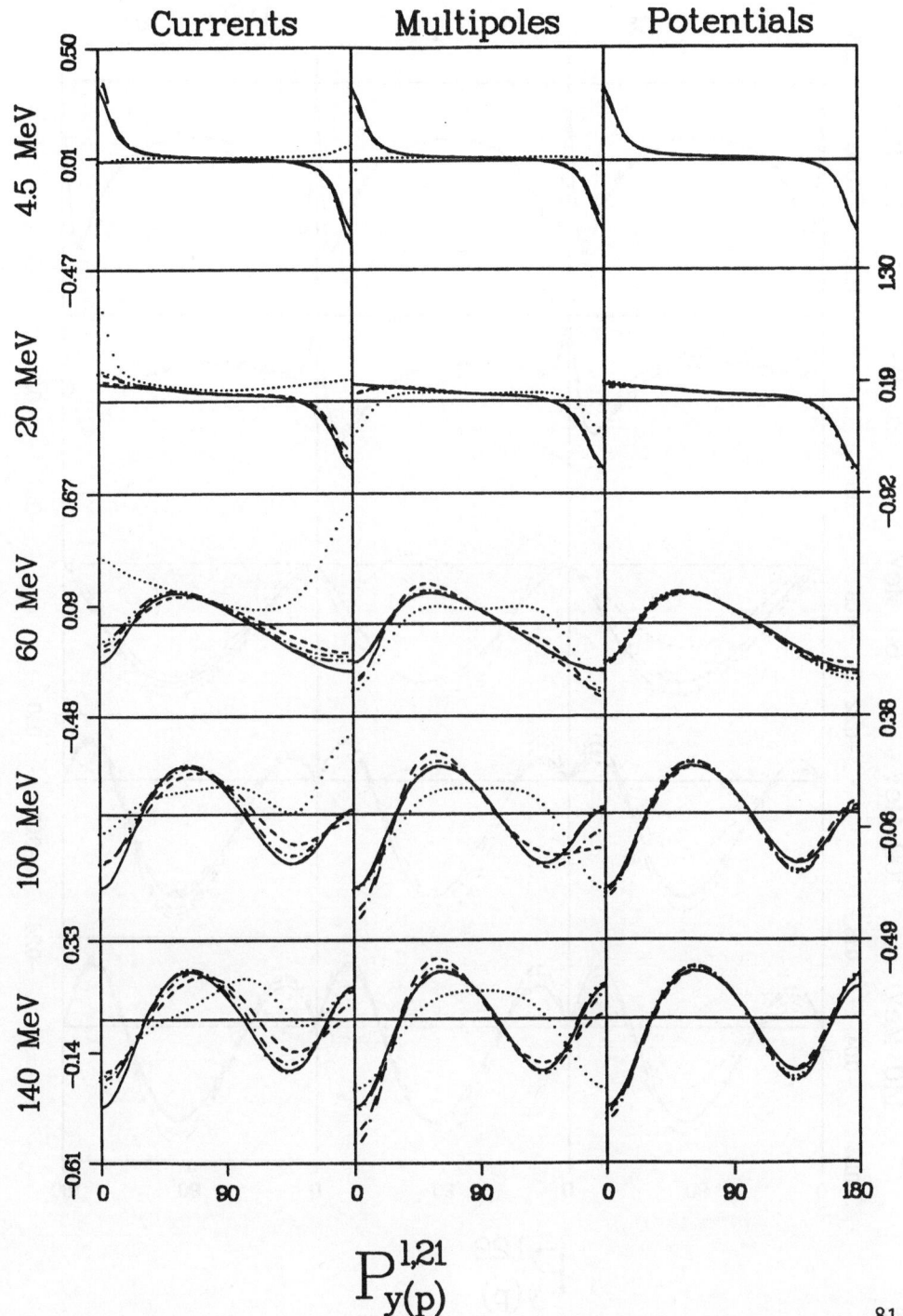

$$P_{y(p)}^{1,21}$$

Proton Polarization $P_y^{l,IM}(p)$

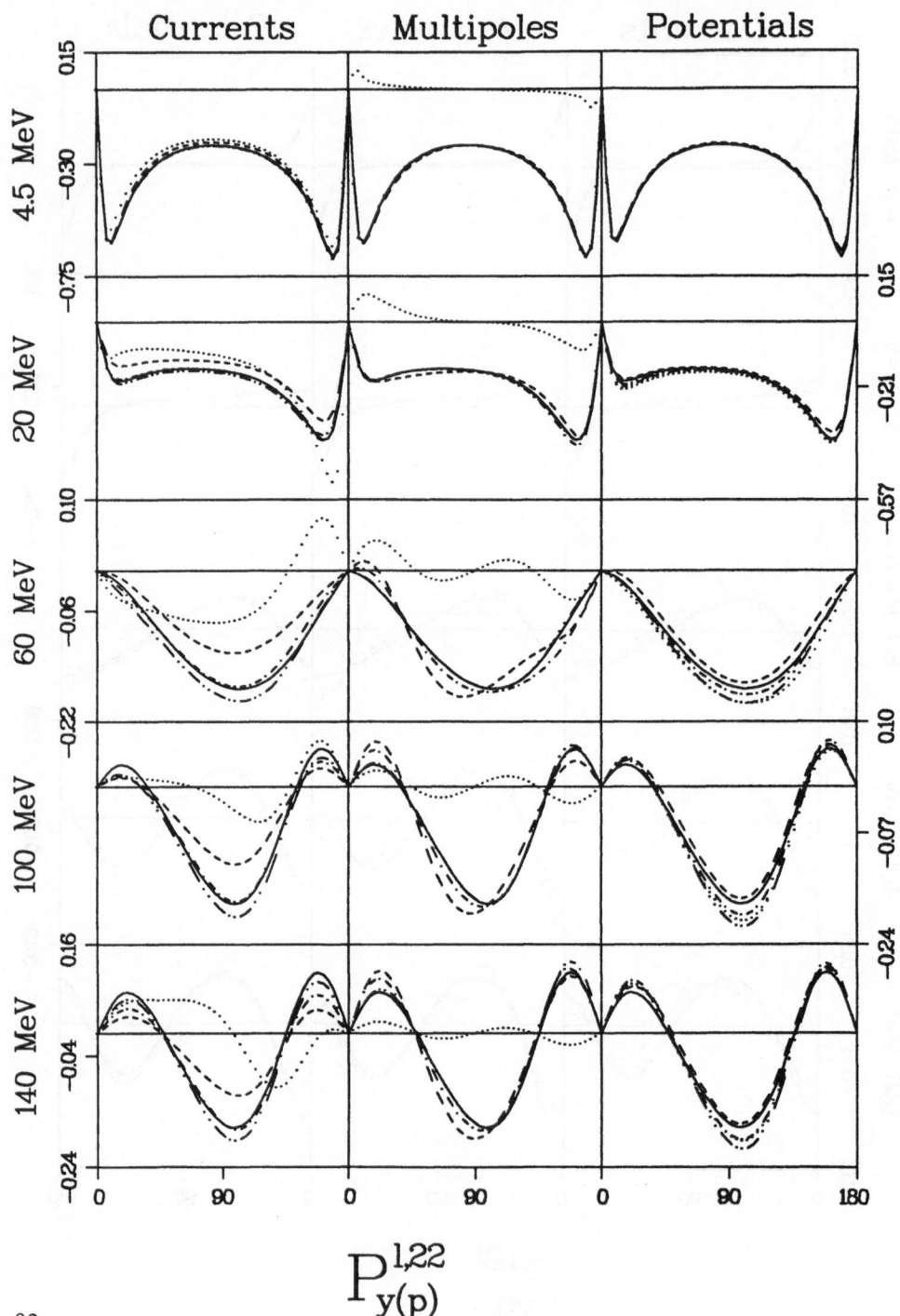

$$P_{y(p)}^{1,22}$$

5.14 Neutron Polarization $P_y^{0,IM}(n)$

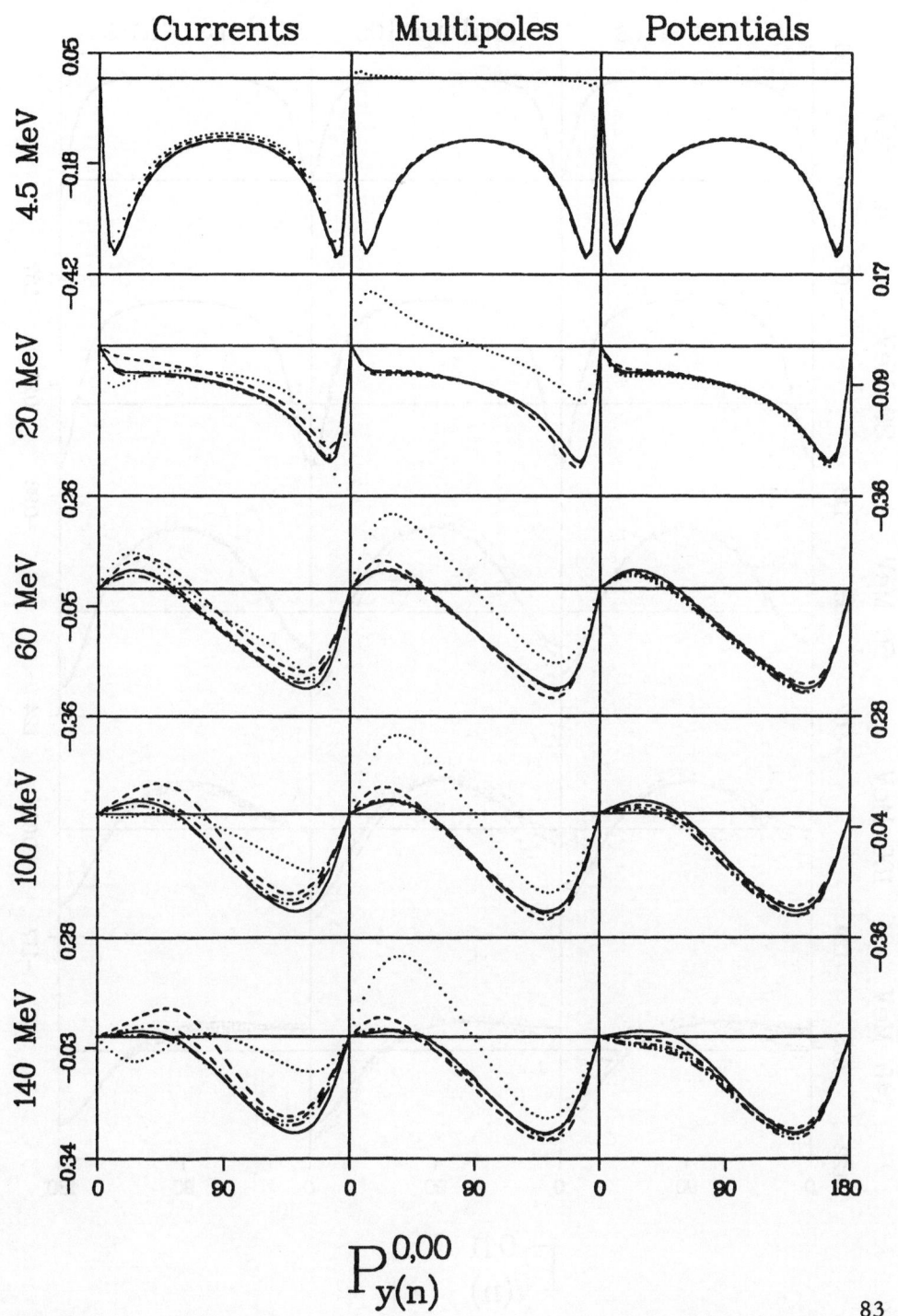

$$P_{y(n)}^{0,00}$$

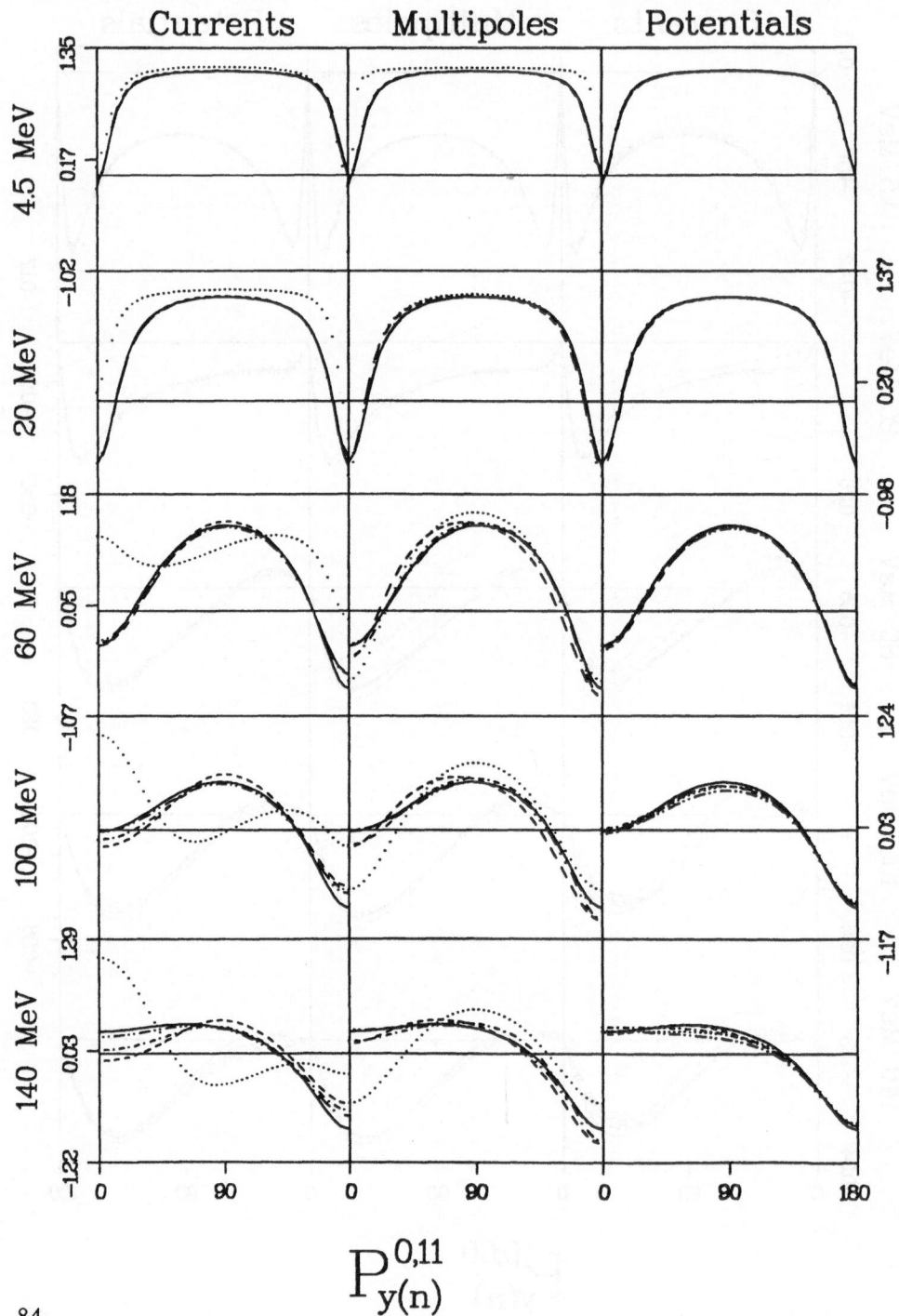

$$P_{y(n)}^{0,11}$$

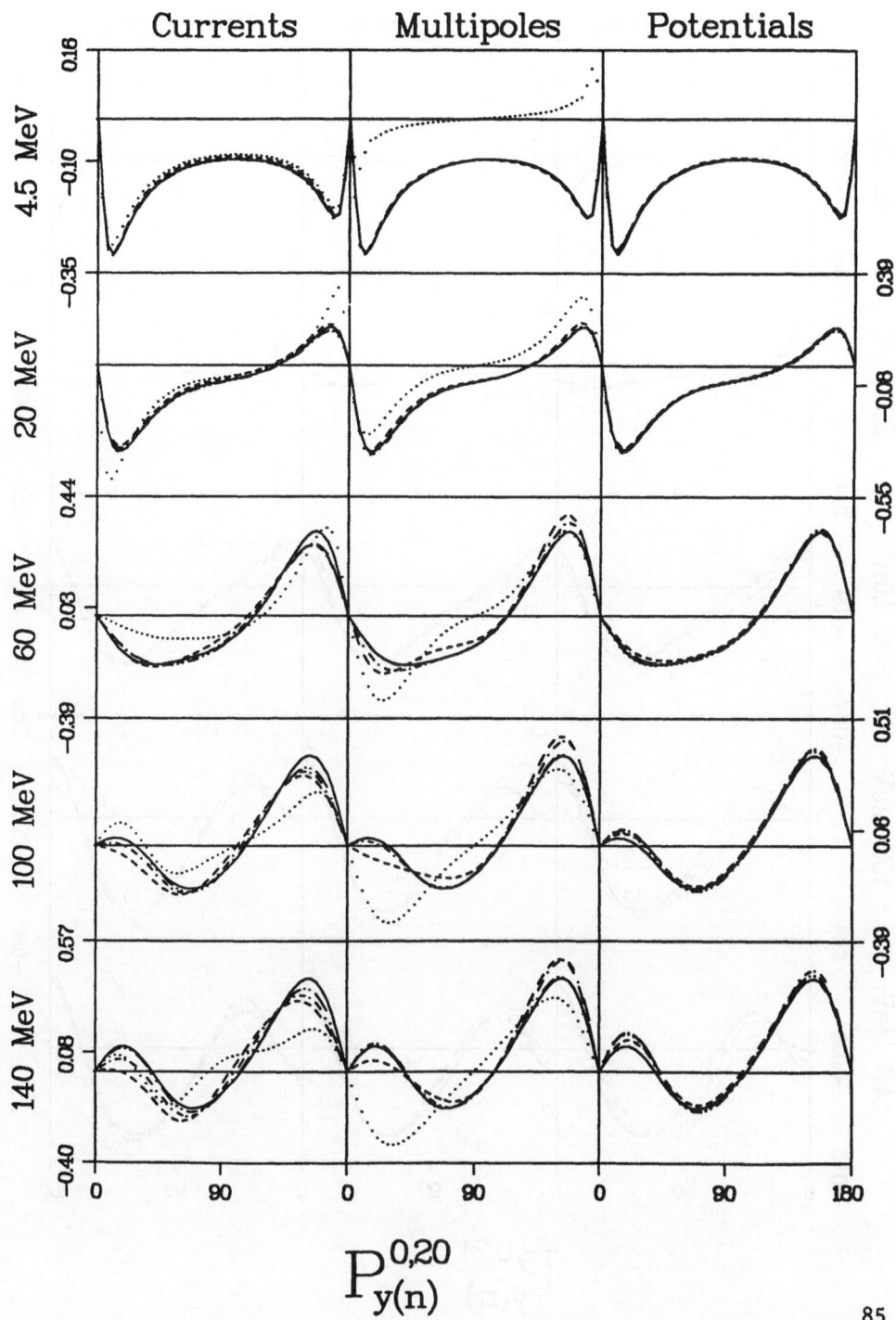

$$P_{y(n)}^{0,20}$$

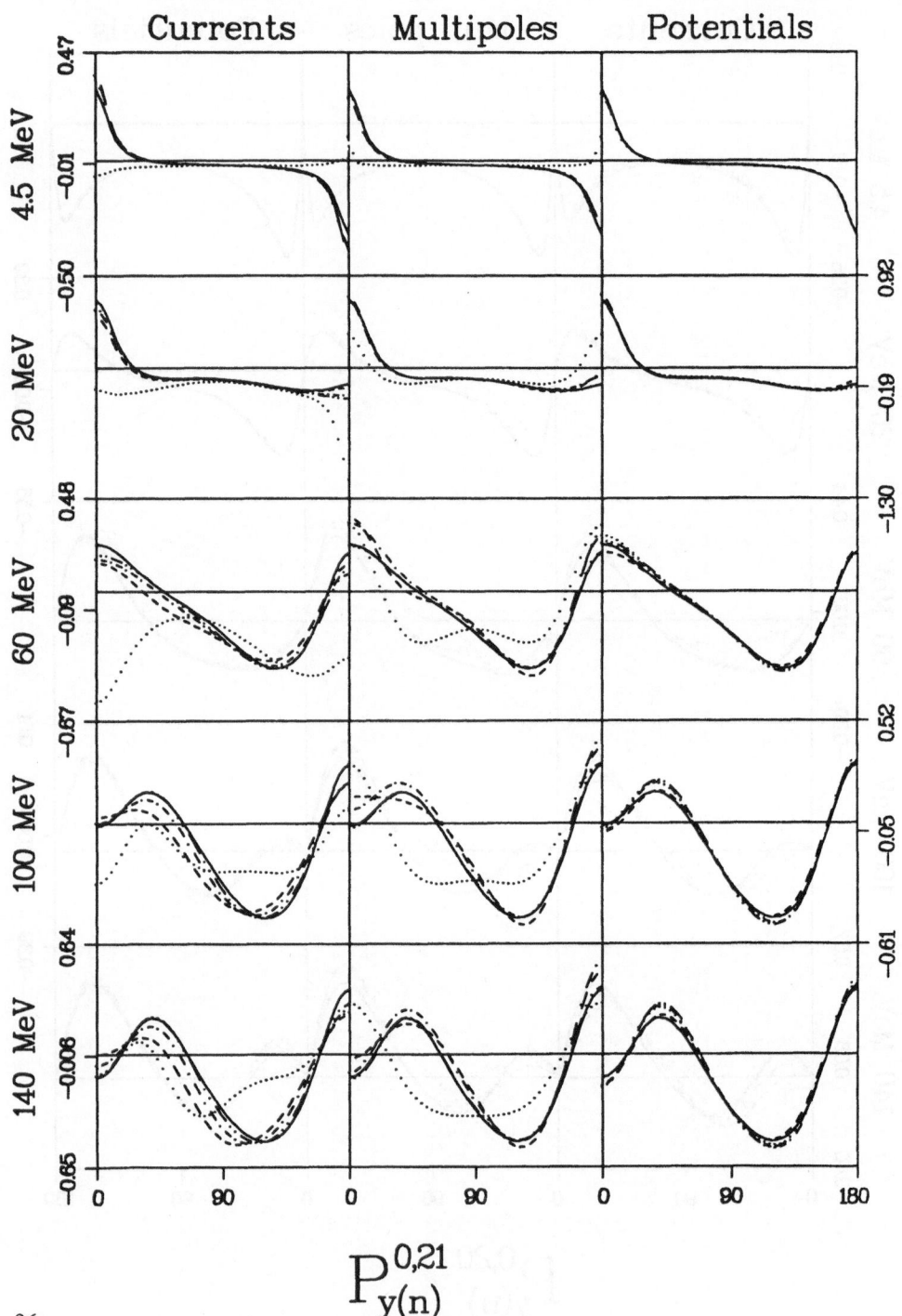

$$P_{y(n)}^{0,21}$$

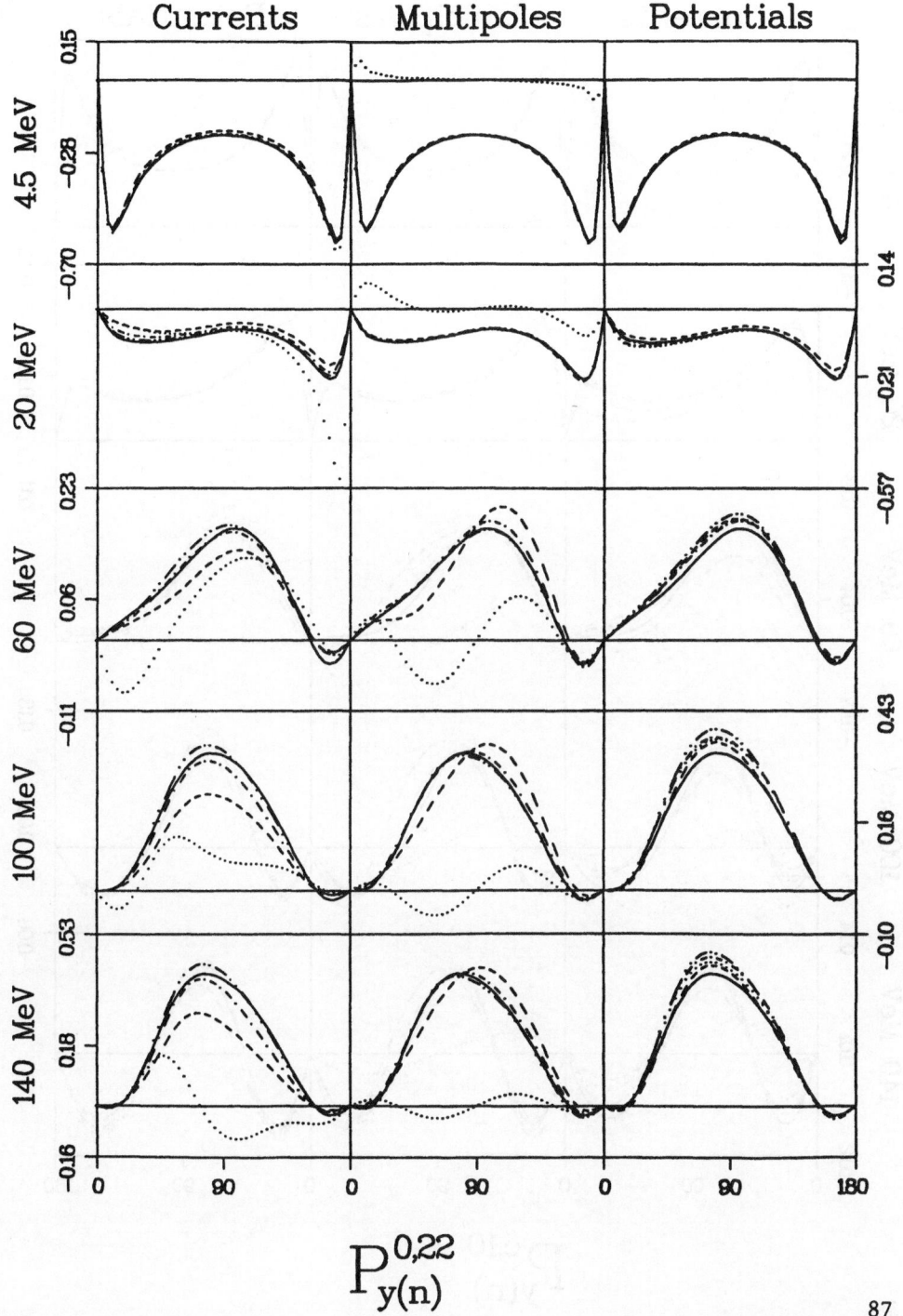

$$P_{y(n)}^{0,22}$$

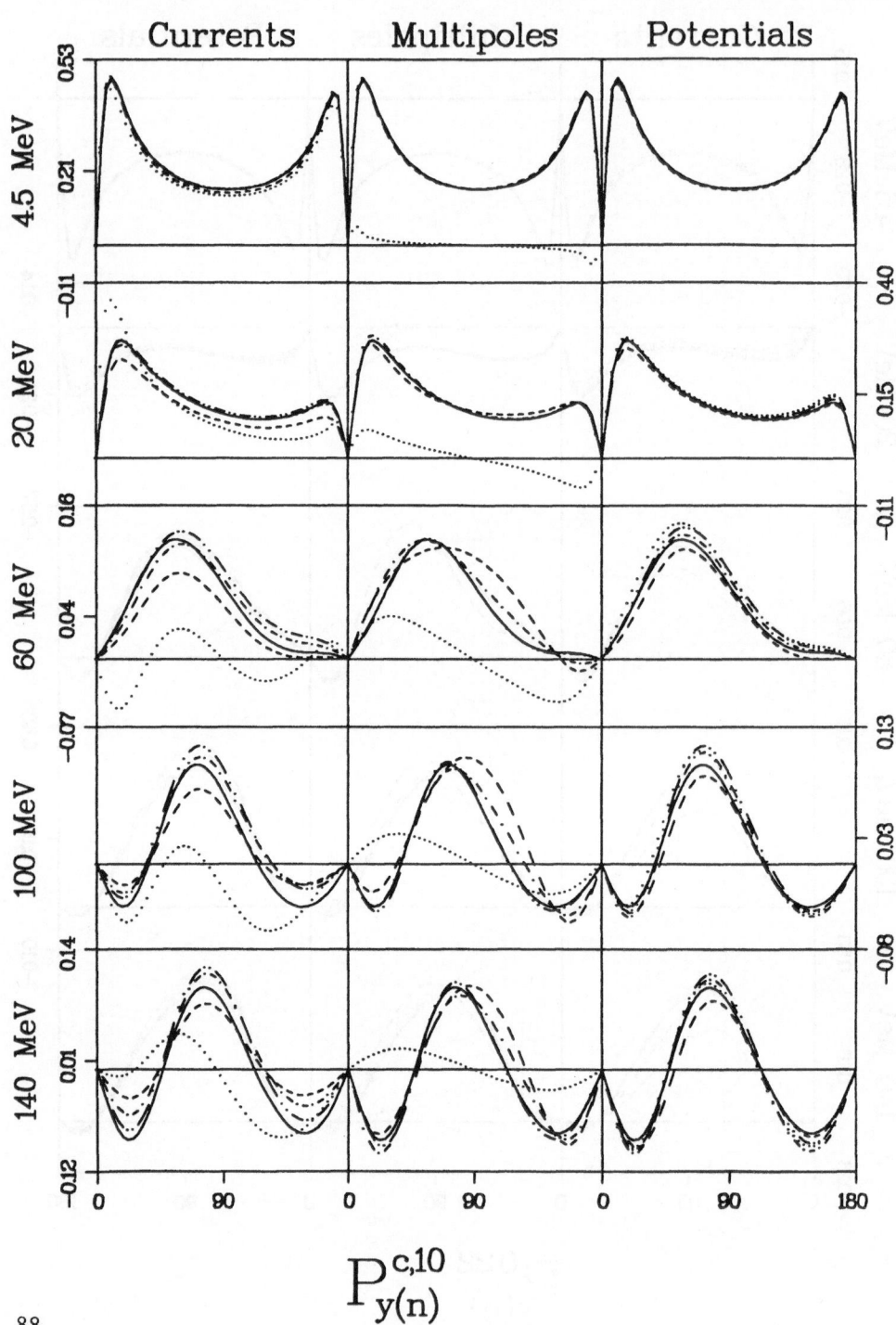

$$P_{y(n)}^{c,10}$$

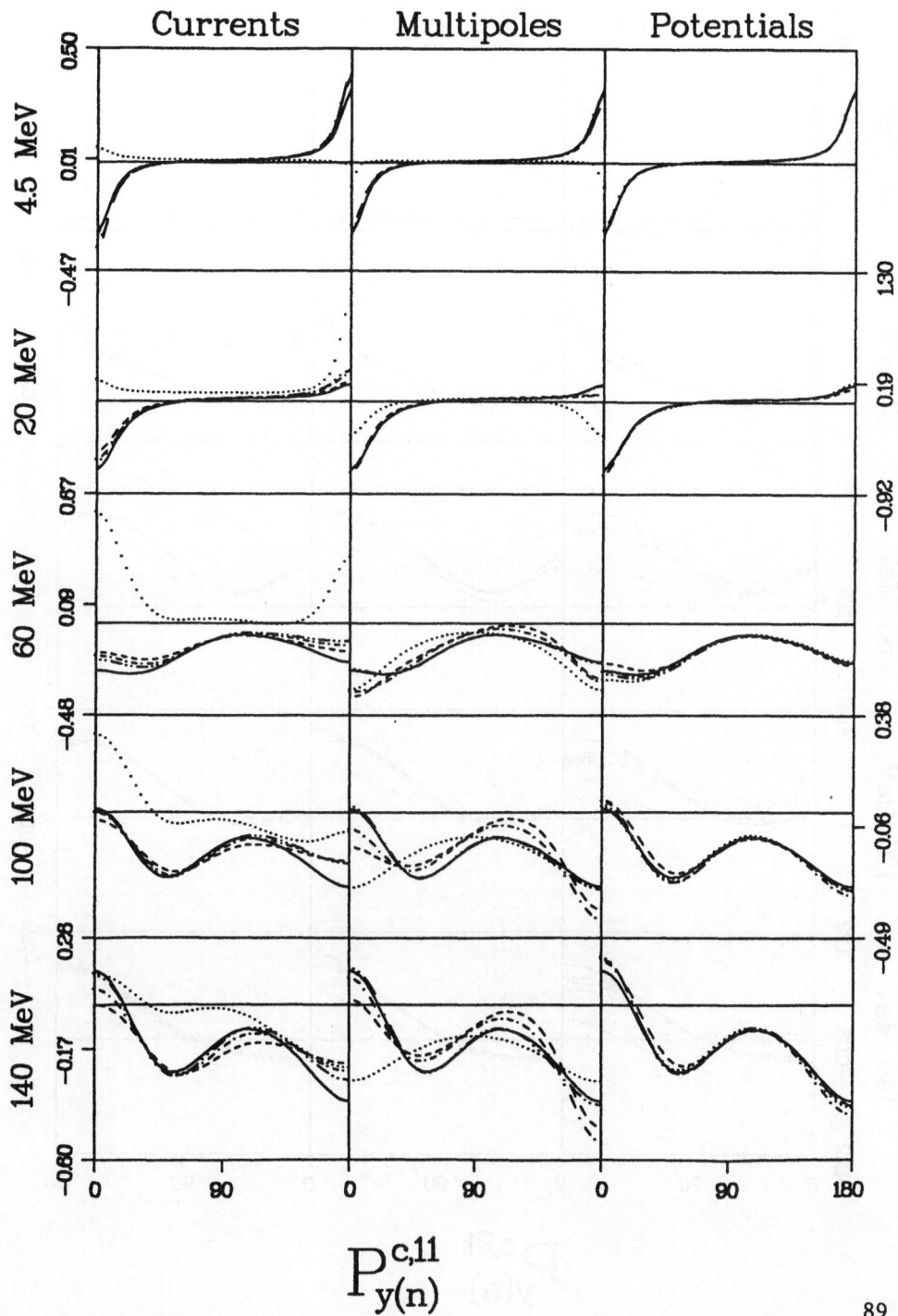

$$P_{y(n)}^{c,11}$$

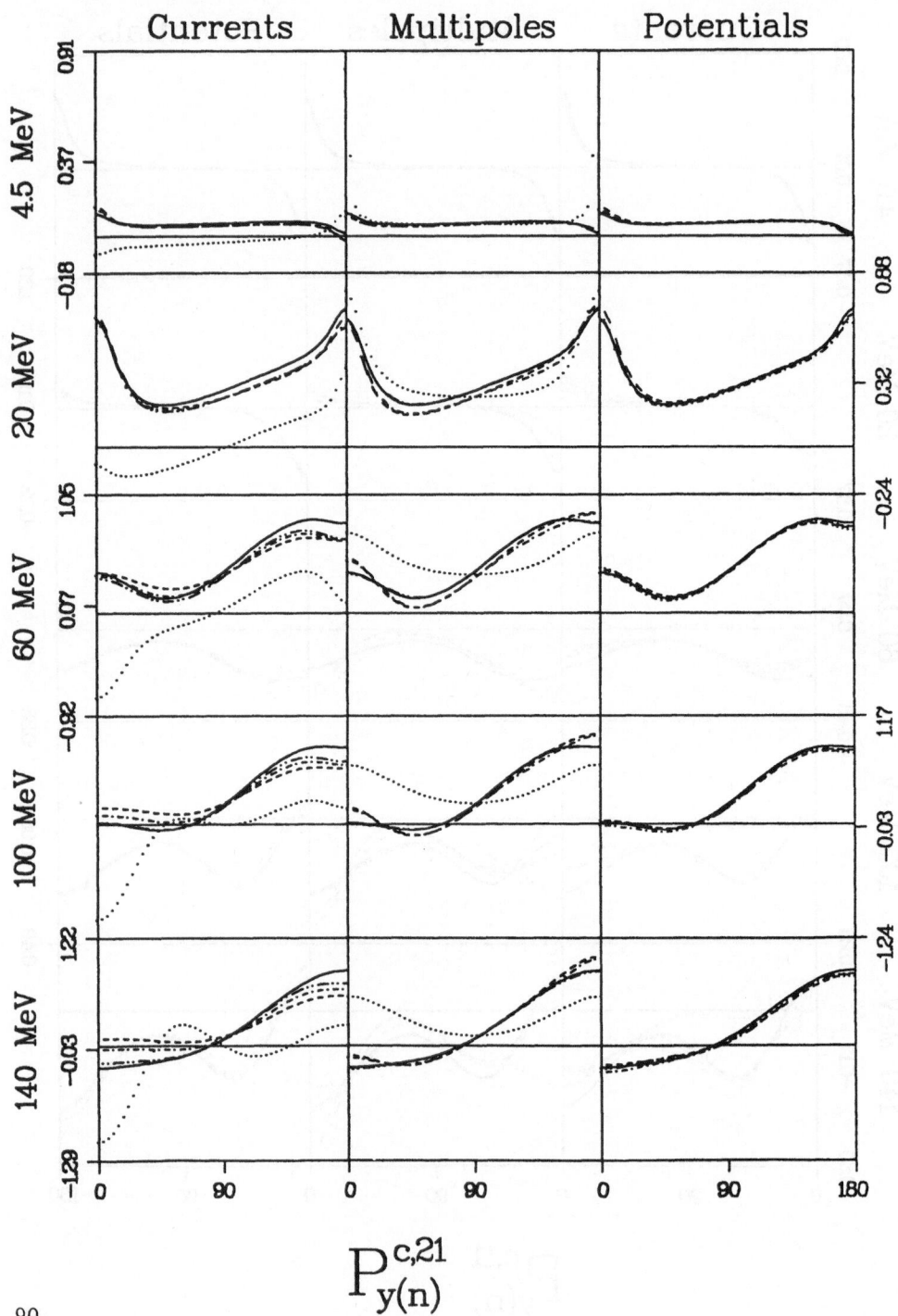

$$\text{P}_{\text{y(n)}}^{\text{c,21}}$$

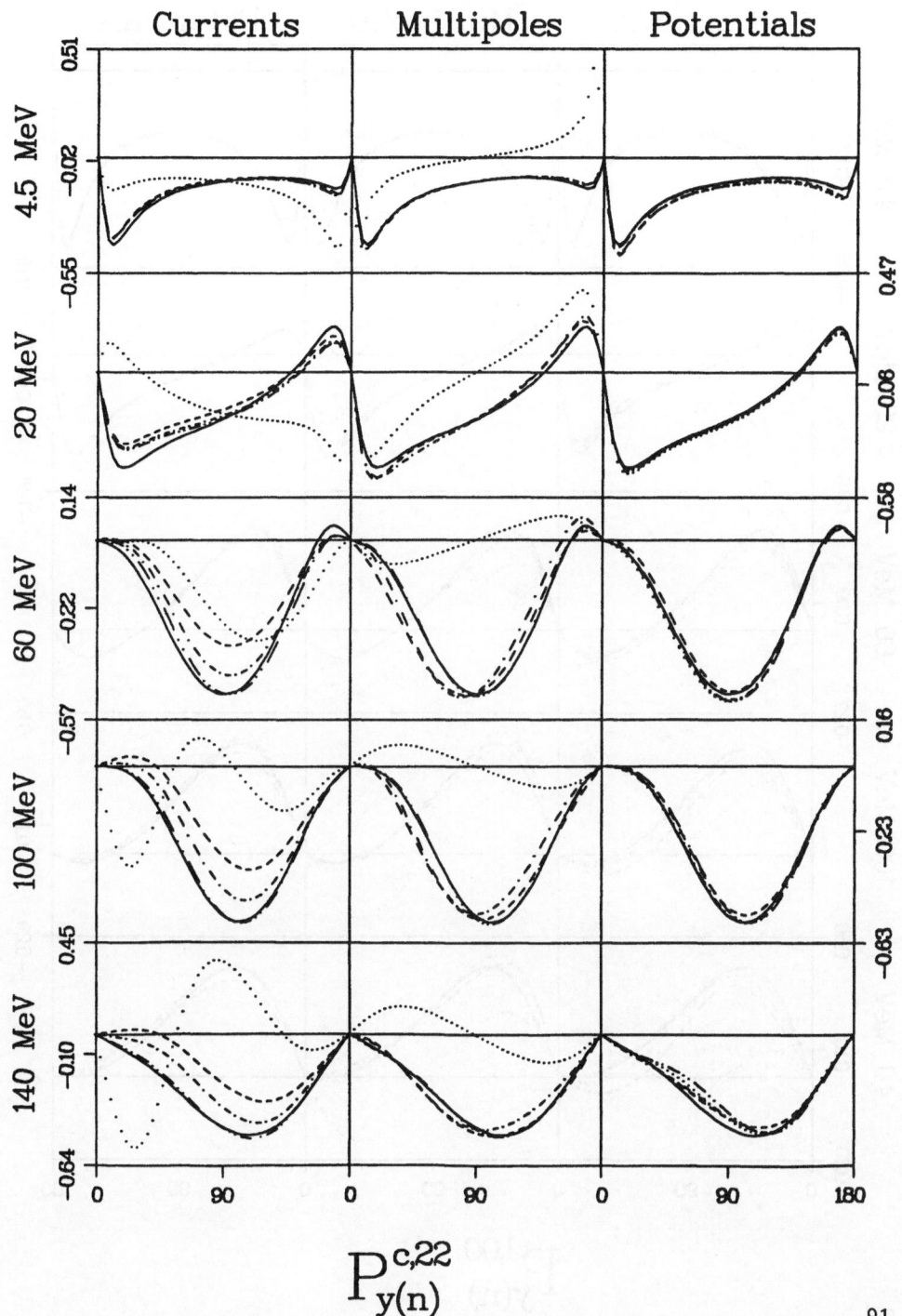

$$P_{y(n)}^{c,22}$$

5.16 Neutron Polarization $P_y^{l,IM}(n)$

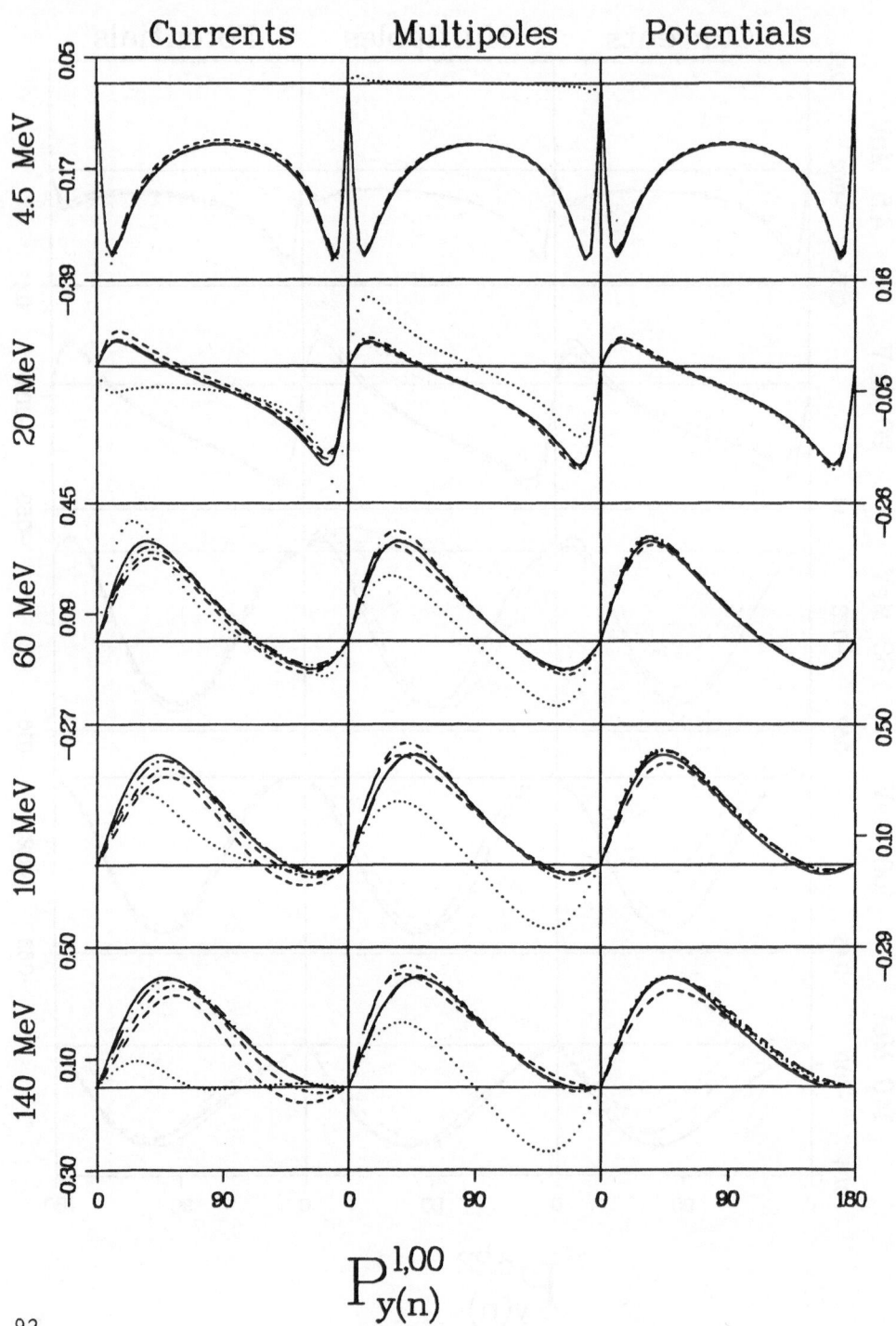

$$P_{y(n)}^{1,00}$$

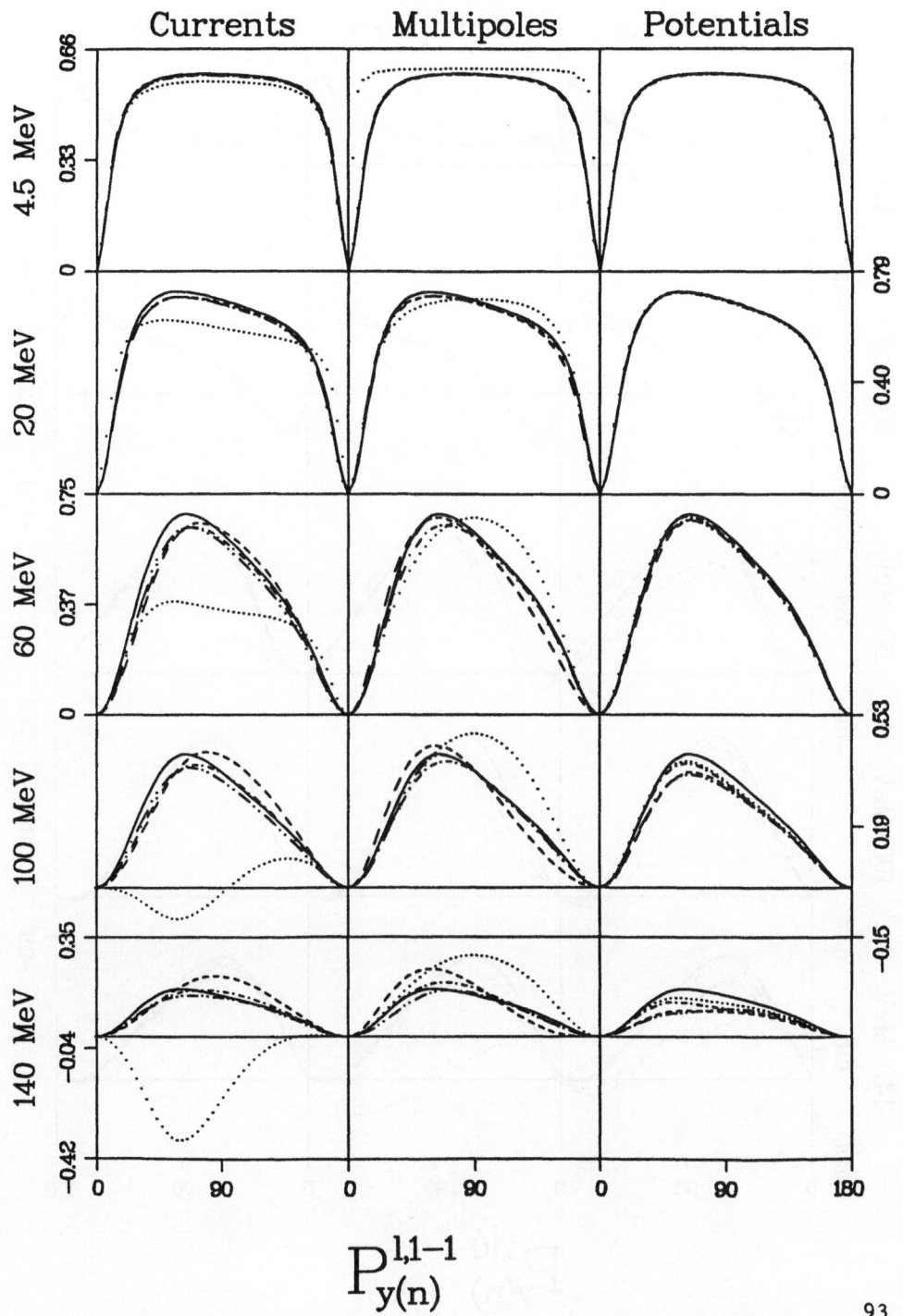

$$P_{y(n)}^{1,1-1}$$

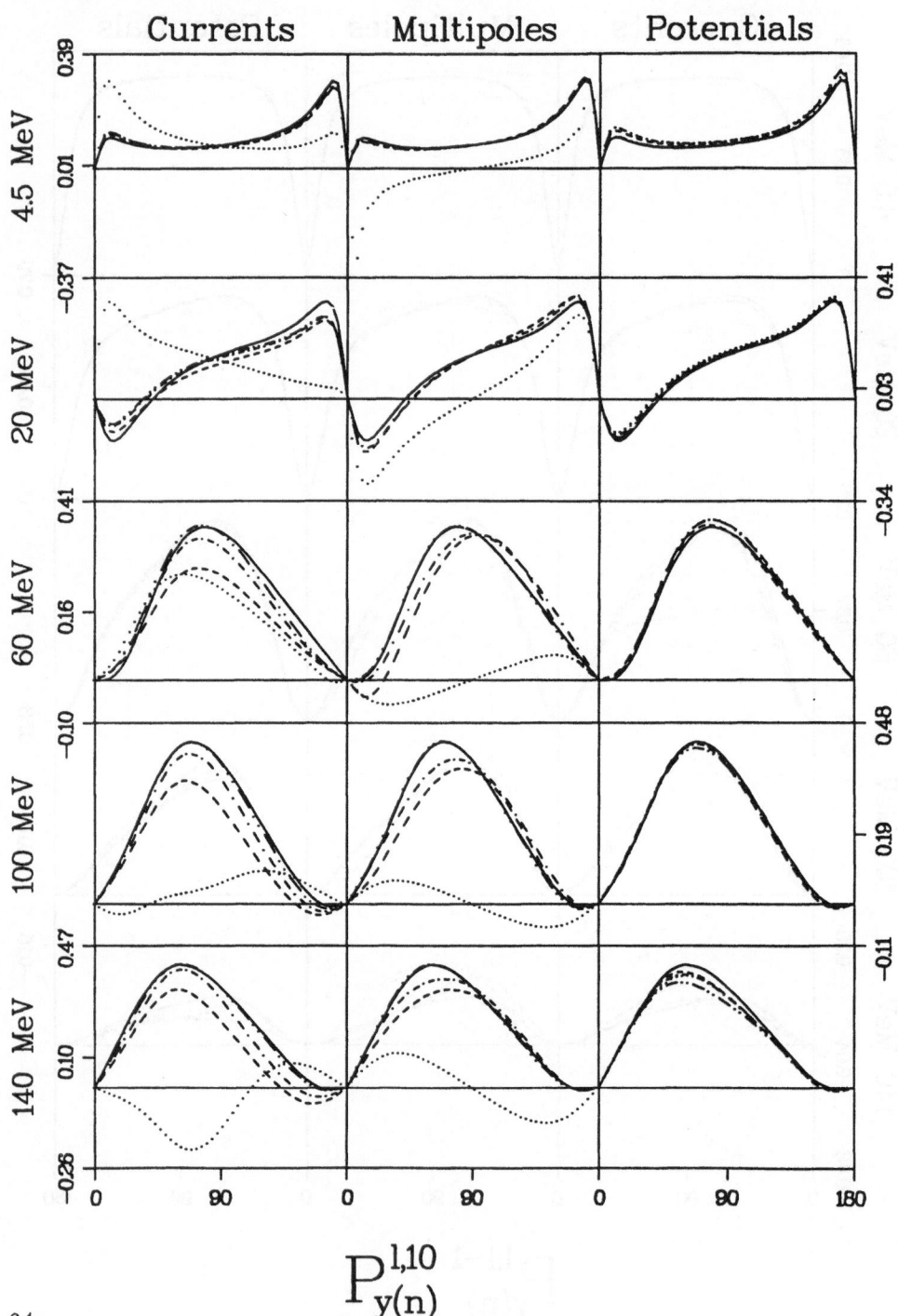

$$P_{y(n)}^{l,10}$$

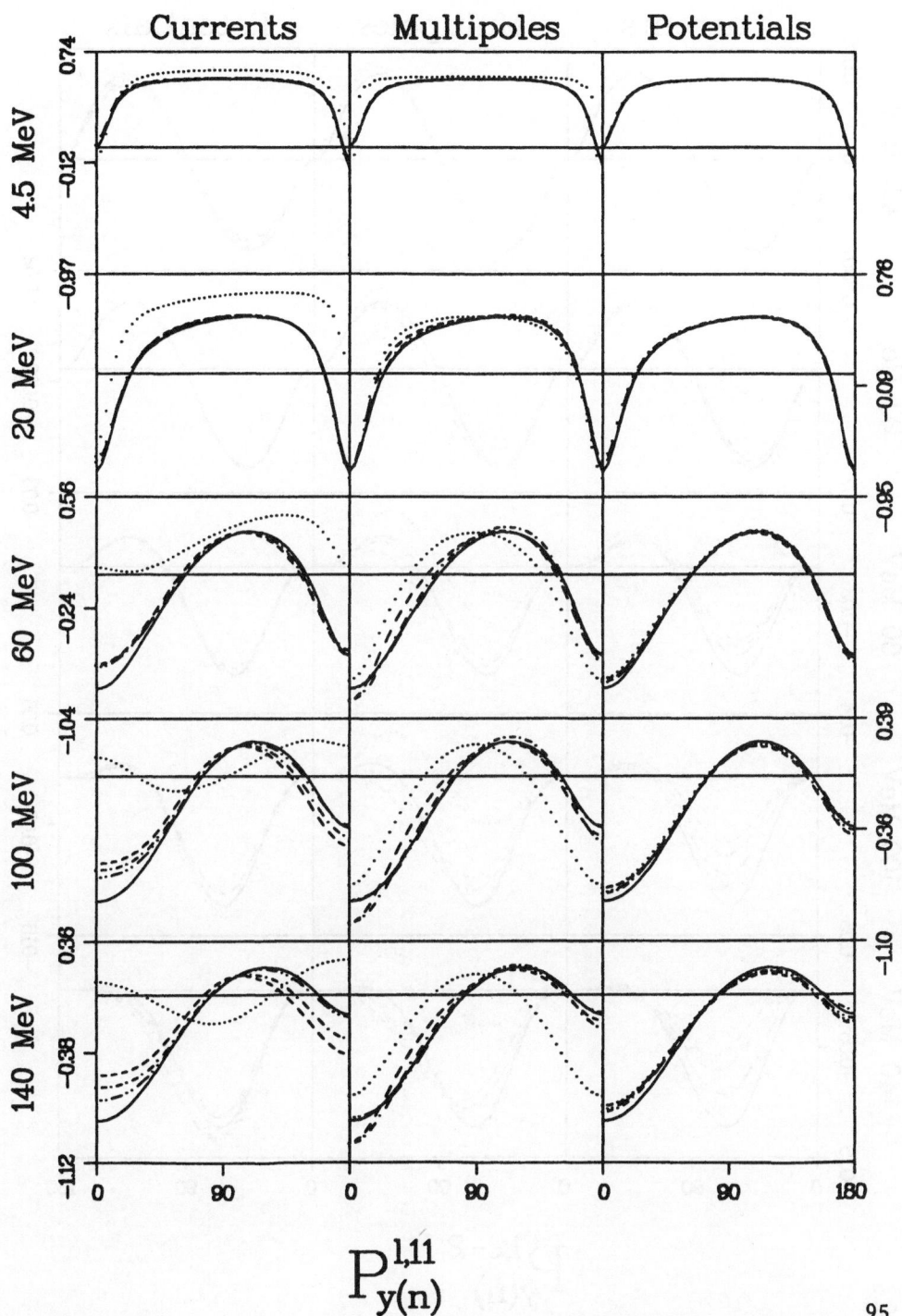

Currents   Multipoles   Potentials

$$P_{y(n)}^{l,11}$$

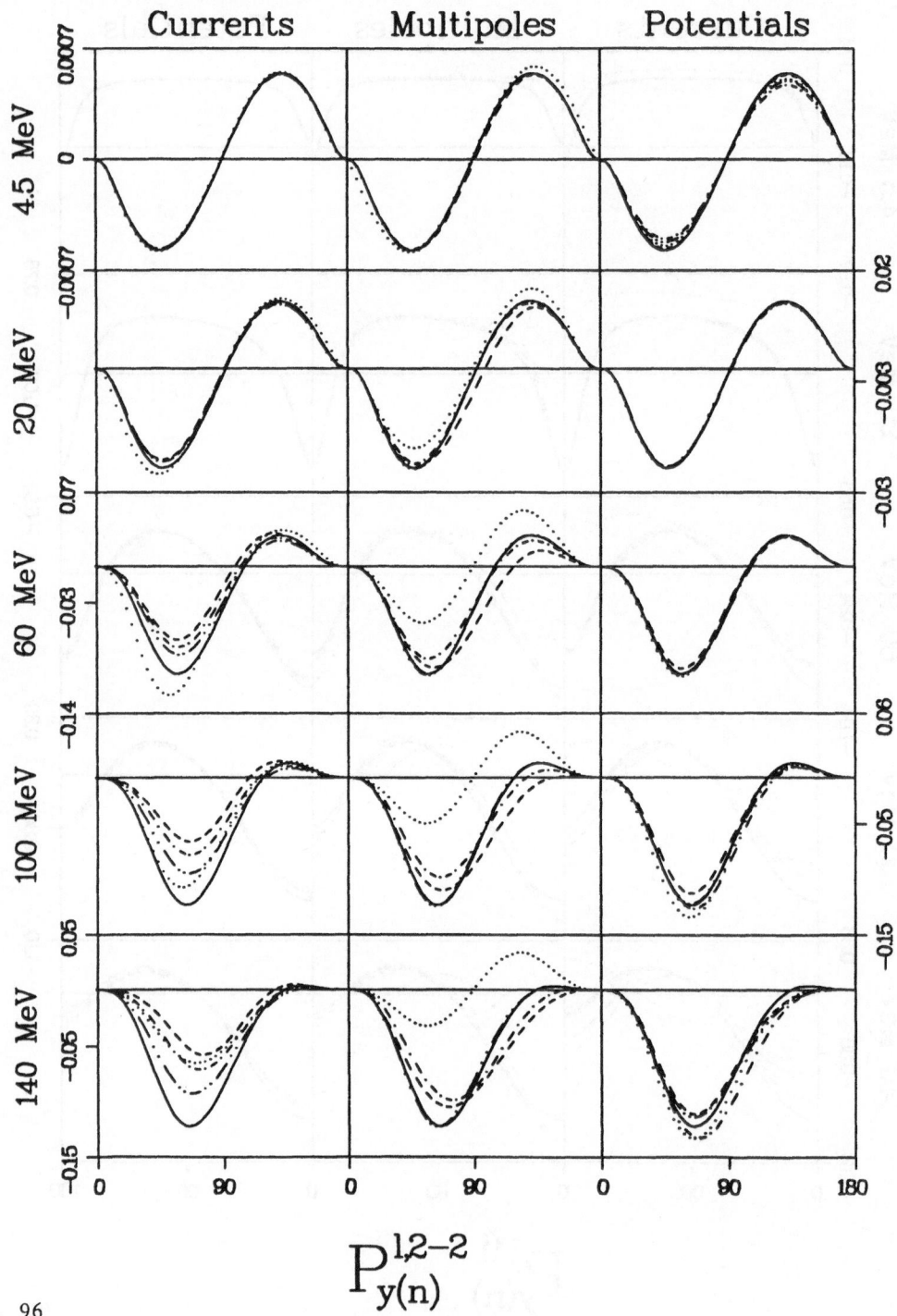

Currents   Multipoles   Potentials

4.5 MeV

20 MeV

60 MeV

100 MeV

140 MeV

$$P_{y(n)}^{1,2-2}$$

Neutron Polarization $P_y^{l,IM}(n)$

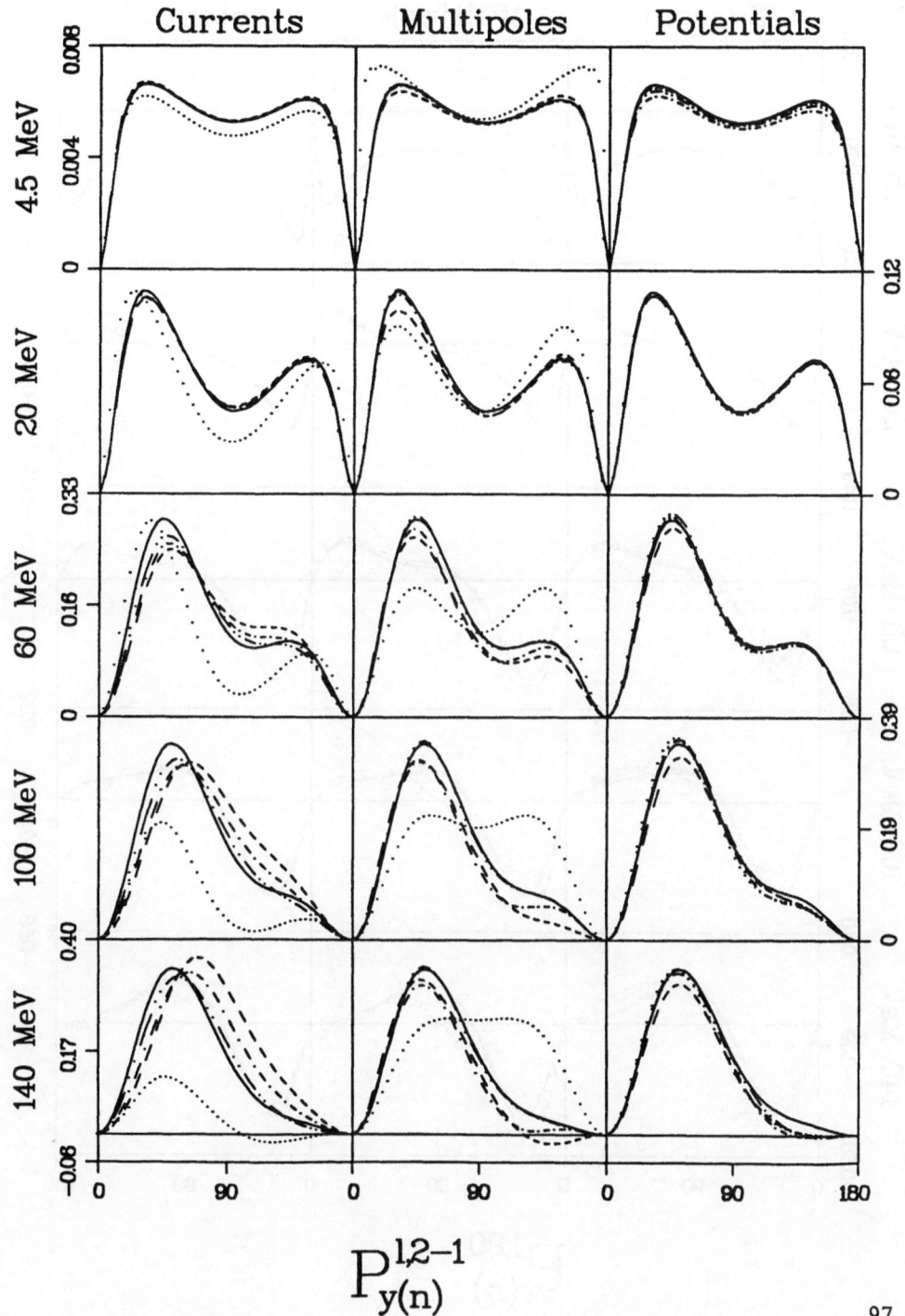

$$P_{y(n)}^{1,2-1}$$

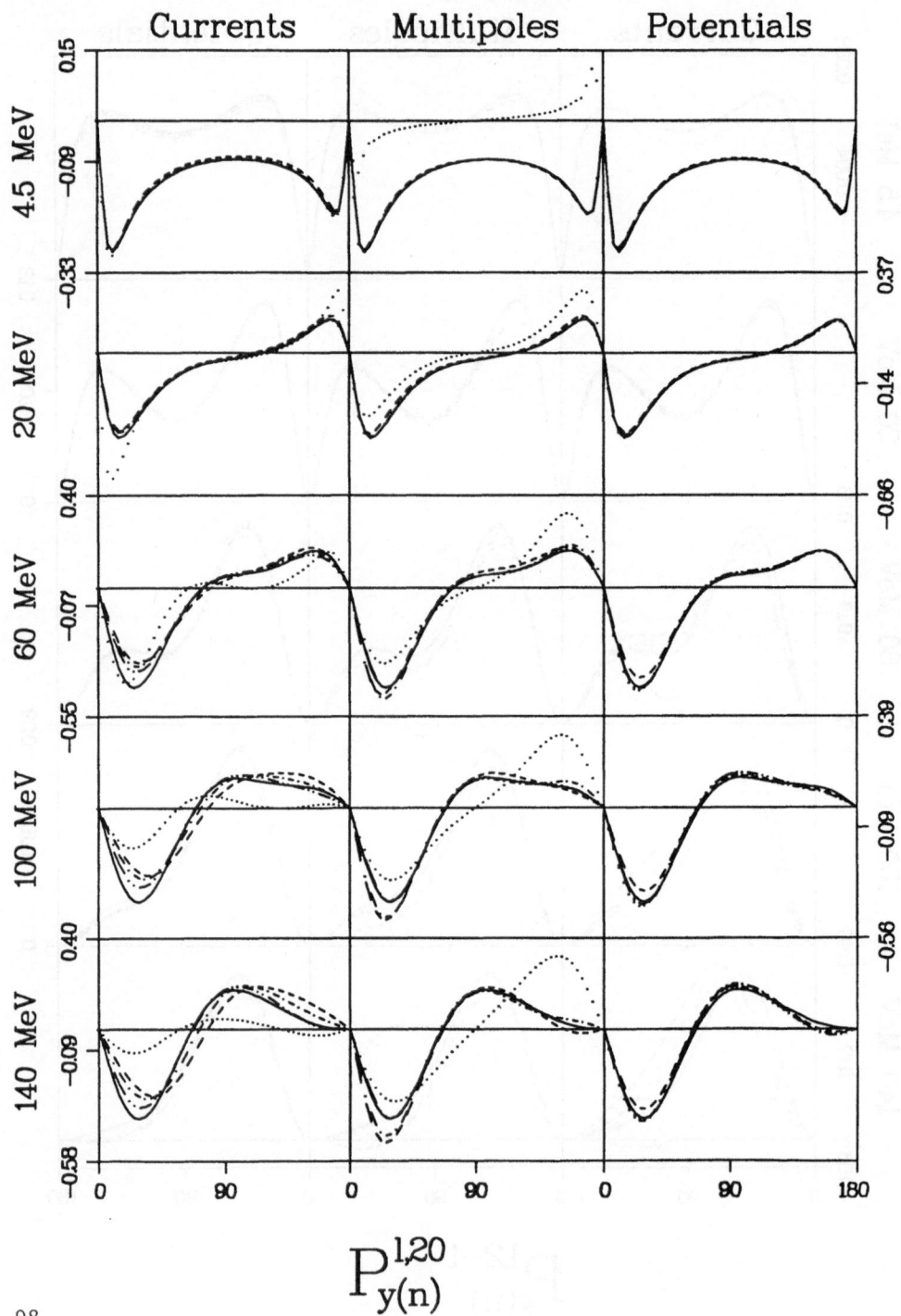

$$P_{y(n)}^{1,20}$$

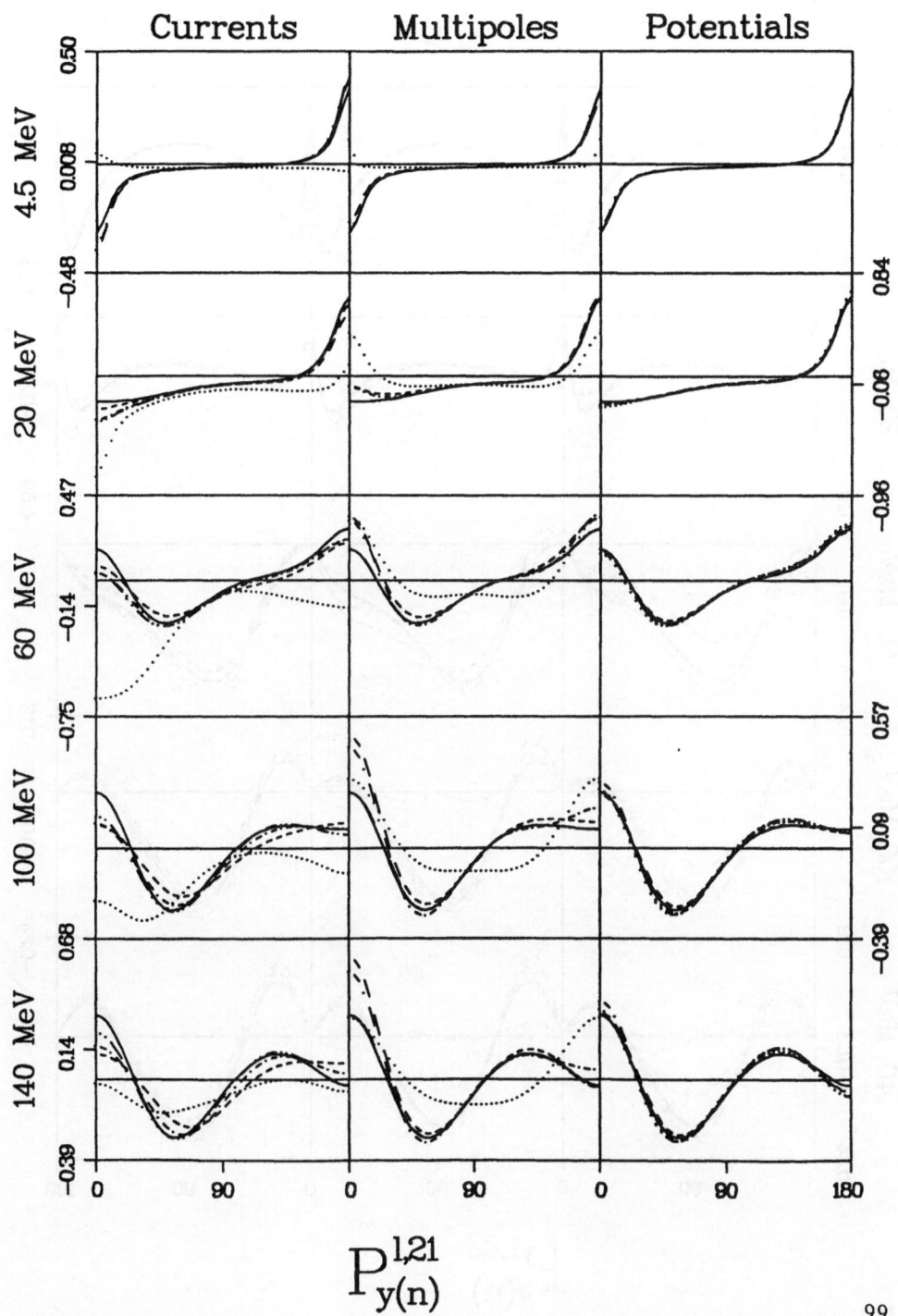

$$P_{y(n)}^{1,21}$$

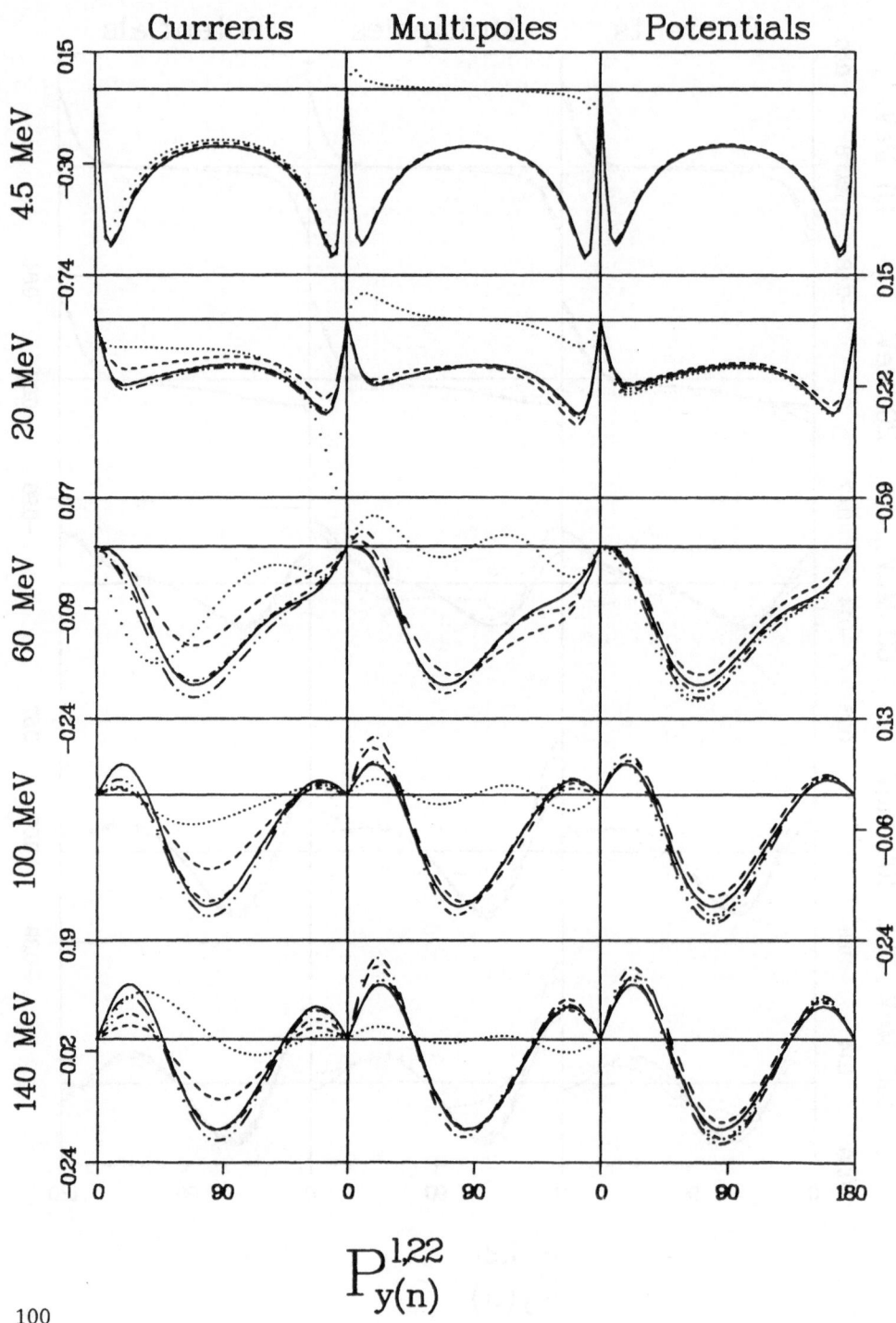

$$P_{y(n)}^{1,22}$$

5.17 Proton Polarization $P_z^{0,IM}(p)$

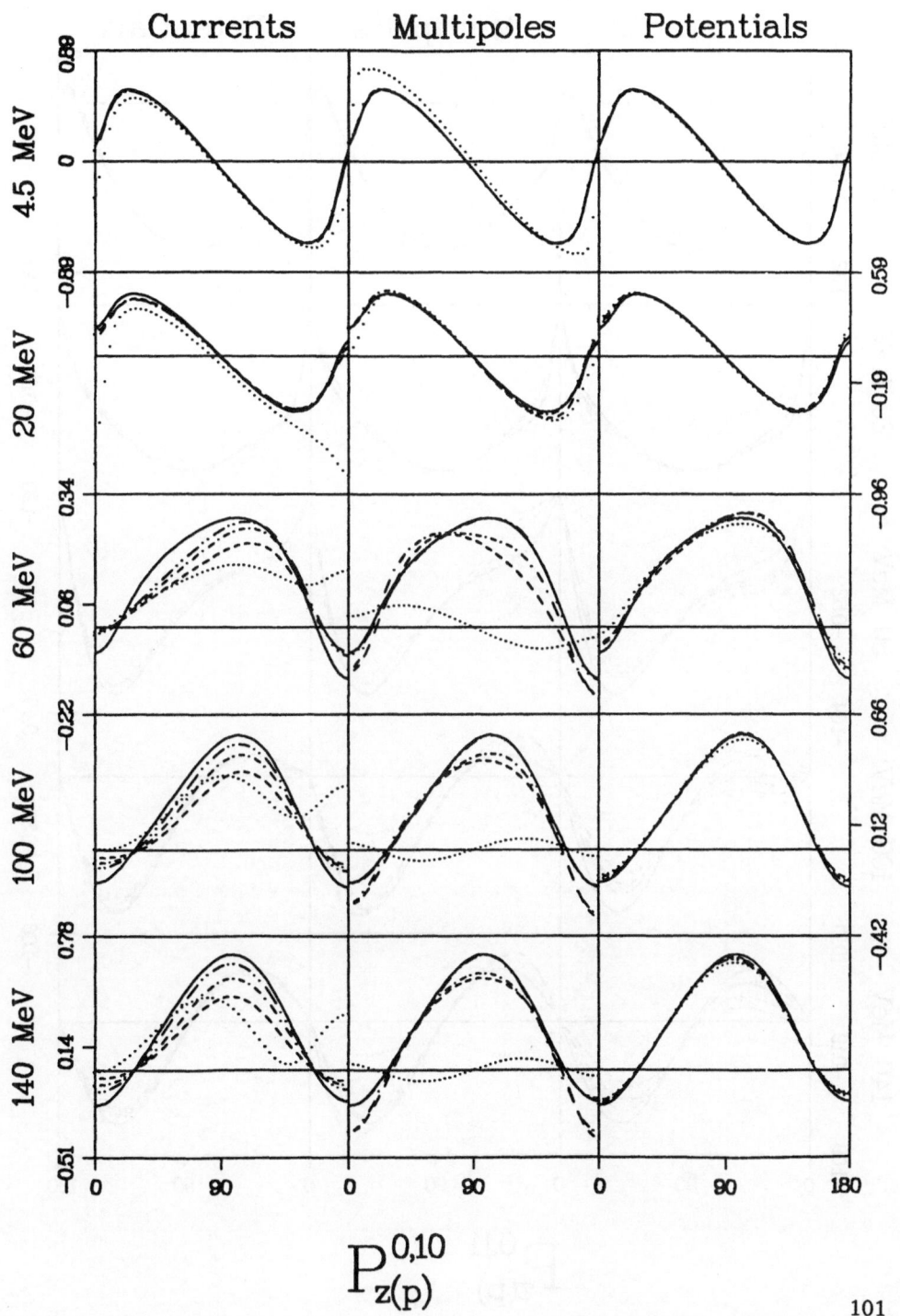

$$P_{z(p)}^{0,10}$$

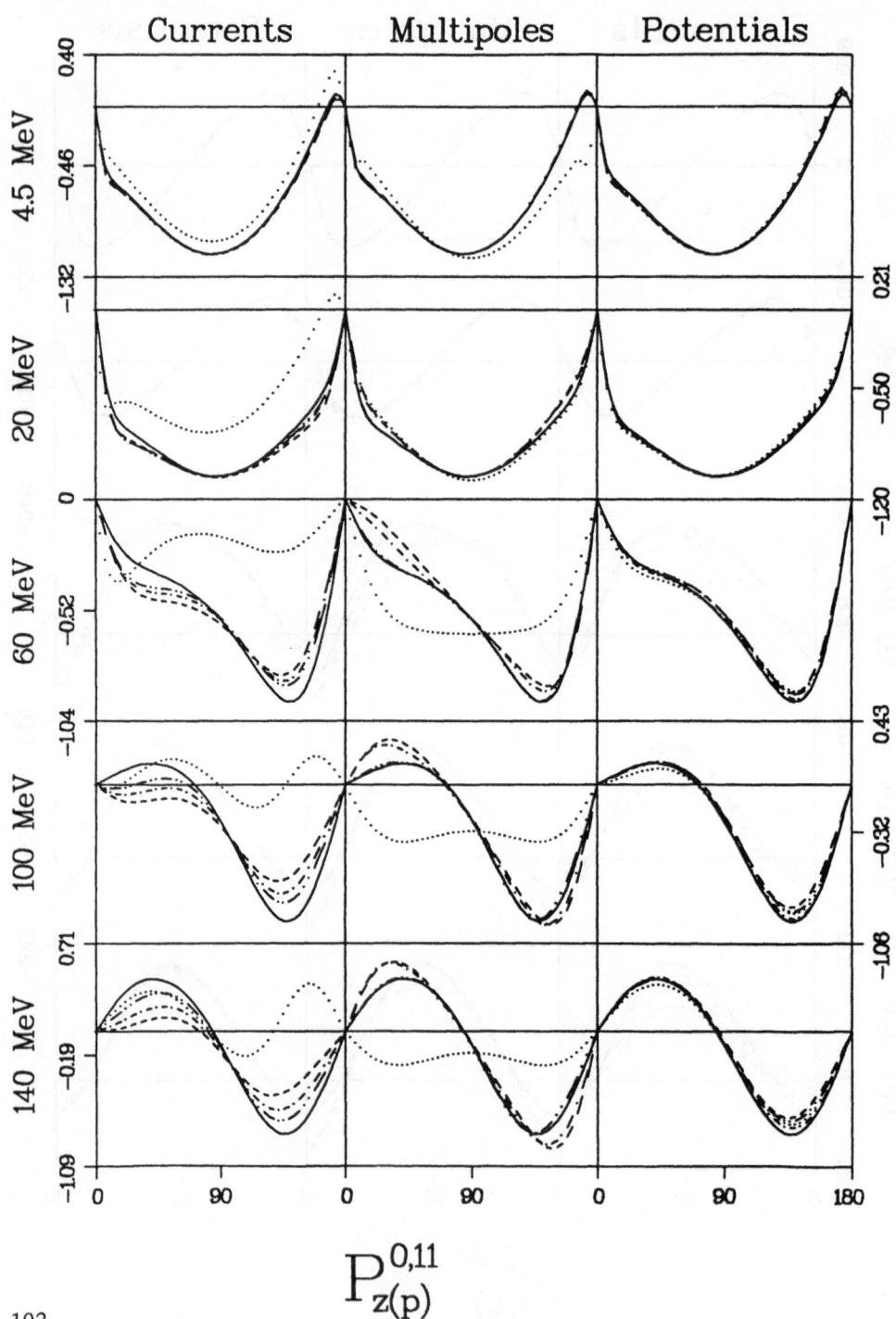

$$P_{z(p)}^{0,11}$$

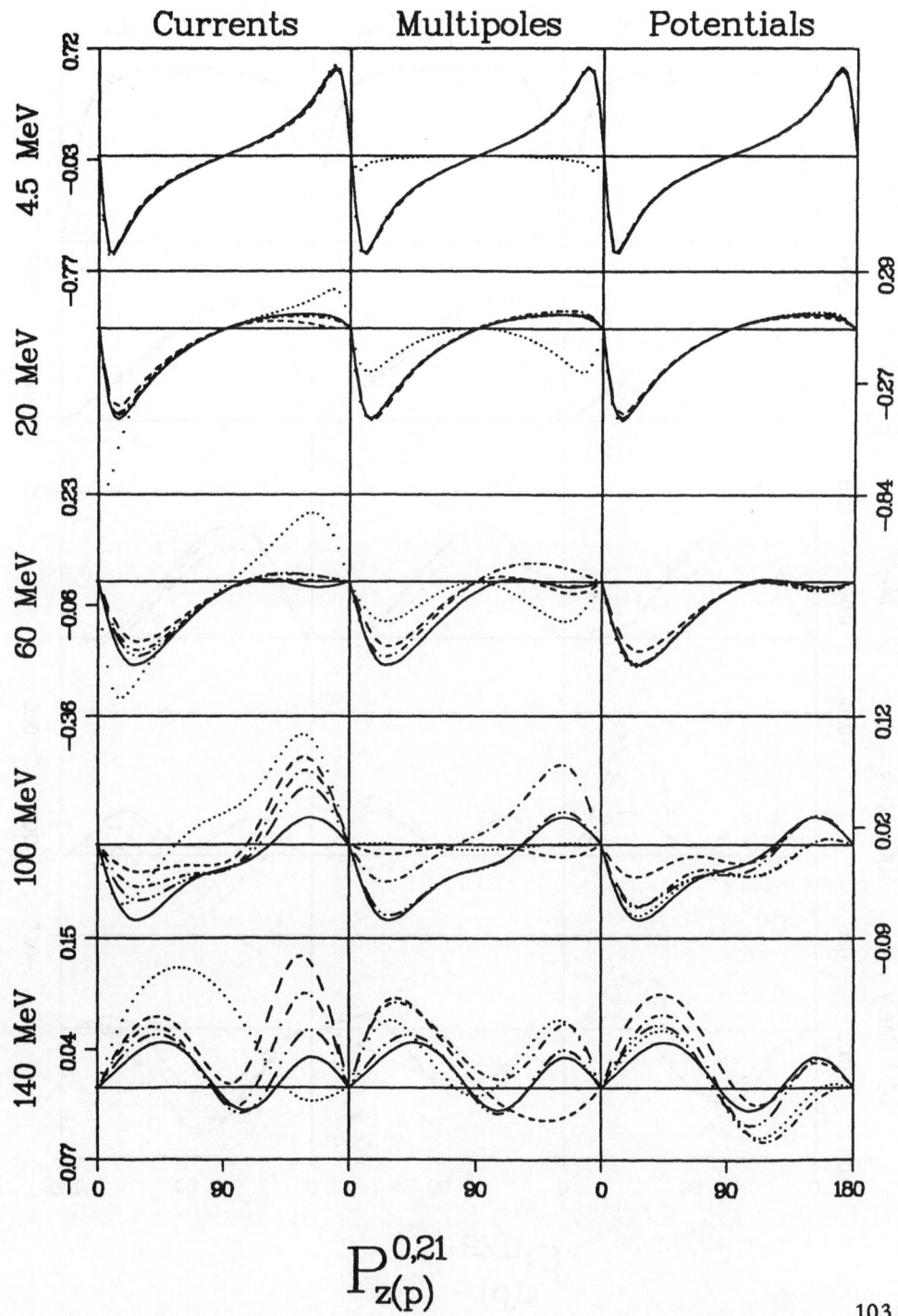

$$P_{z(p)}^{0,21}$$

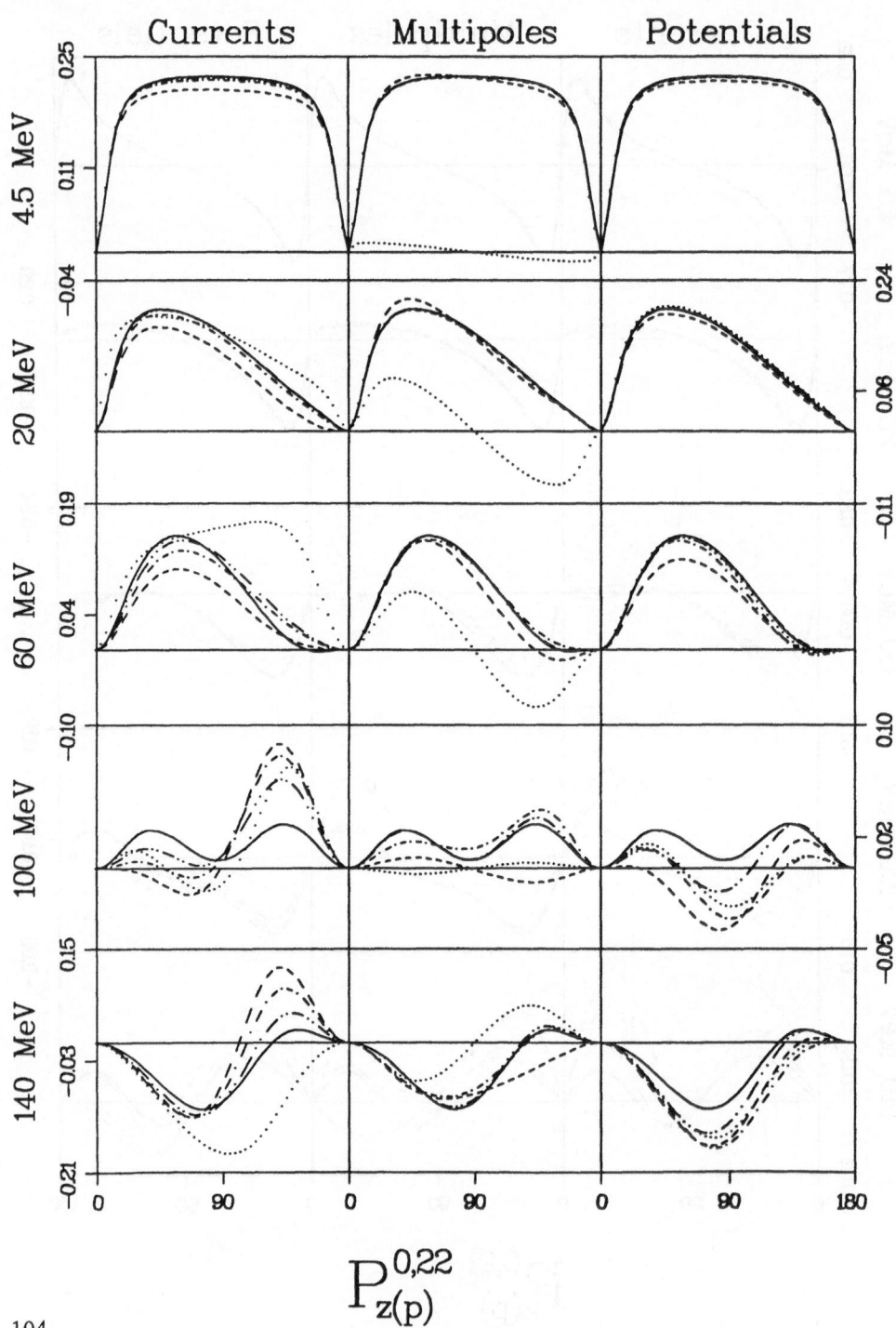

$$P_{z(p)}^{0,22}$$

5.18 Proton Polarization $P_z^{c,IM}(p)$

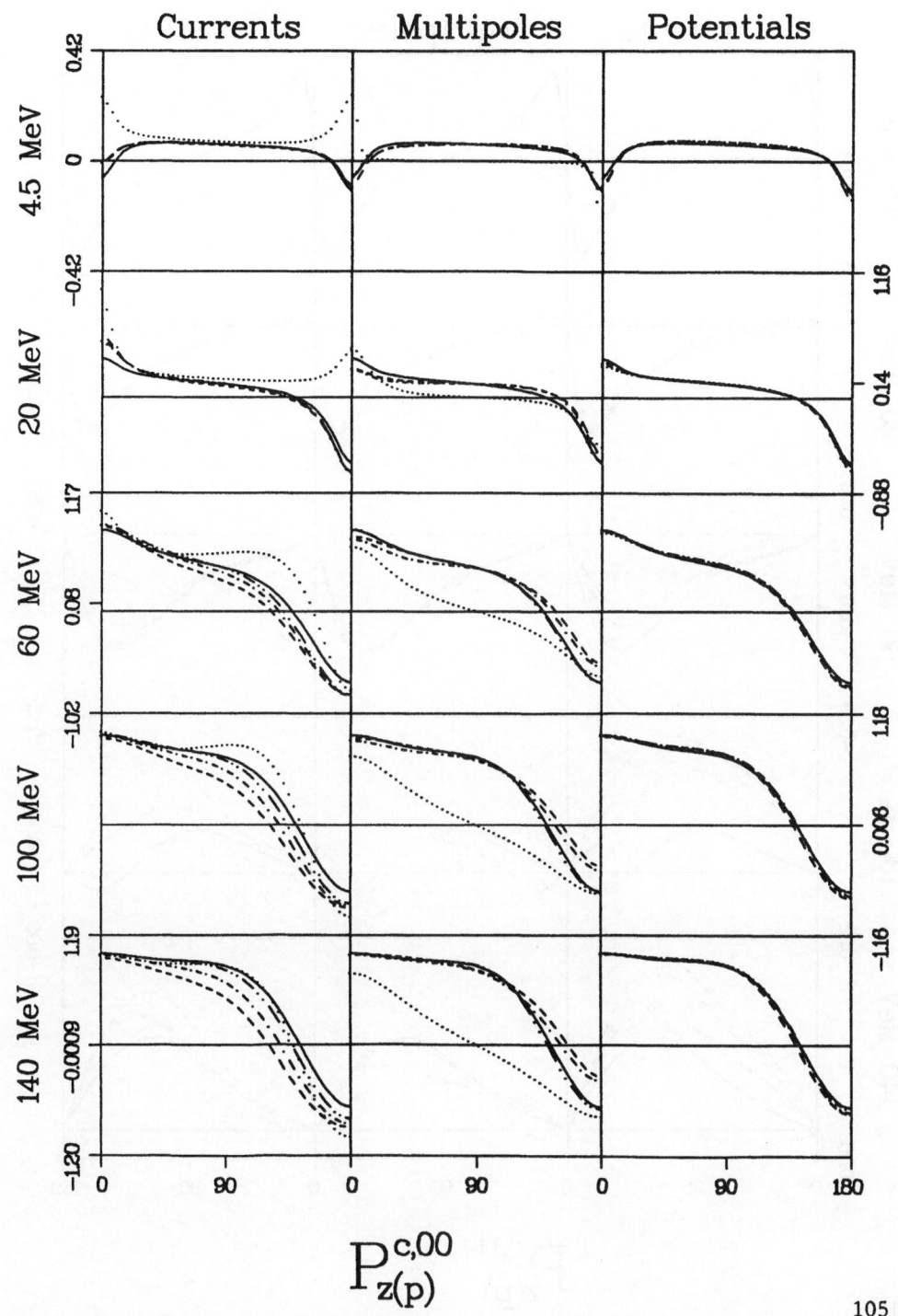

$$P_{z(p)}^{c,00}$$

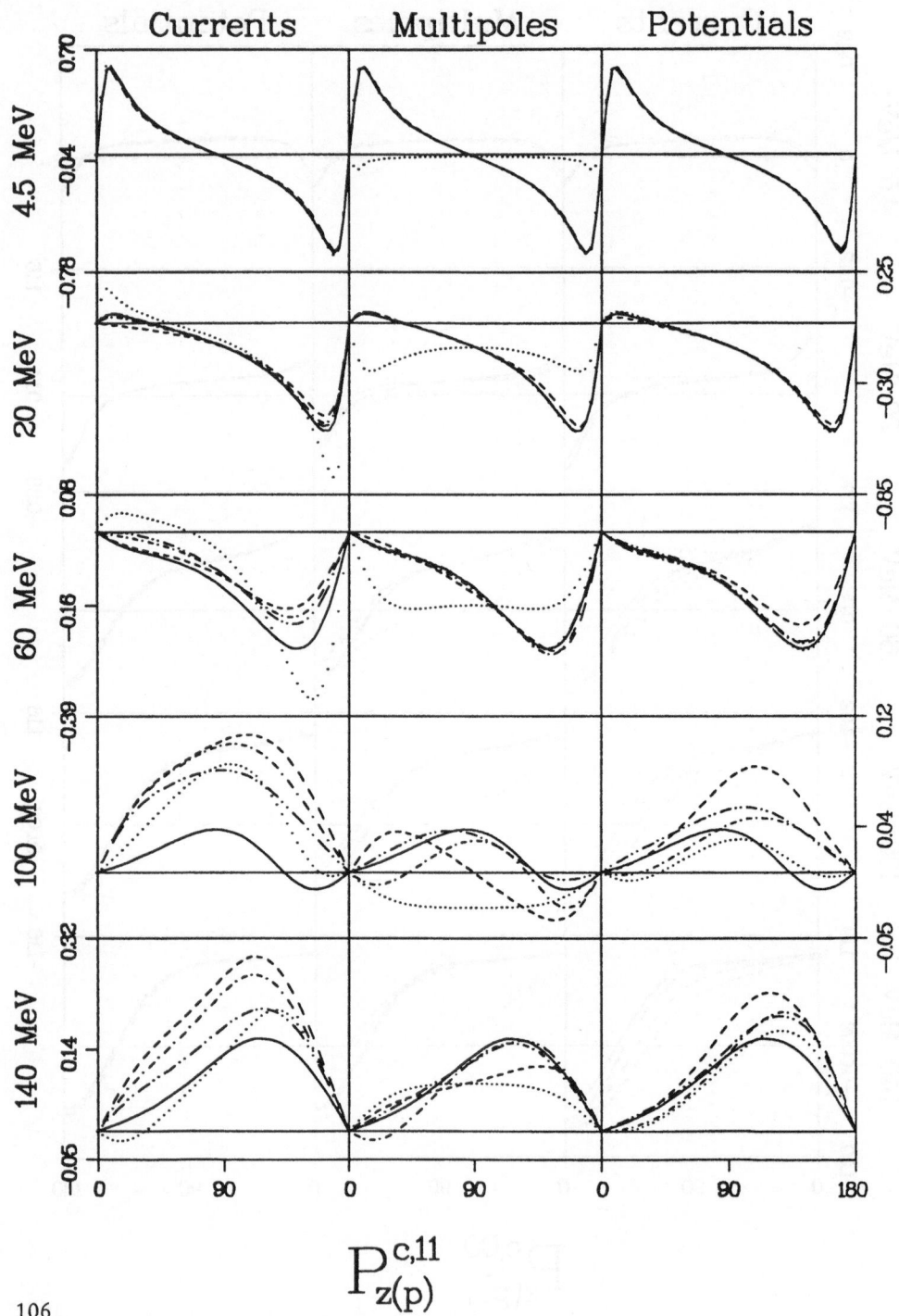

$$P_{z(p)}^{c,11}$$

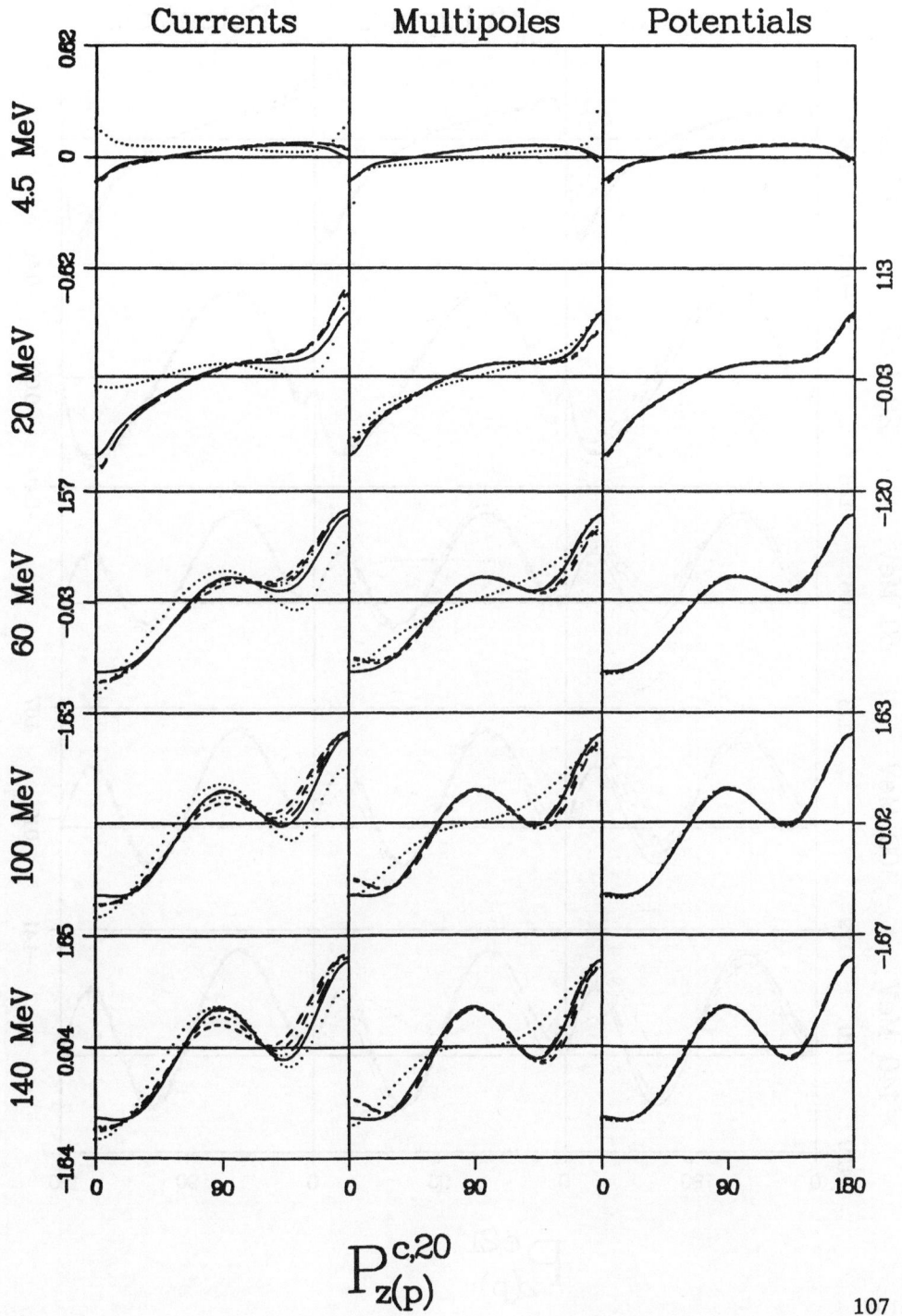

$$P_{z(p)}^{c,20}$$

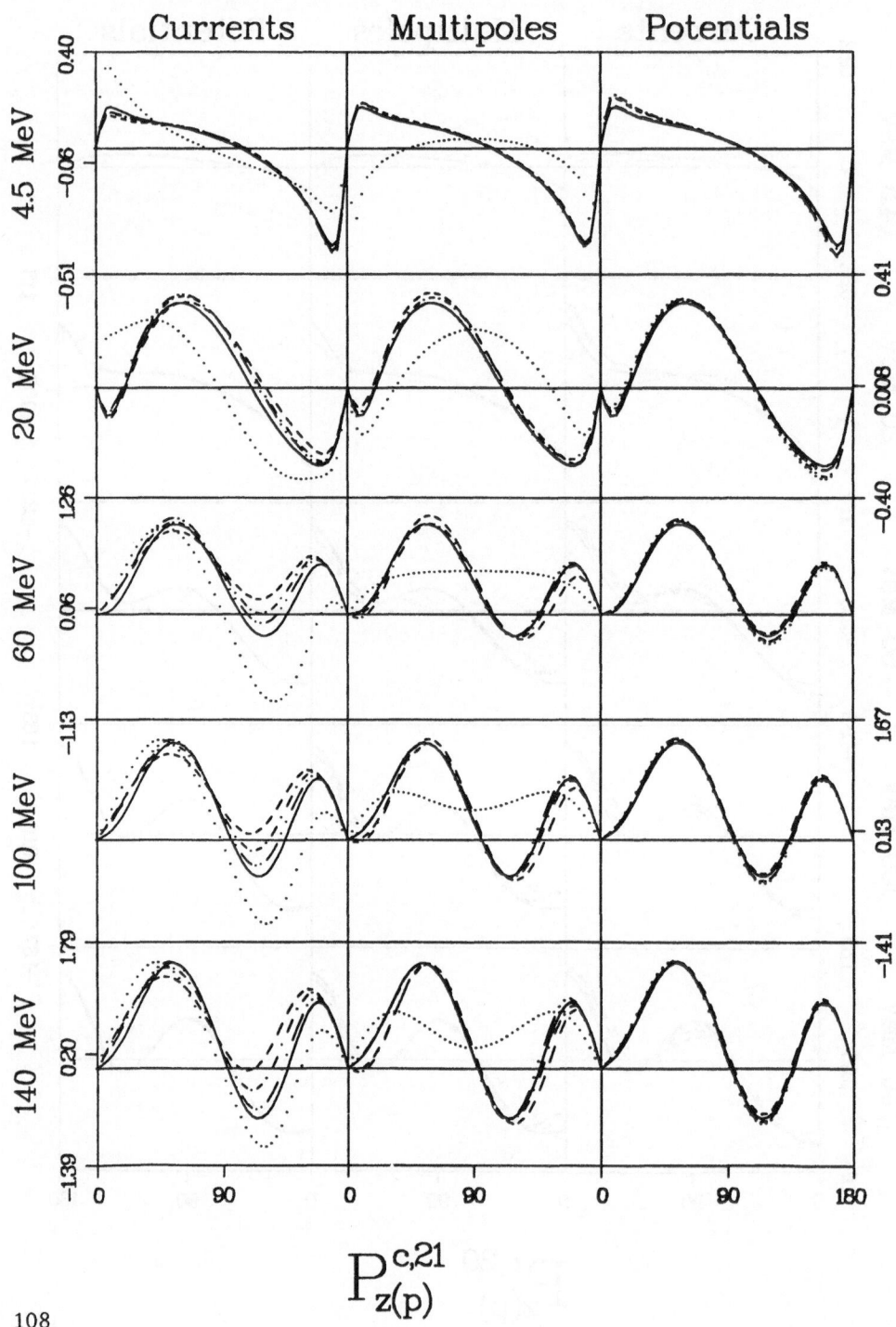

$$P_{z(p)}^{c,21}$$

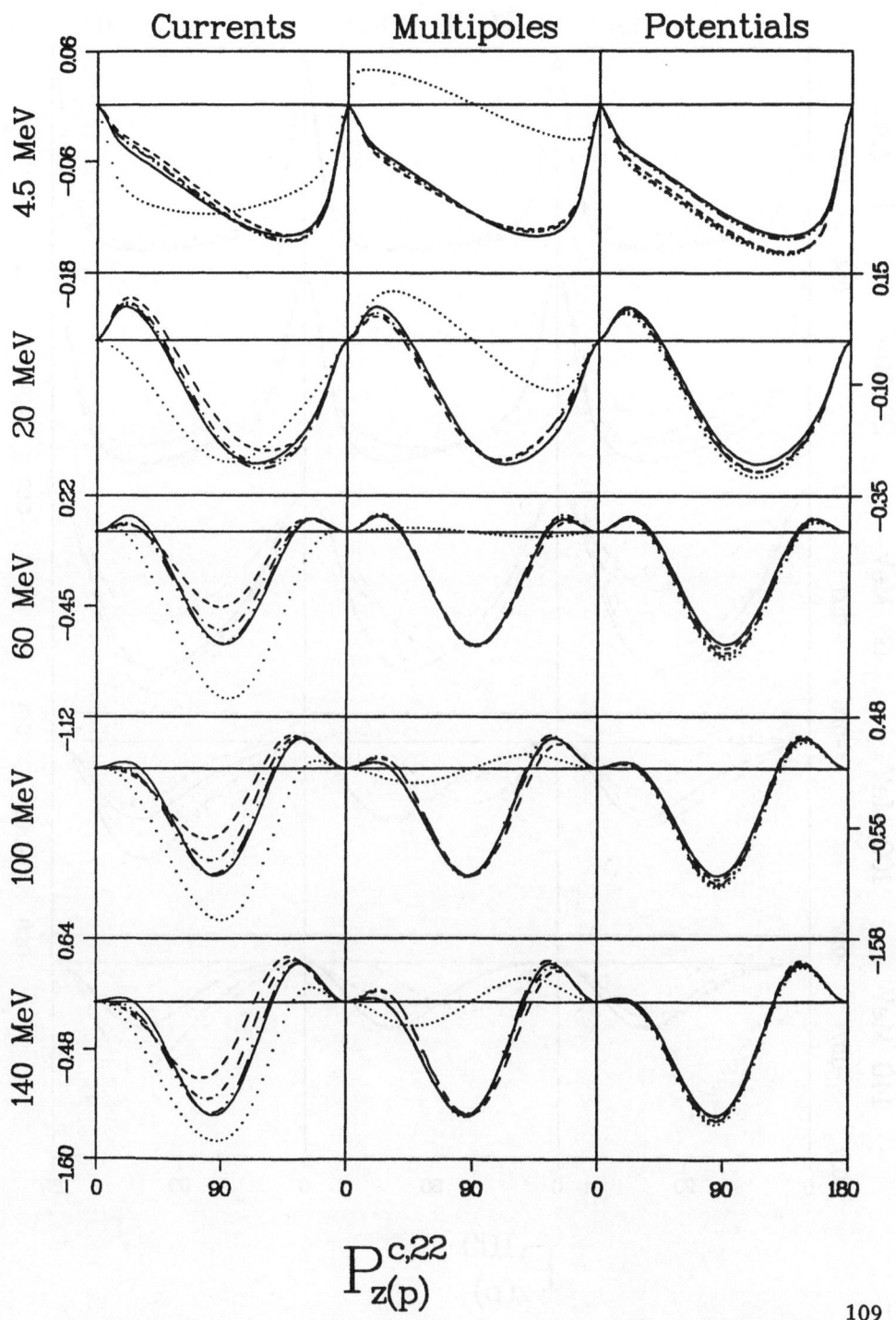

$$P_{z(p)}^{c,22}$$

5.19 Proton Polarization $P_z^{l,IM}(p)$

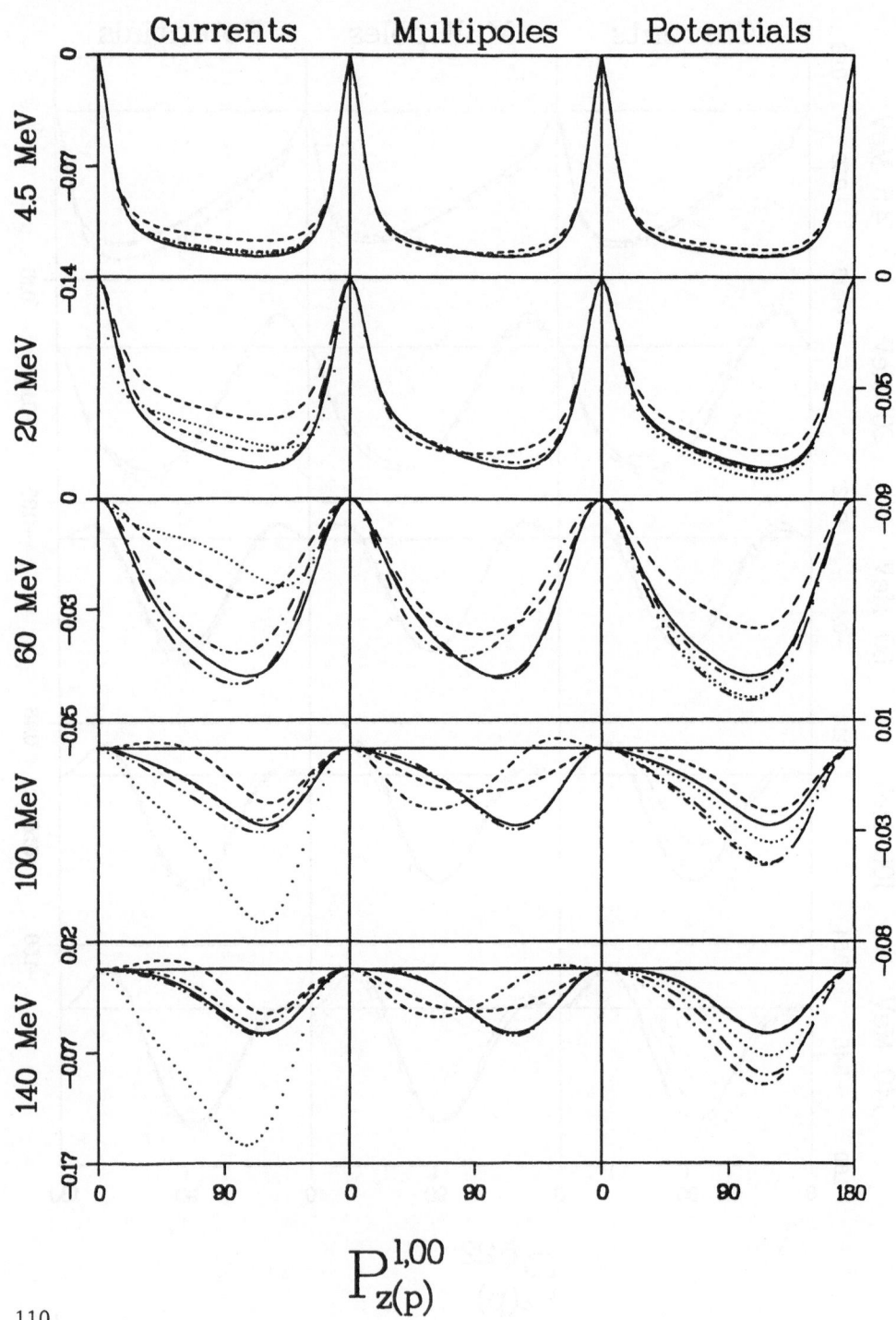

$$P_{z(p)}^{1,00}$$

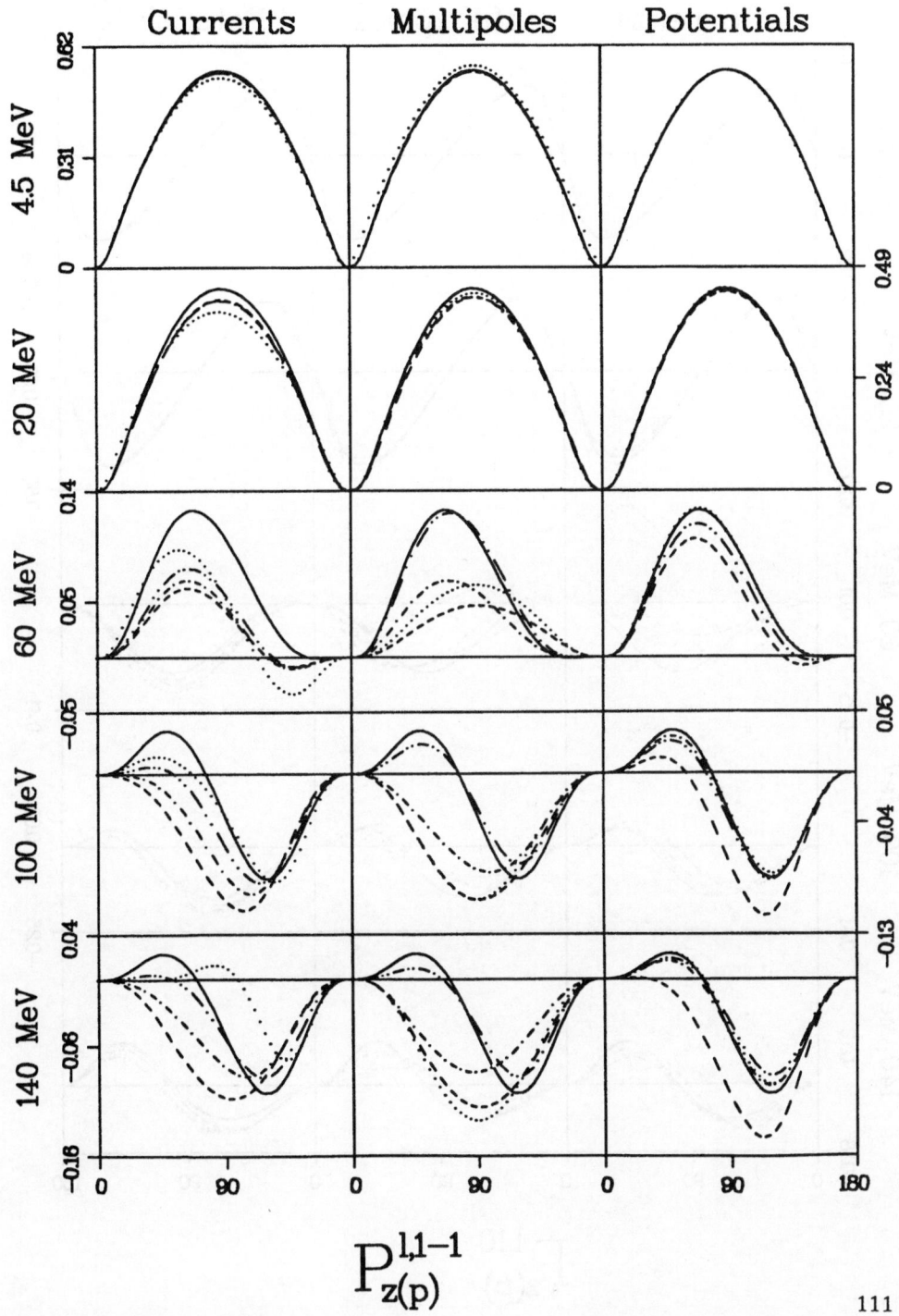

$$P_{z(p)}^{l,1-1}$$

Proton Polarization $P_z^{l,IM}(p)$

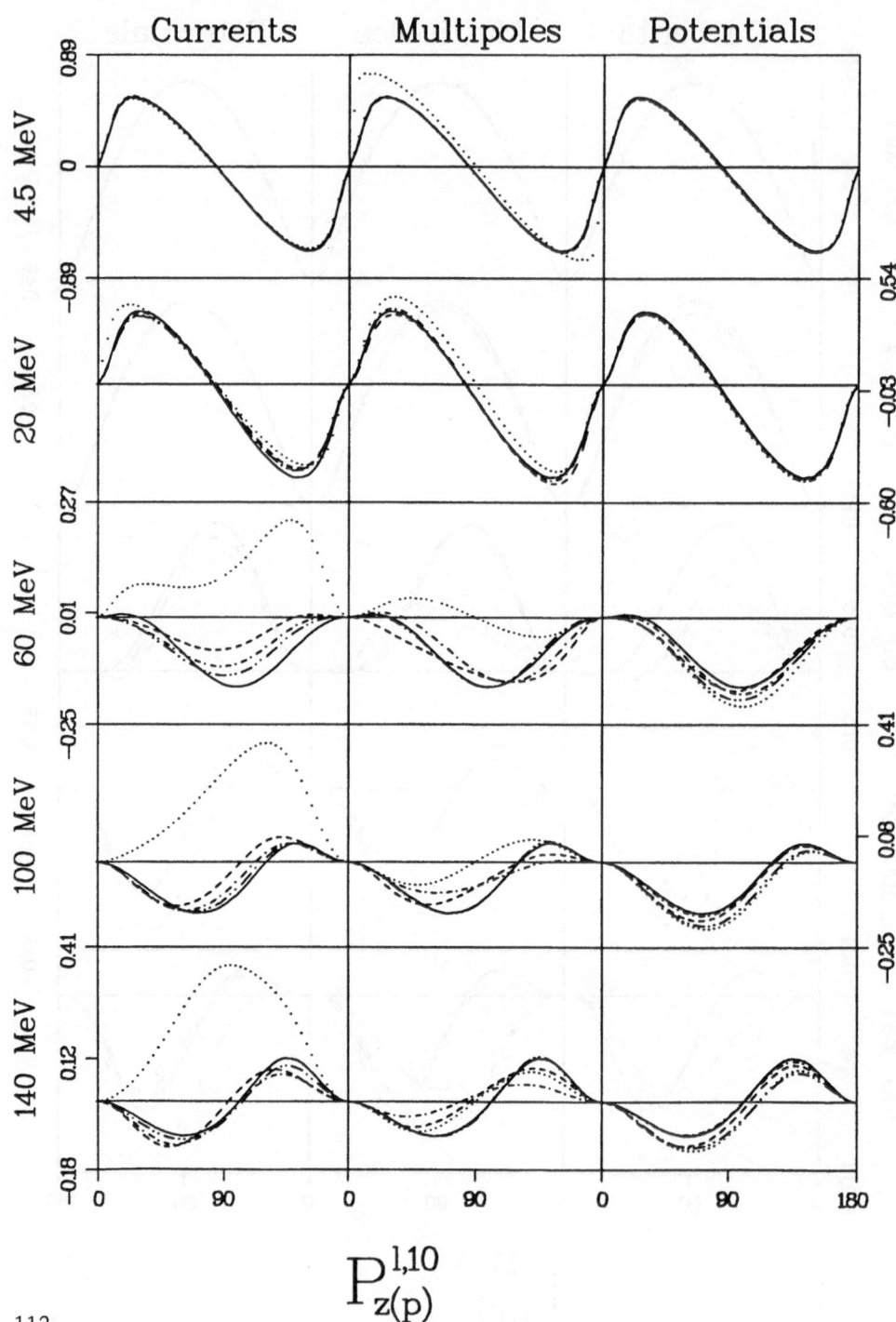

$$P_{z(p)}^{l,10}$$

Proton Polarization $P_z^{l,IM}(p)$

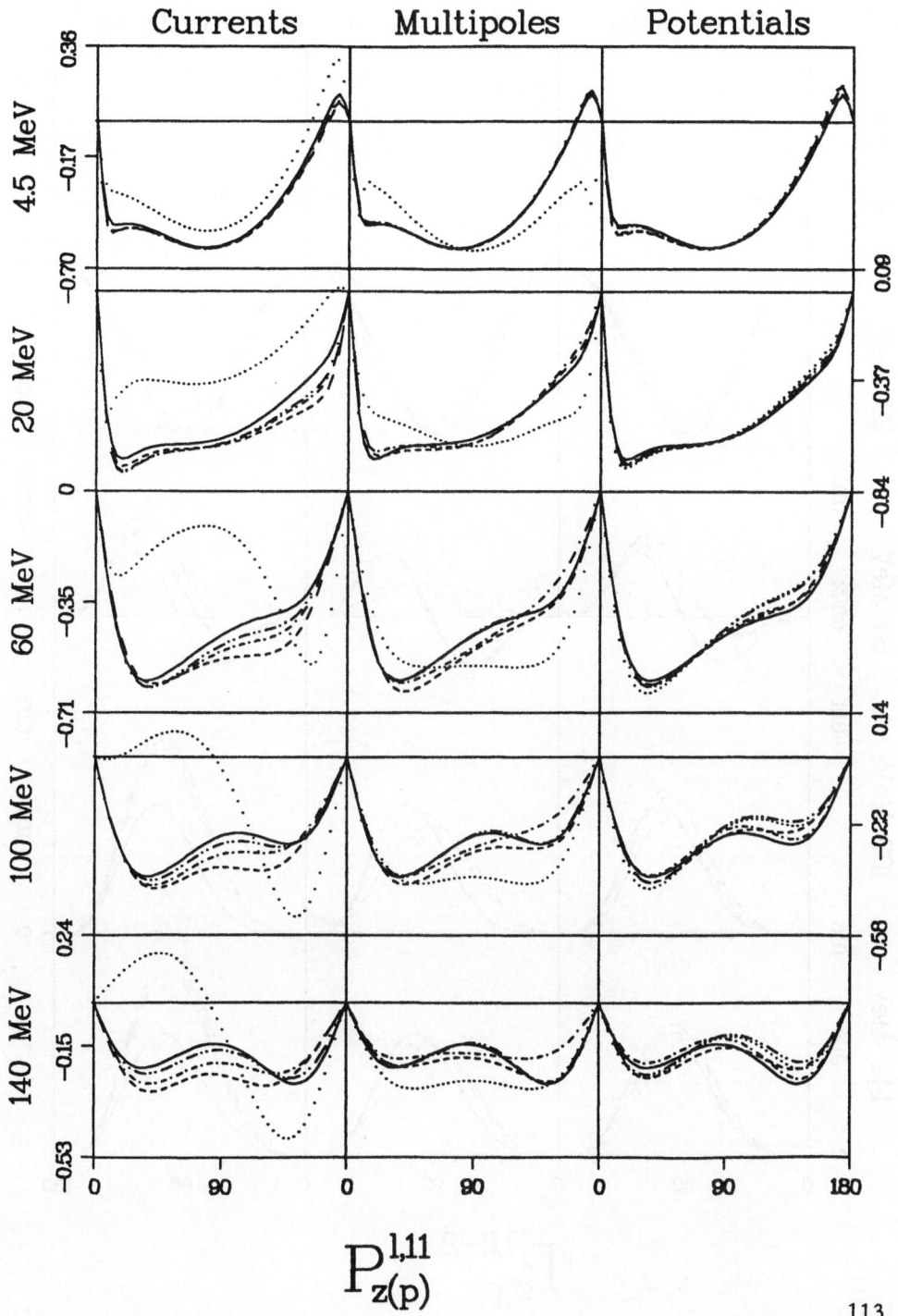

$$P_{z(p)}^{l,11}$$

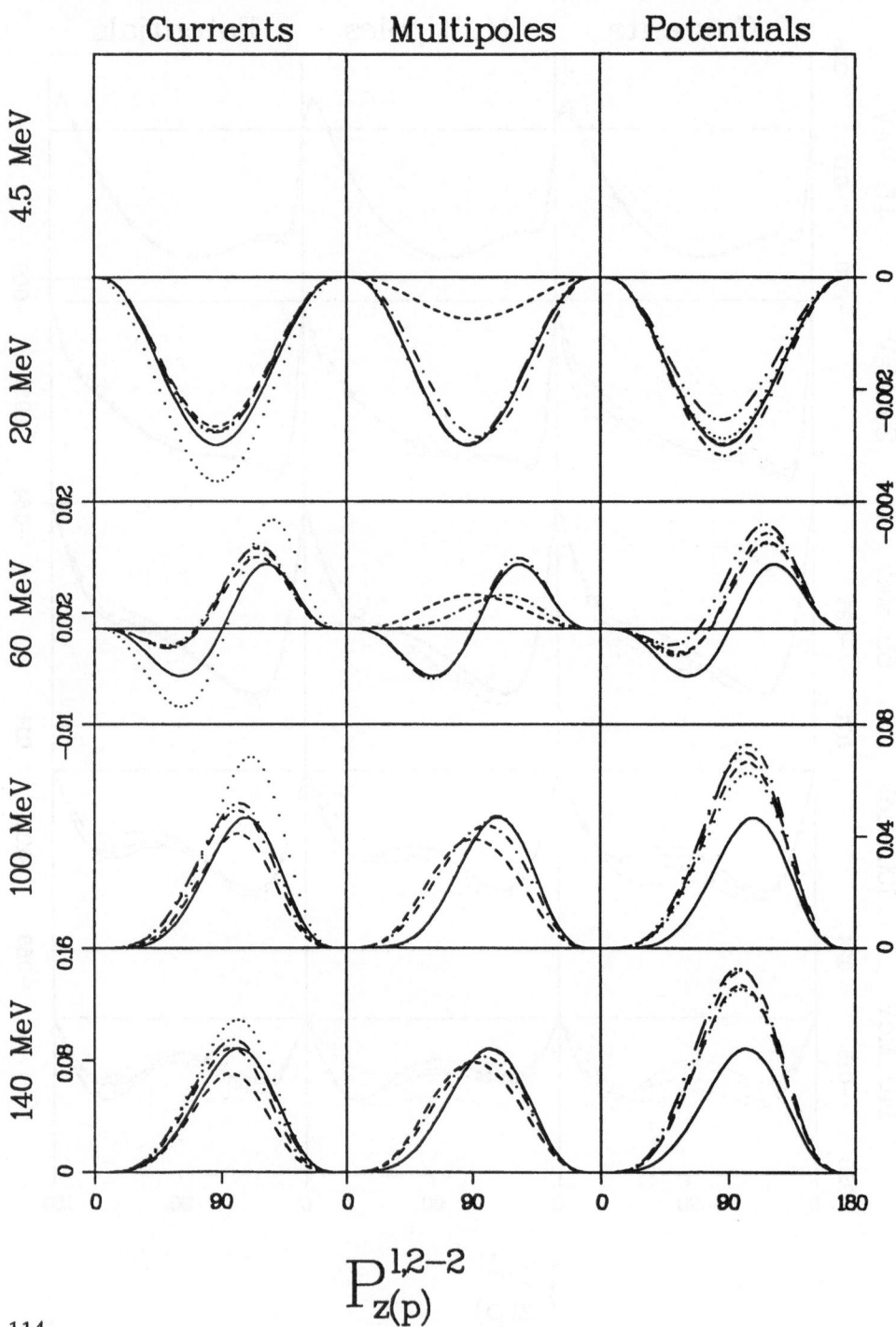

Currents   Multipoles   Potentials

4.5 MeV   20 MeV   60 MeV   100 MeV   140 MeV

$$P_{z(p)}^{1,2-2}$$

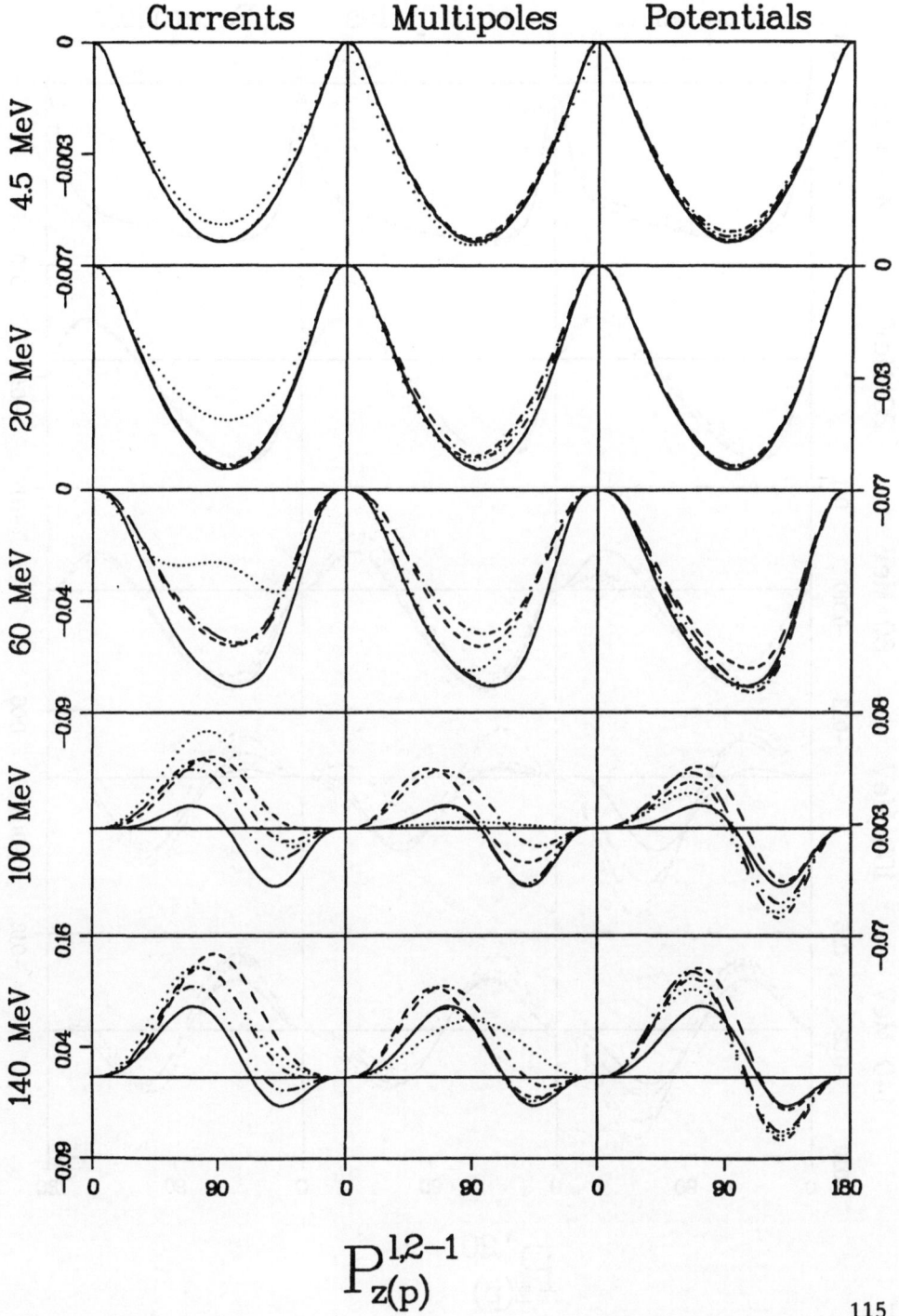

$$P_{z(p)}^{l,2-1}$$

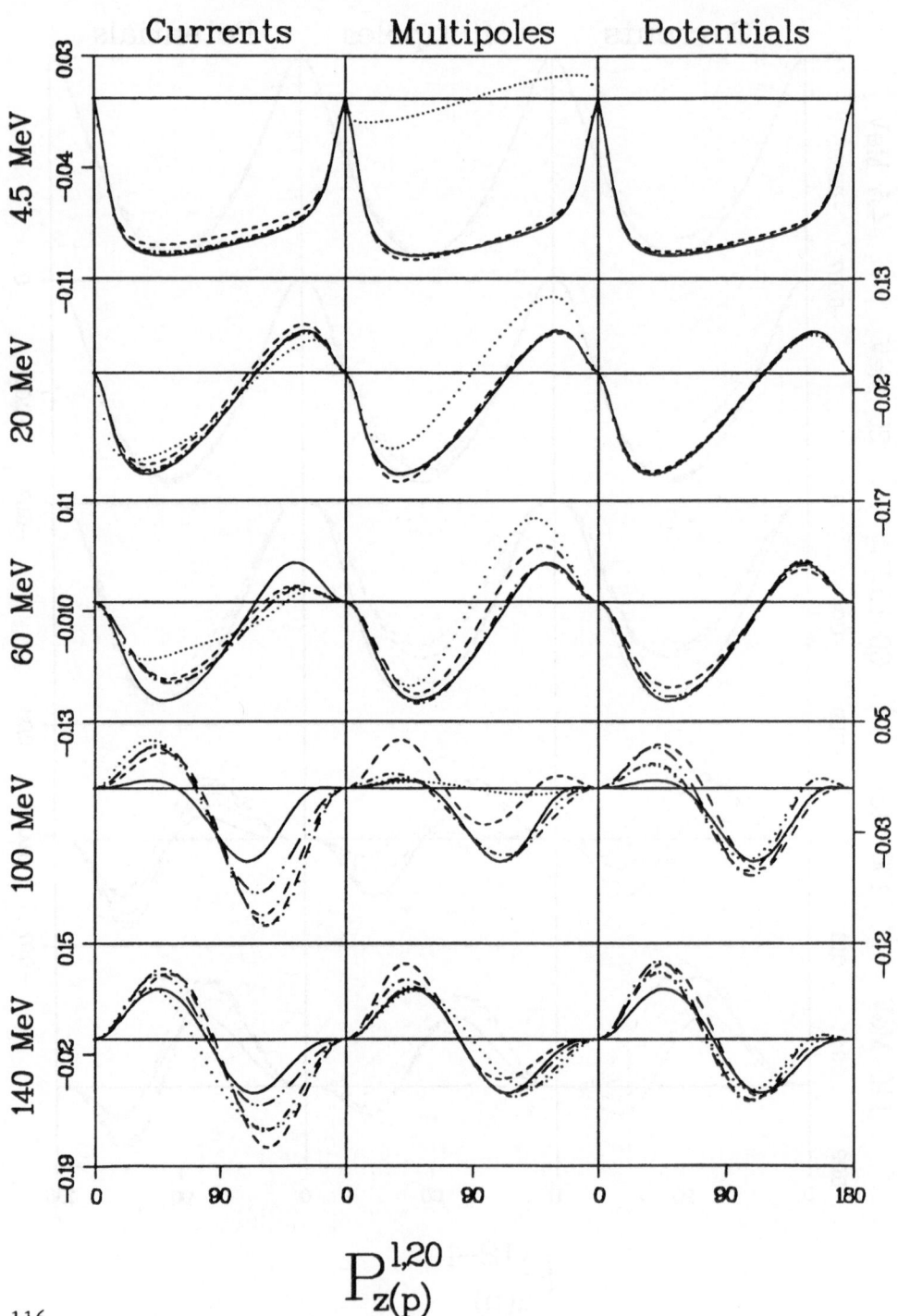

$$P_{z(p)}^{1,20}$$

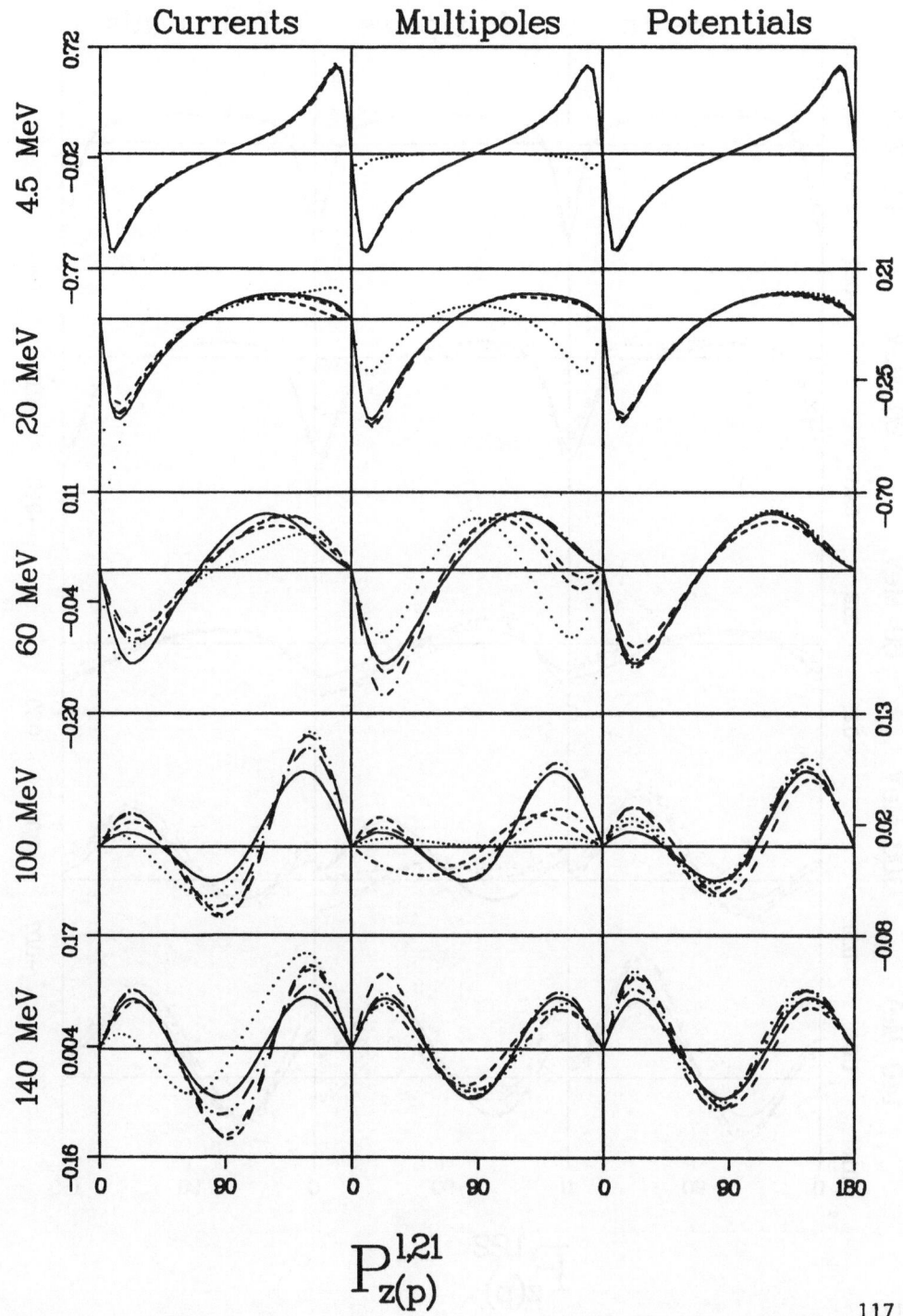

$$P_{z(p)}^{1,21}$$

Proton Polarization $P_z^{l,IM}(p)$

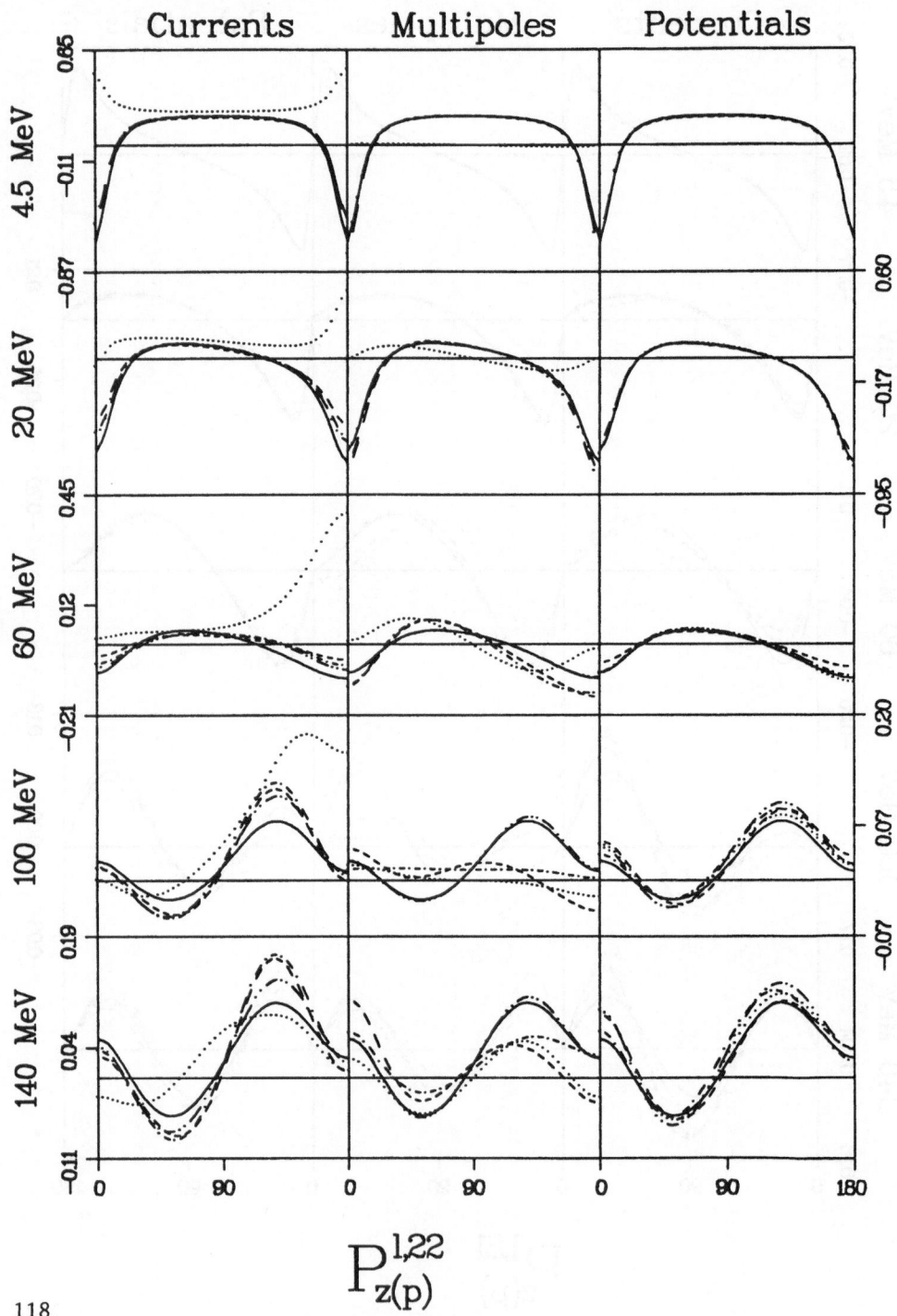

$$P_{z(p)}^{1,22}$$

5.20 Neutron Polarization $P_z^{0,IM}(n)$

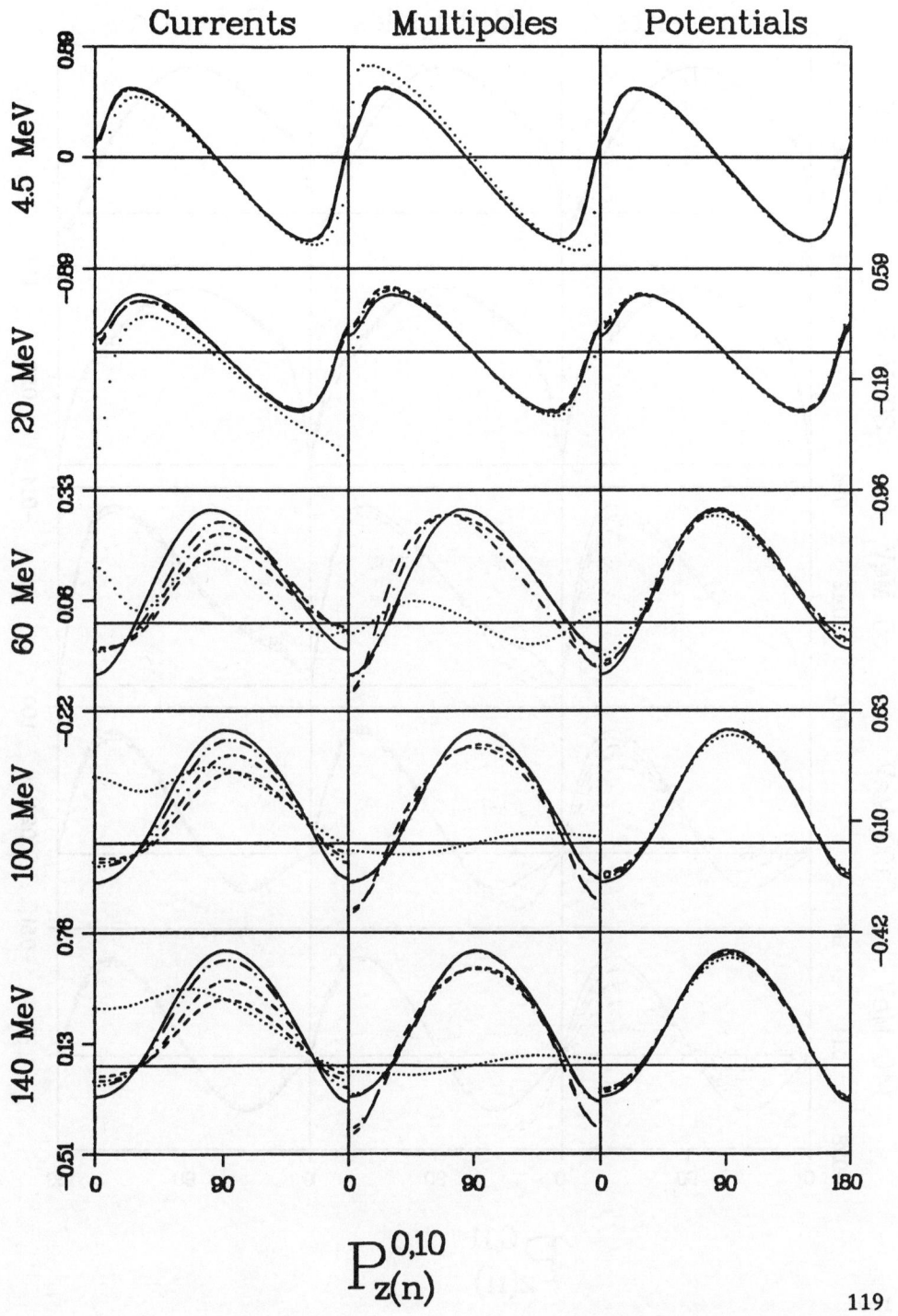

$$P_{z(n)}^{0,10}$$

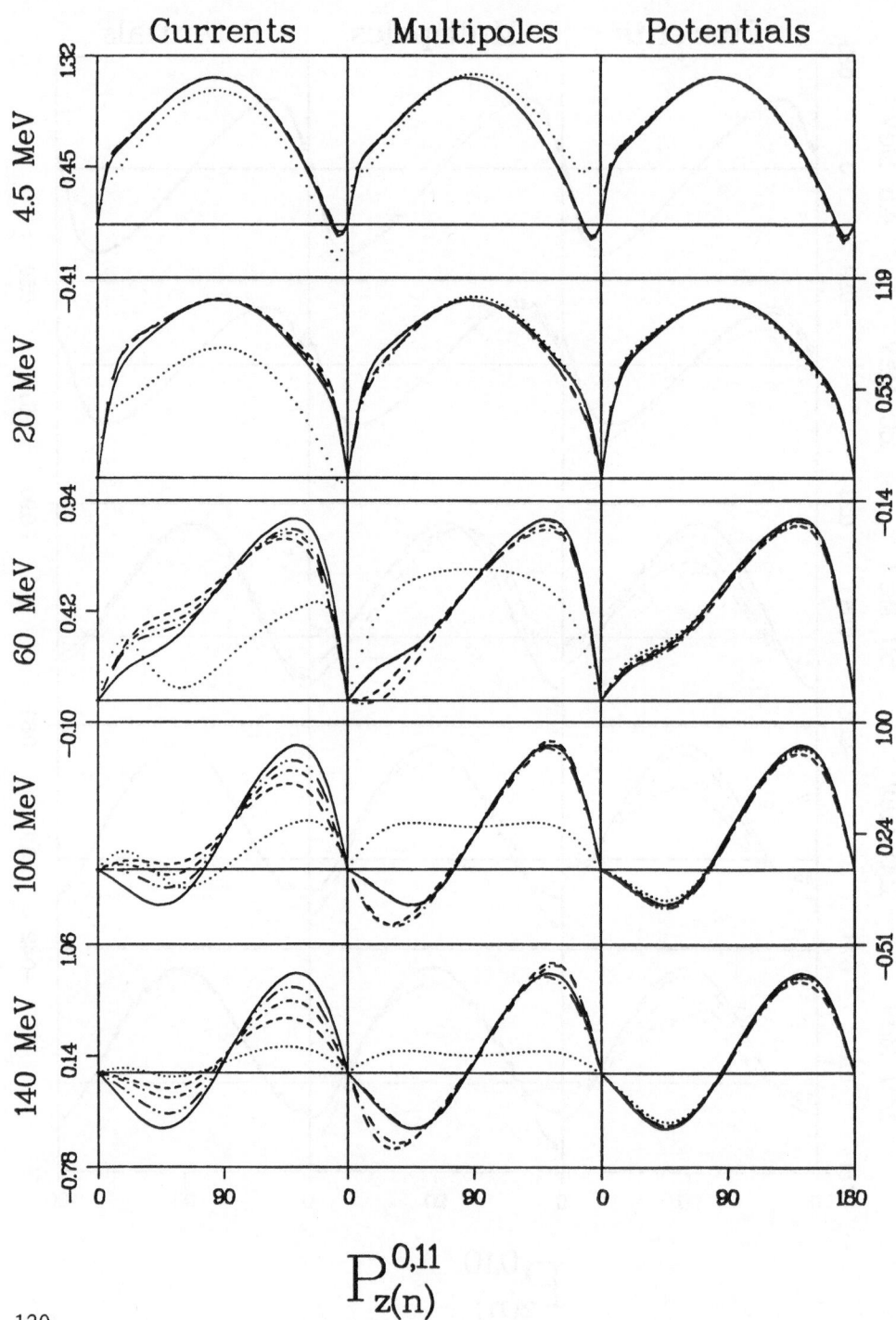

$$P_{z(n)}^{0,11}$$

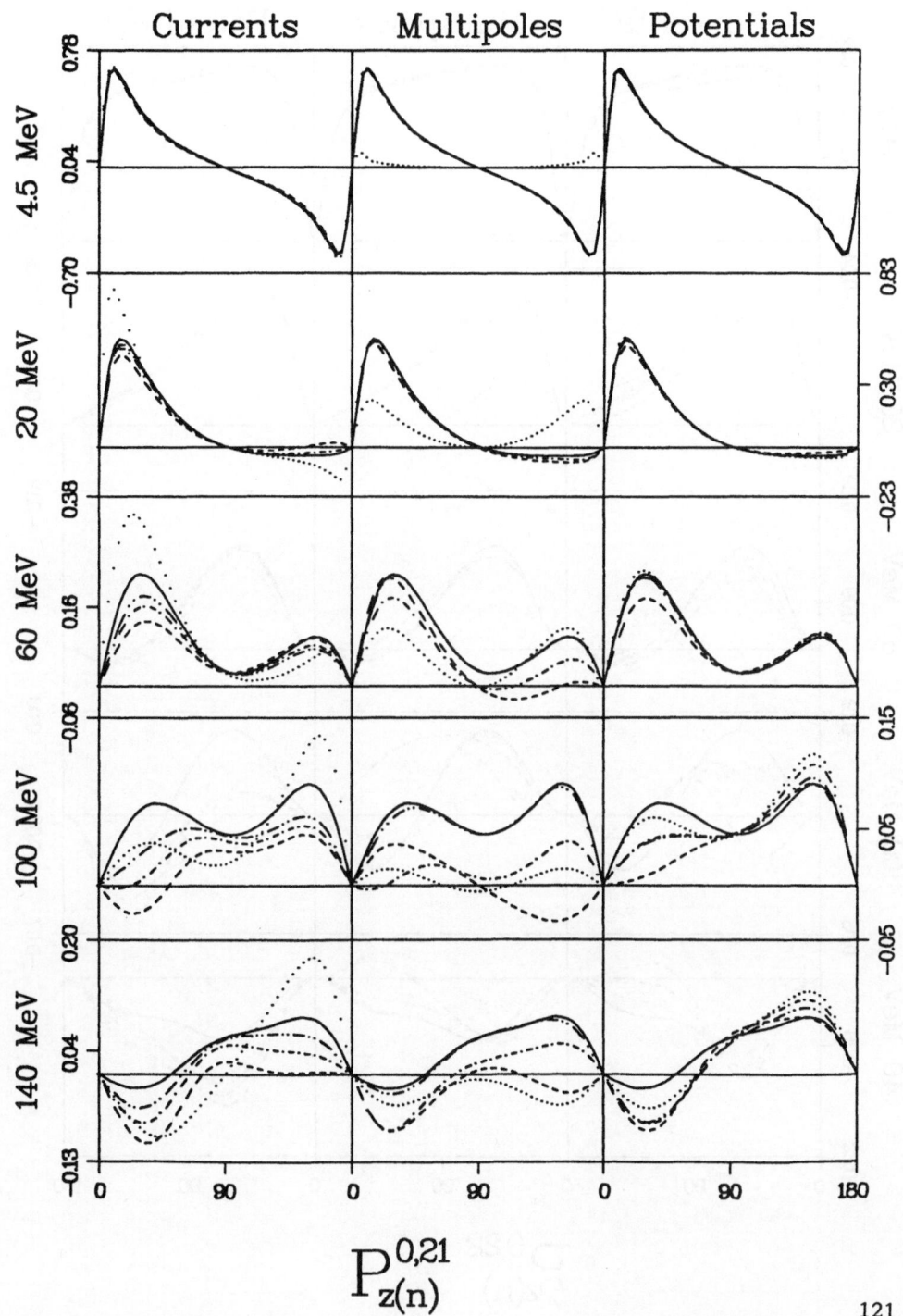

$$P_{z(n)}^{0,21}$$

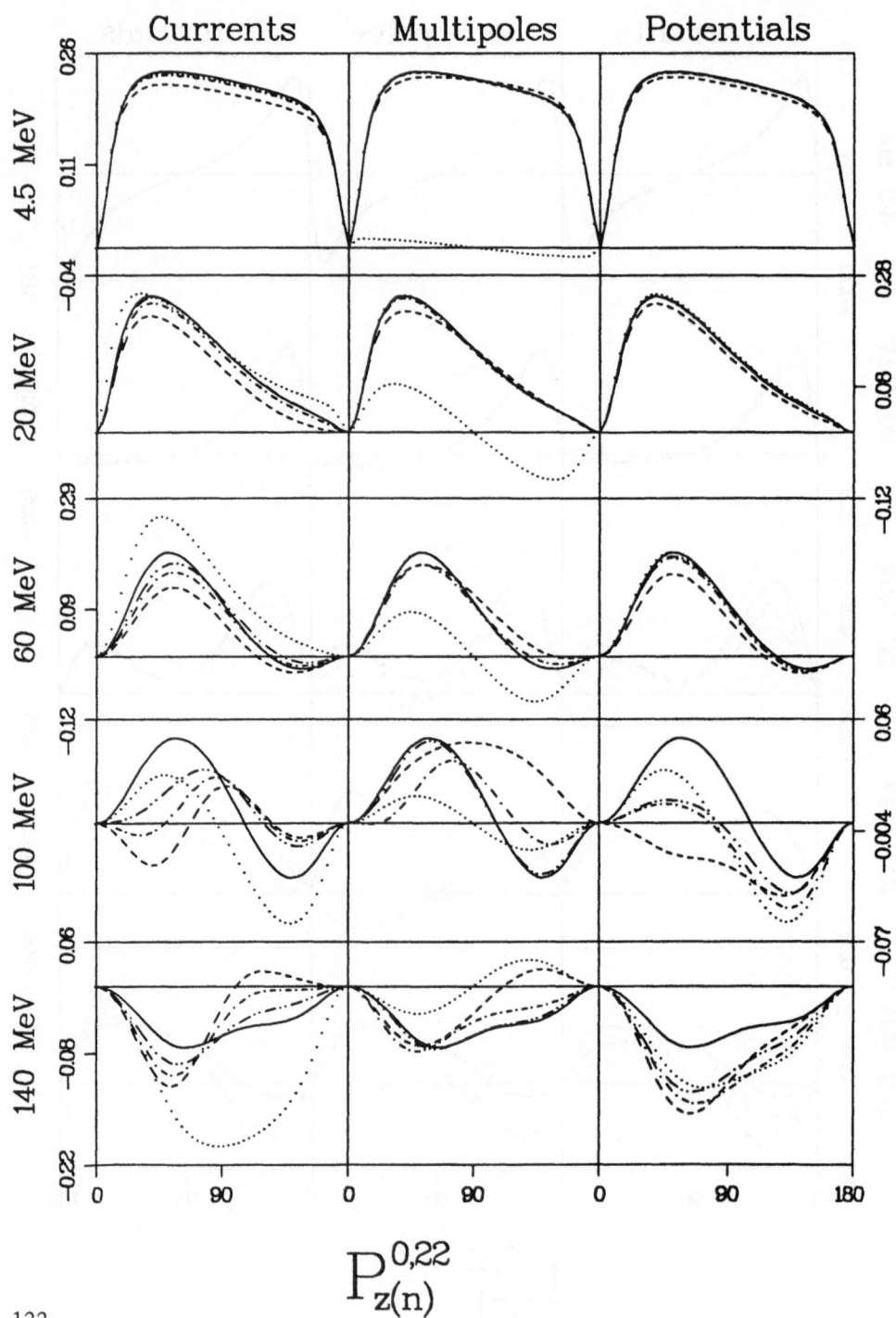

$$P_{z(n)}^{0,22}$$

5.21 Neutron Polarization $P_z^{c,IM}(n)$

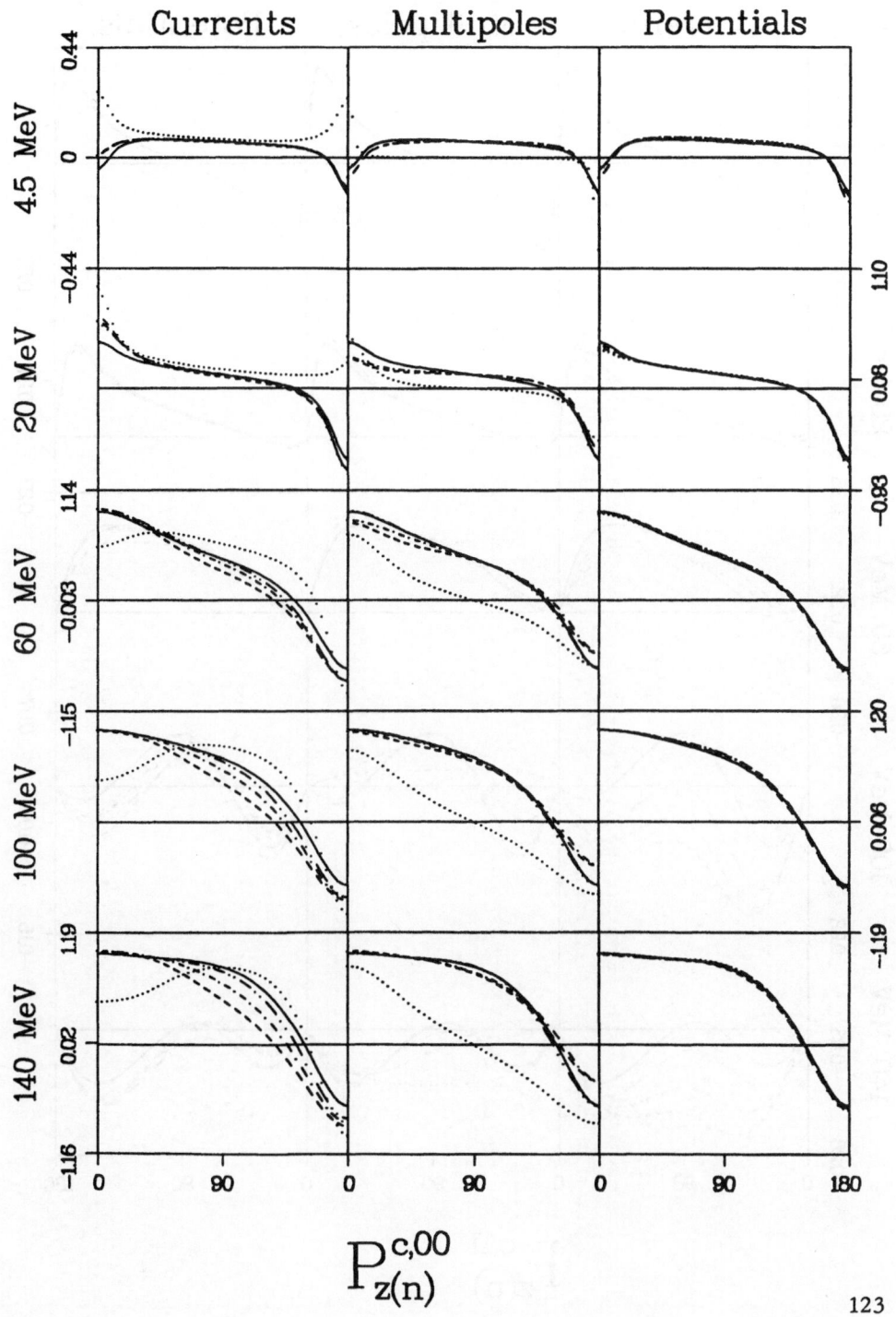

$$P_{z(n)}^{c,00}$$

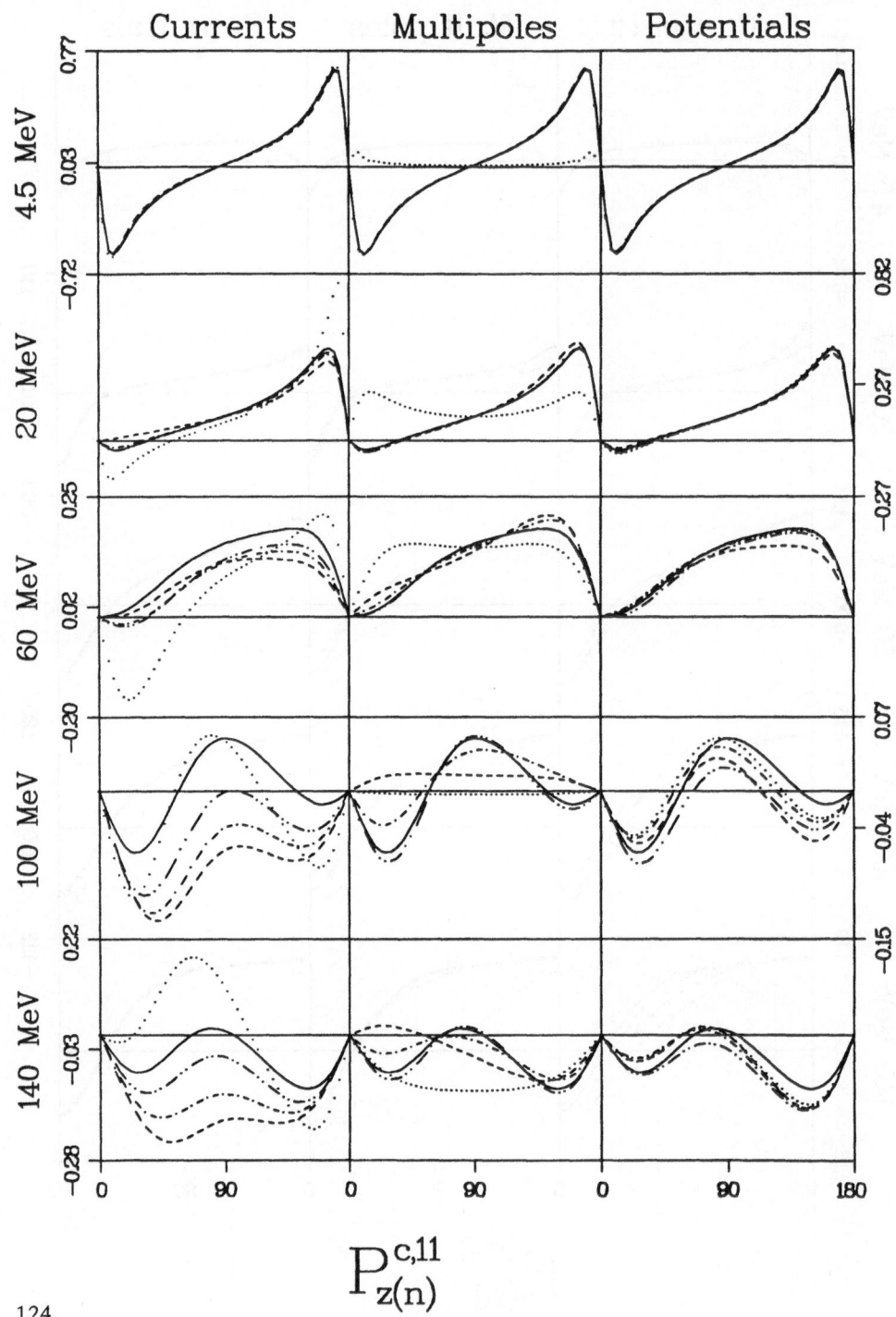

$$P_{z(n)}^{c,11}$$

Neutron Polarization $P_z^{c,IM}(n)$

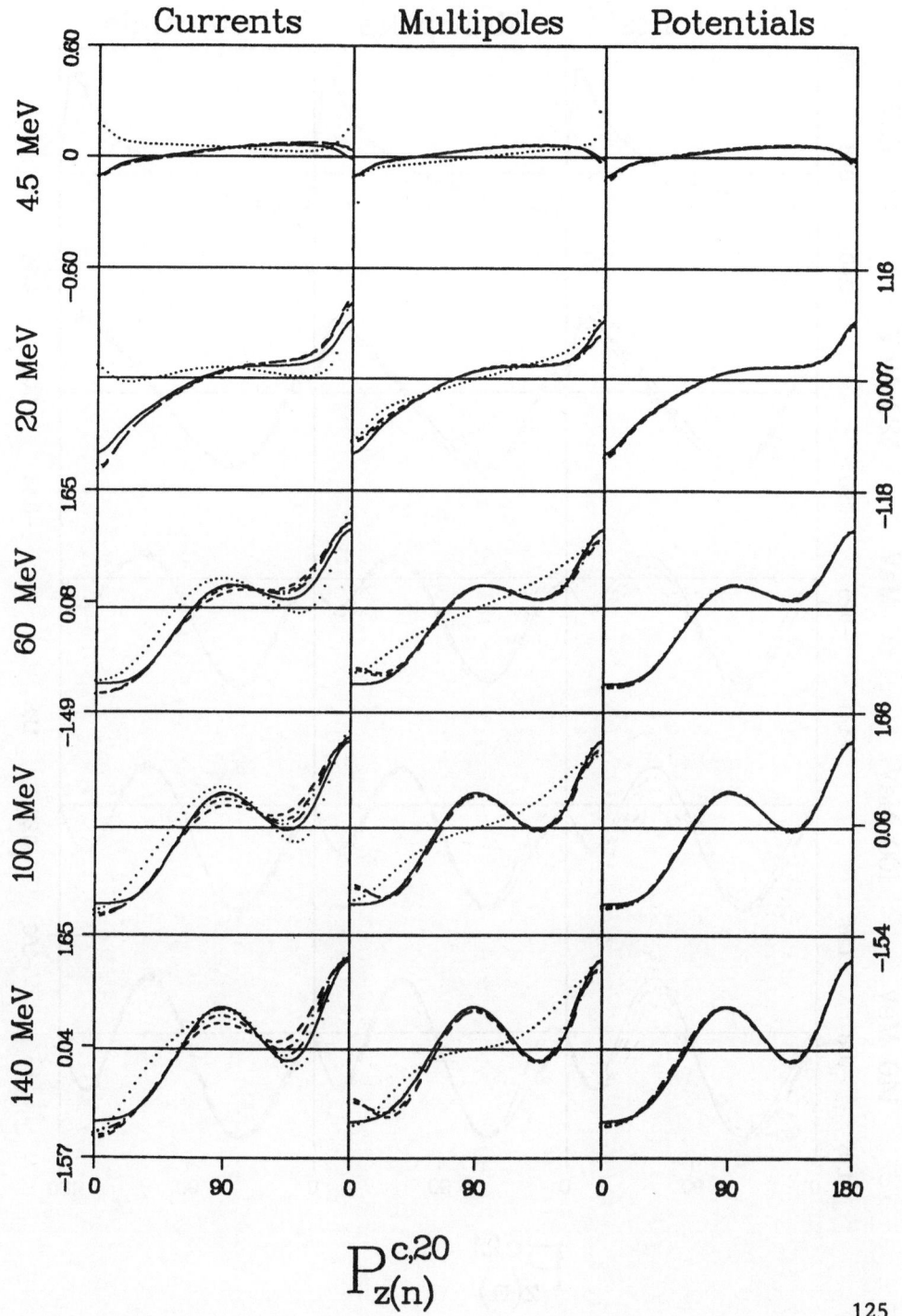

$$P_{z(n)}^{c,20}$$

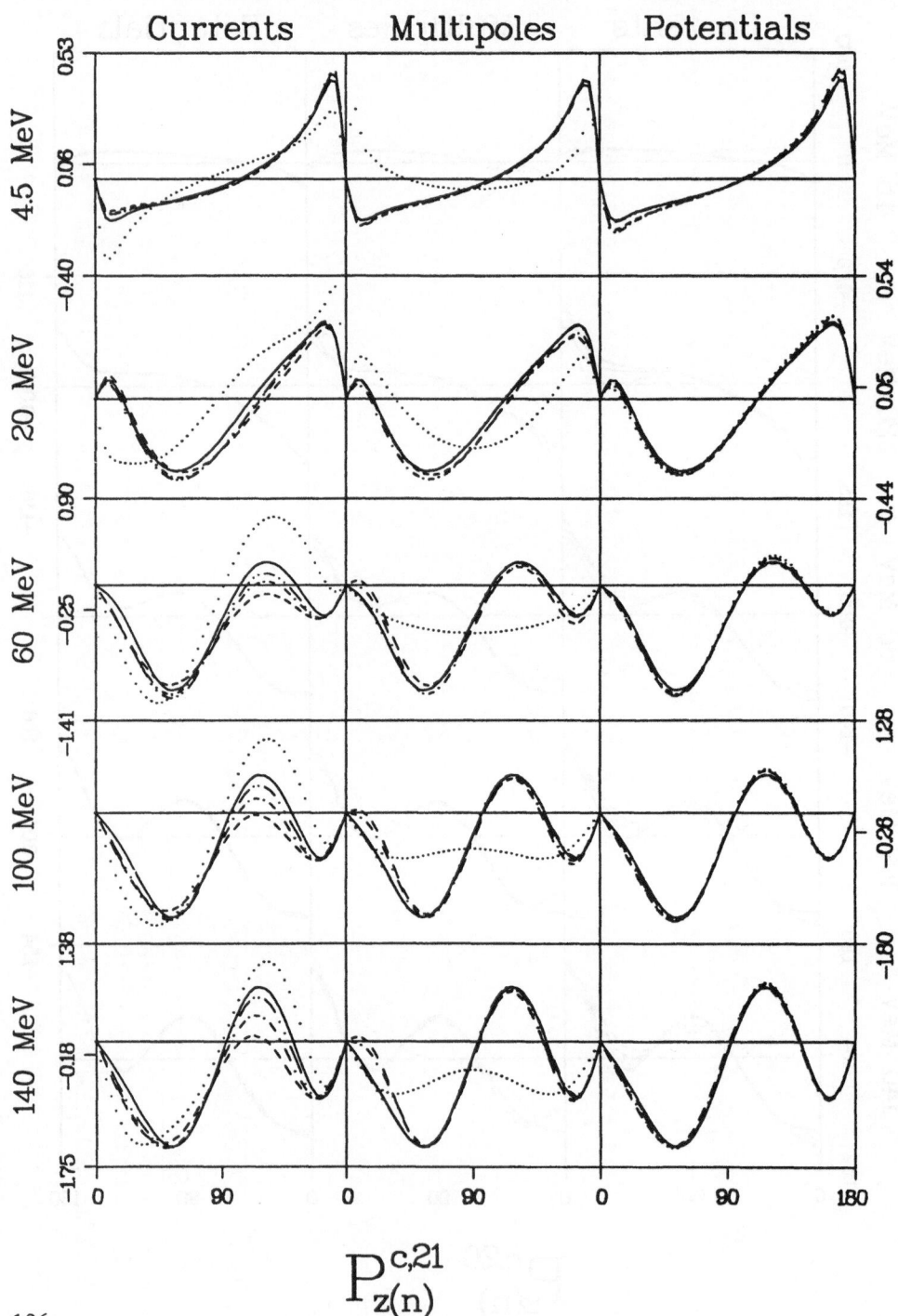

$$P_{z(n)}^{c,21}$$

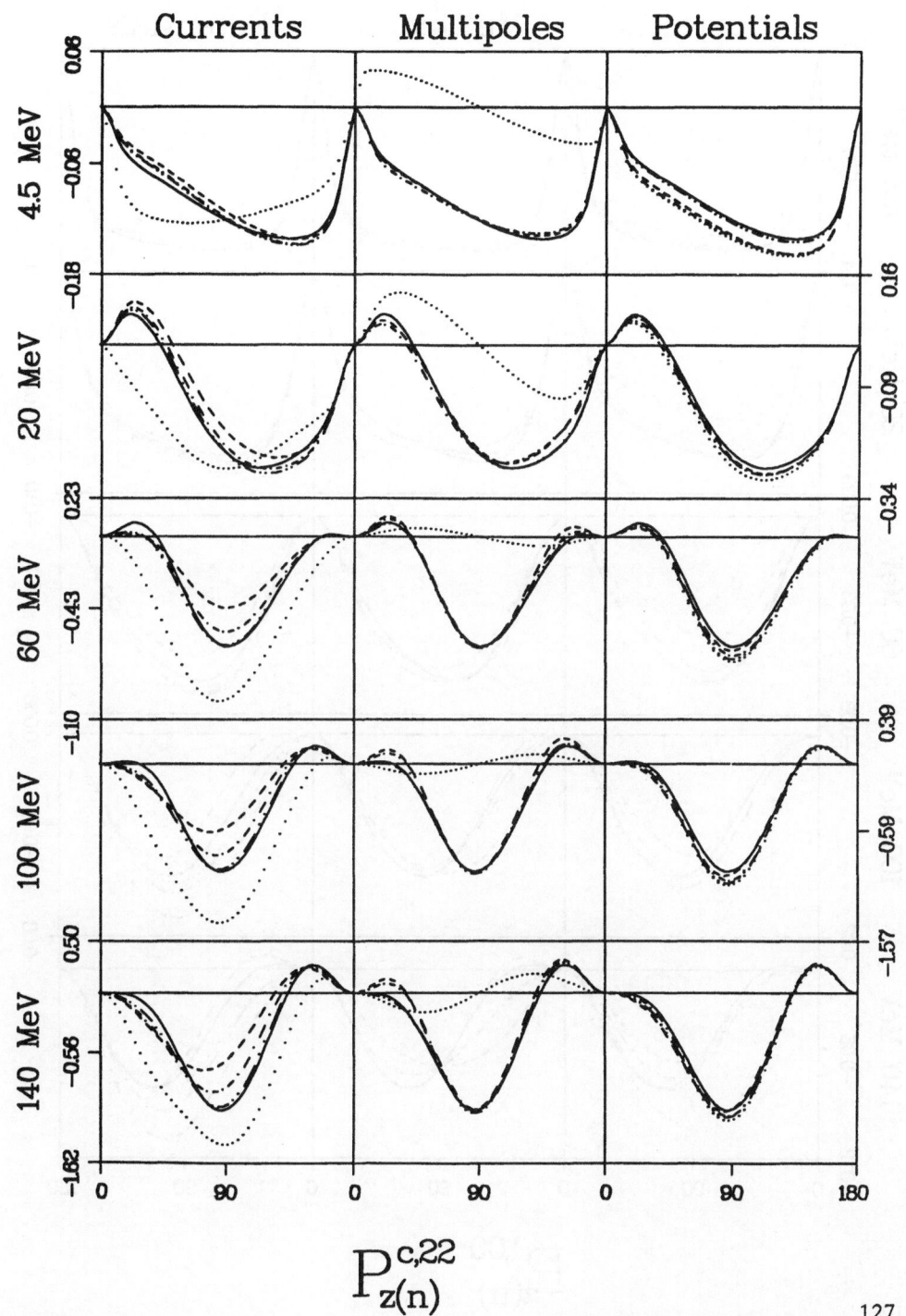

$$P_{z(n)}^{c,22}$$

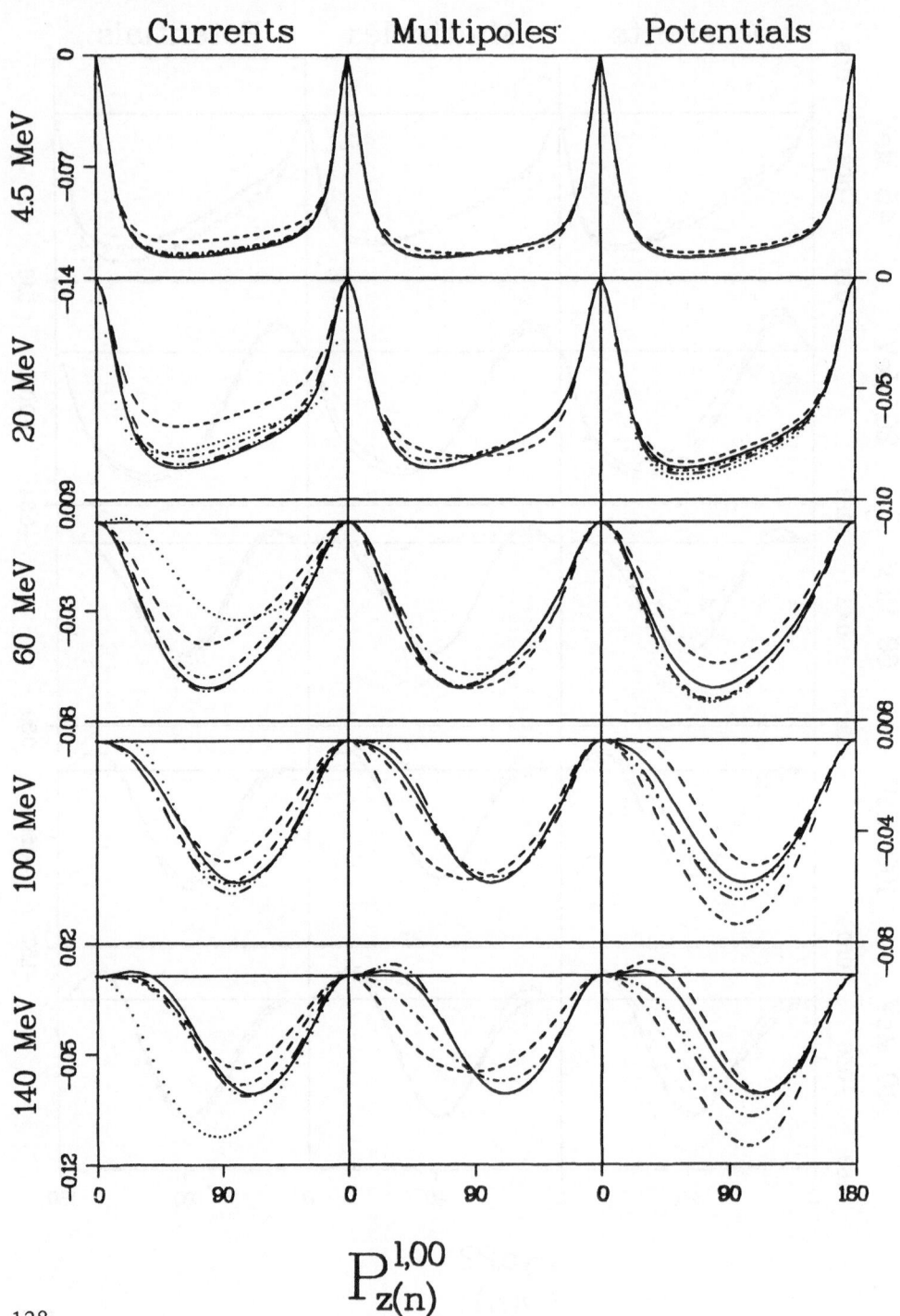

$$P_{z(n)}^{1,00}$$

Neutron Polarization $P_z^{l,IM}(n)$

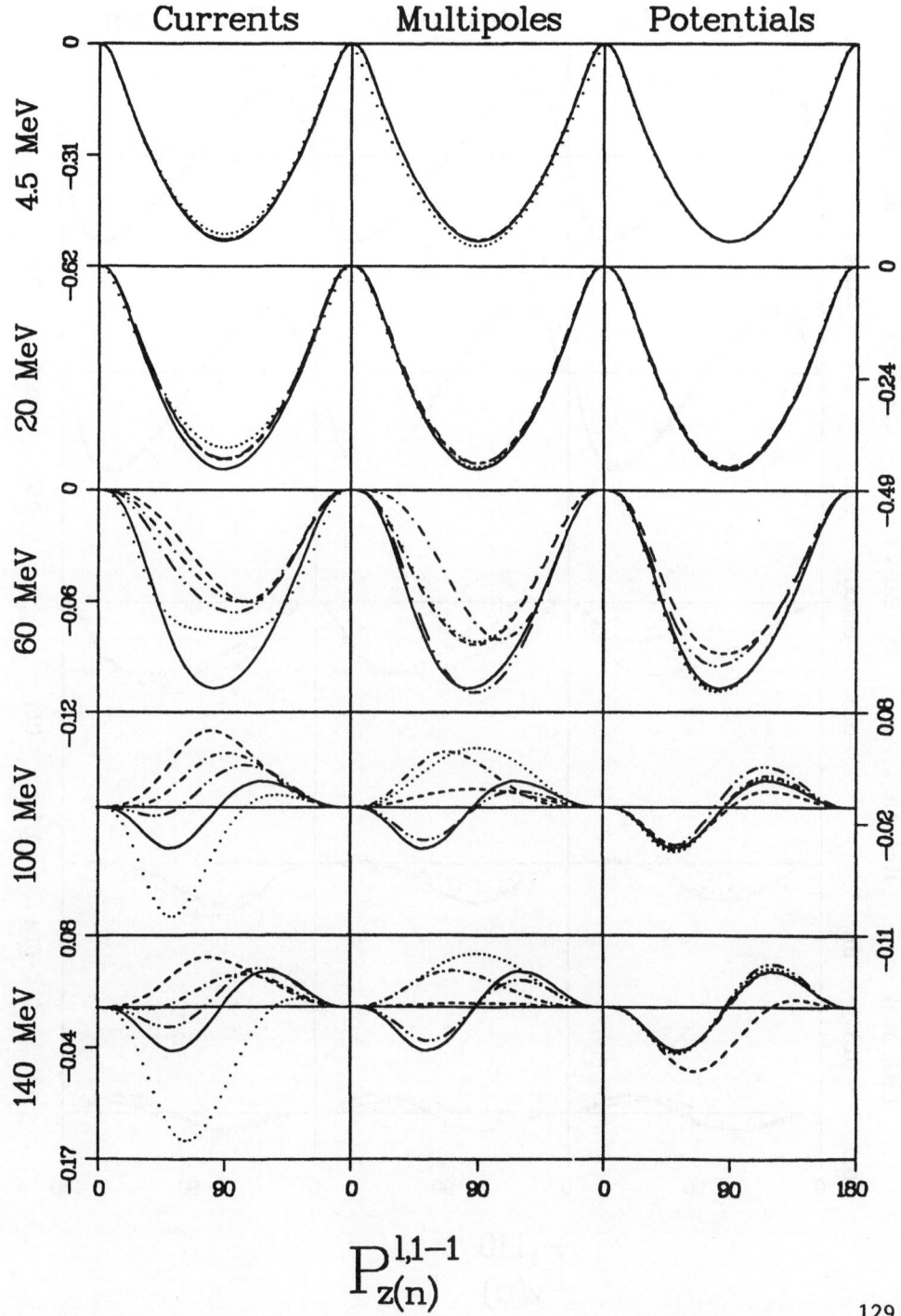

$$P_{z(n)}^{l,1-1}$$

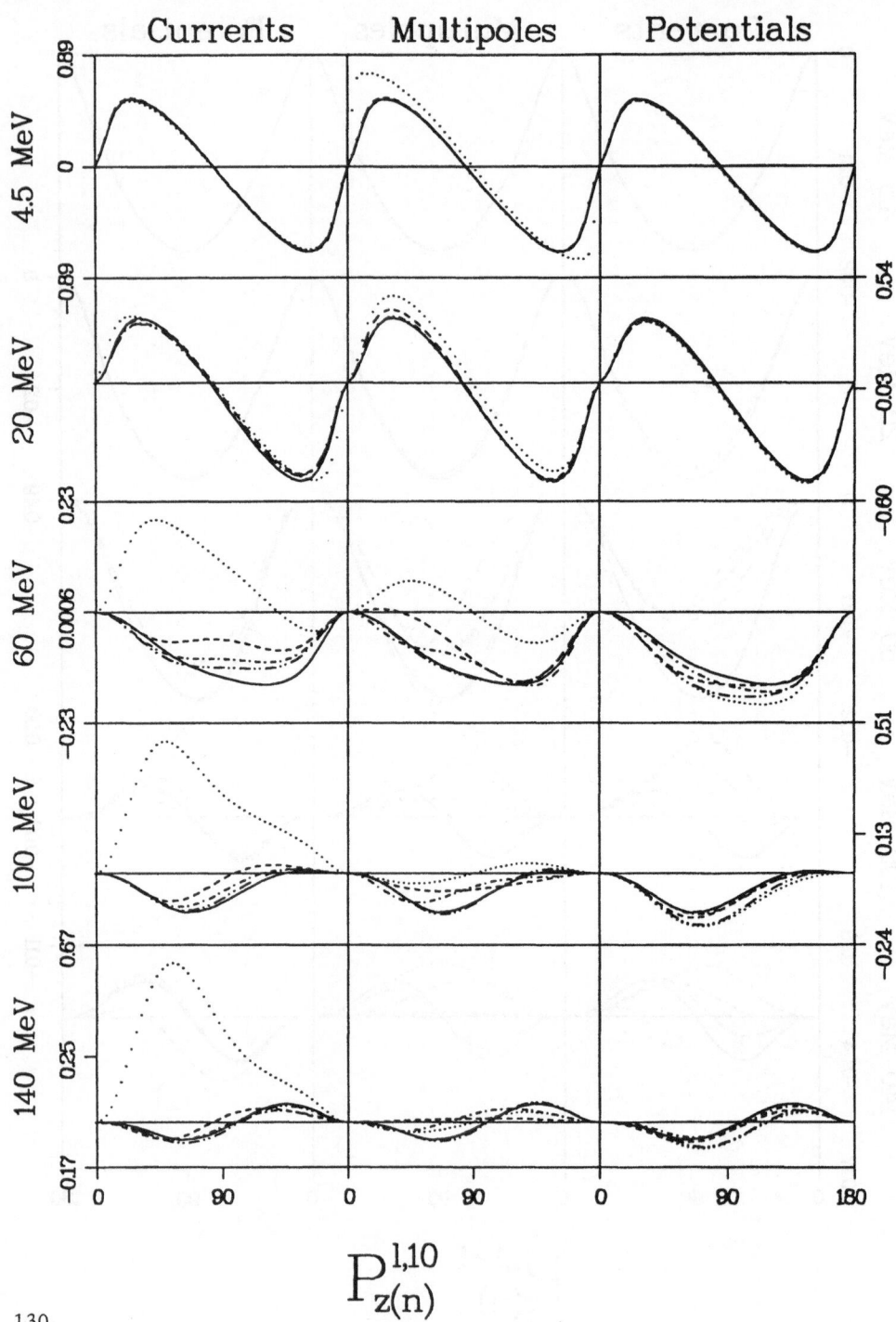

$$P_{z(n)}^{1,10}$$

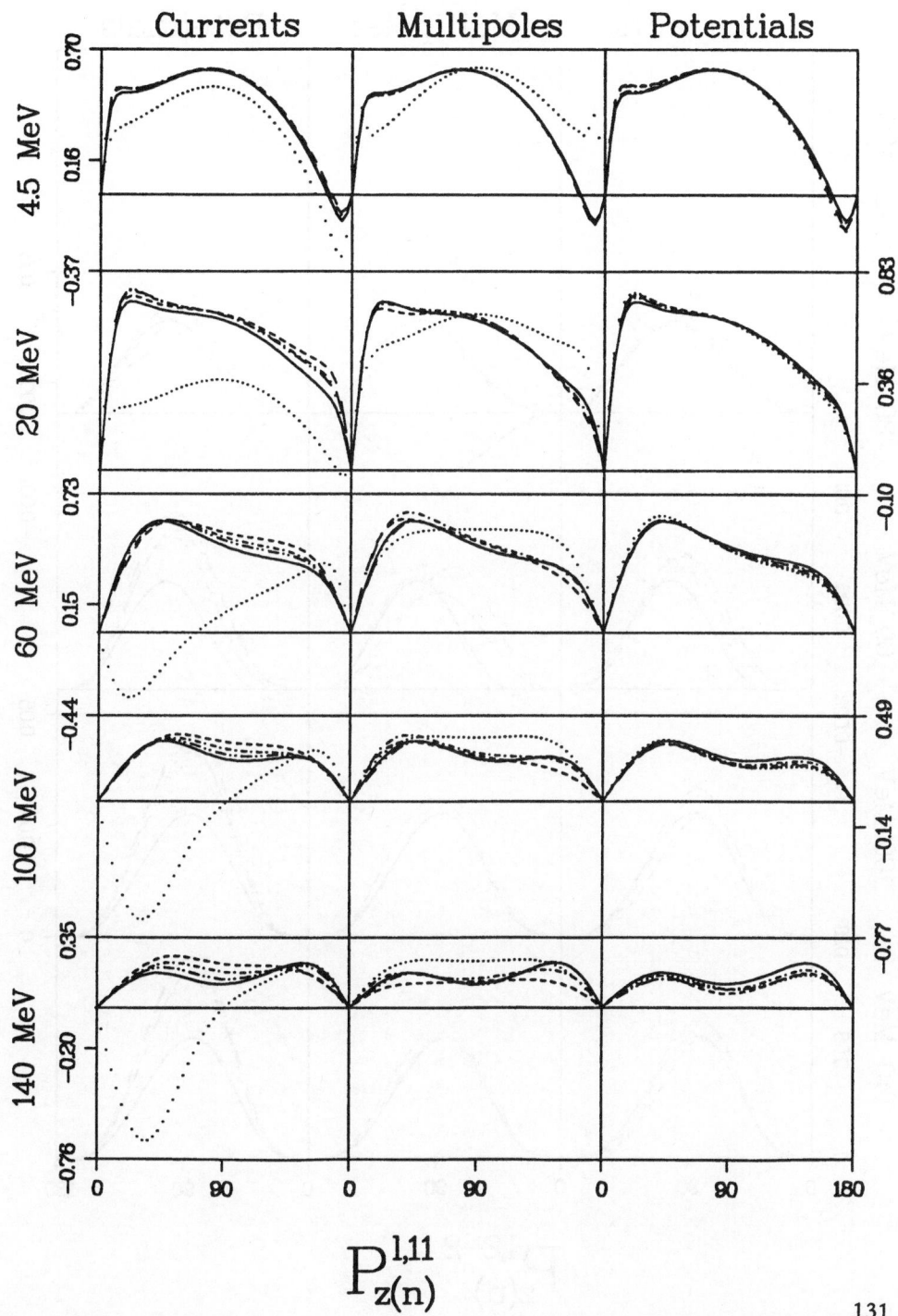

$$P_{z(n)}^{l,11}$$

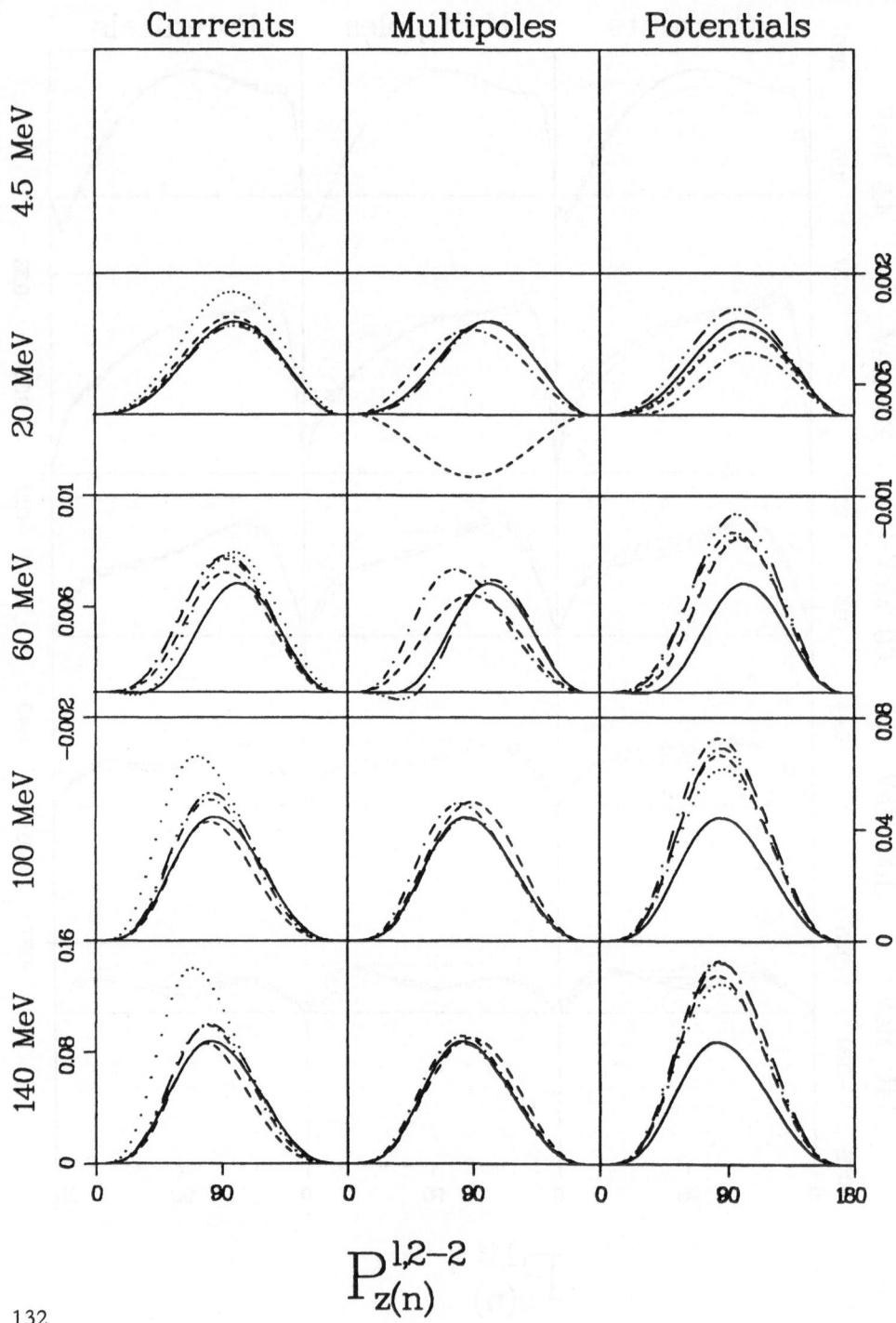

Currents   Multipoles   Potentials

4.5 MeV

20 MeV

60 MeV

100 MeV

140 MeV

$$P_{z(n)}^{l,2-2}$$

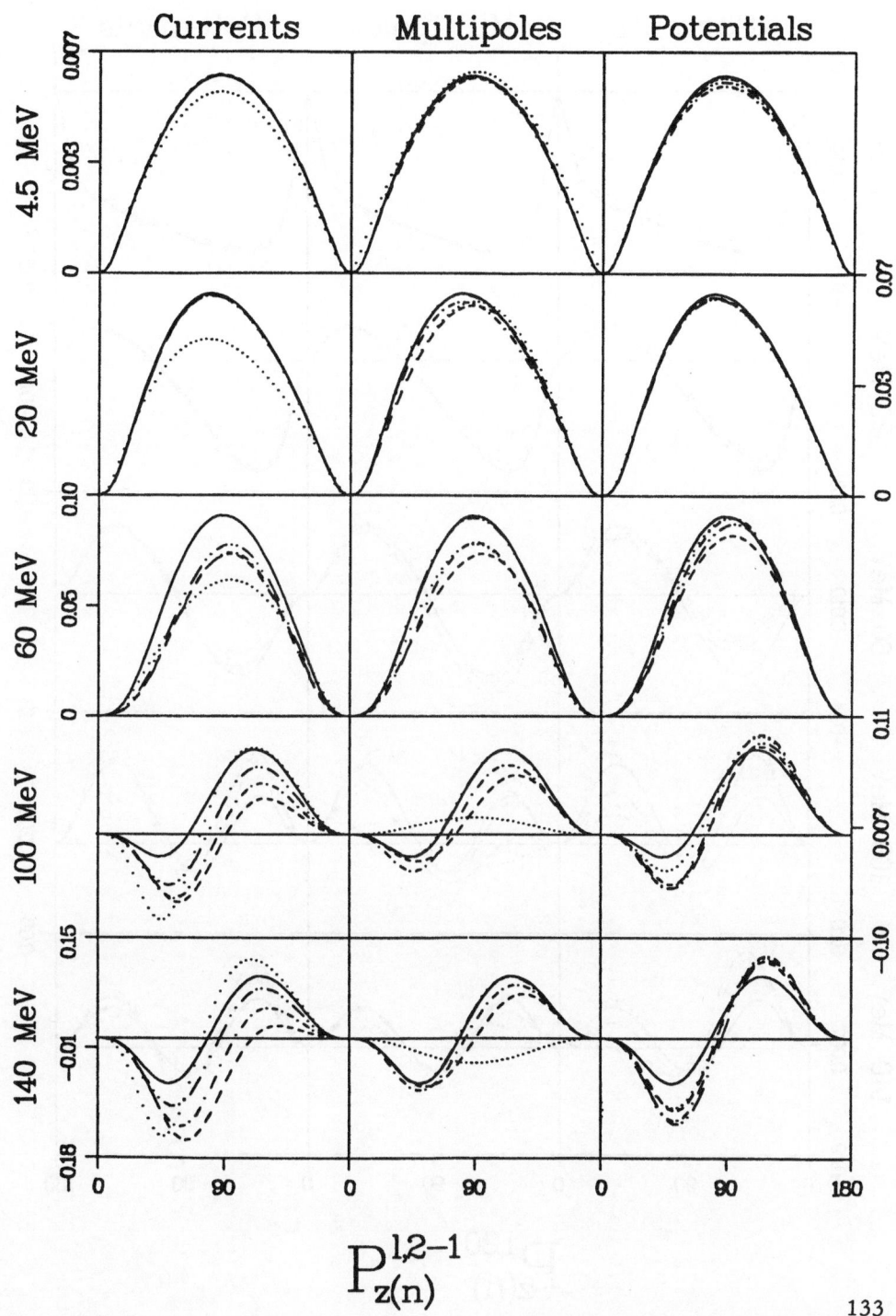

$$P_{z(n)}^{1,2-1}$$

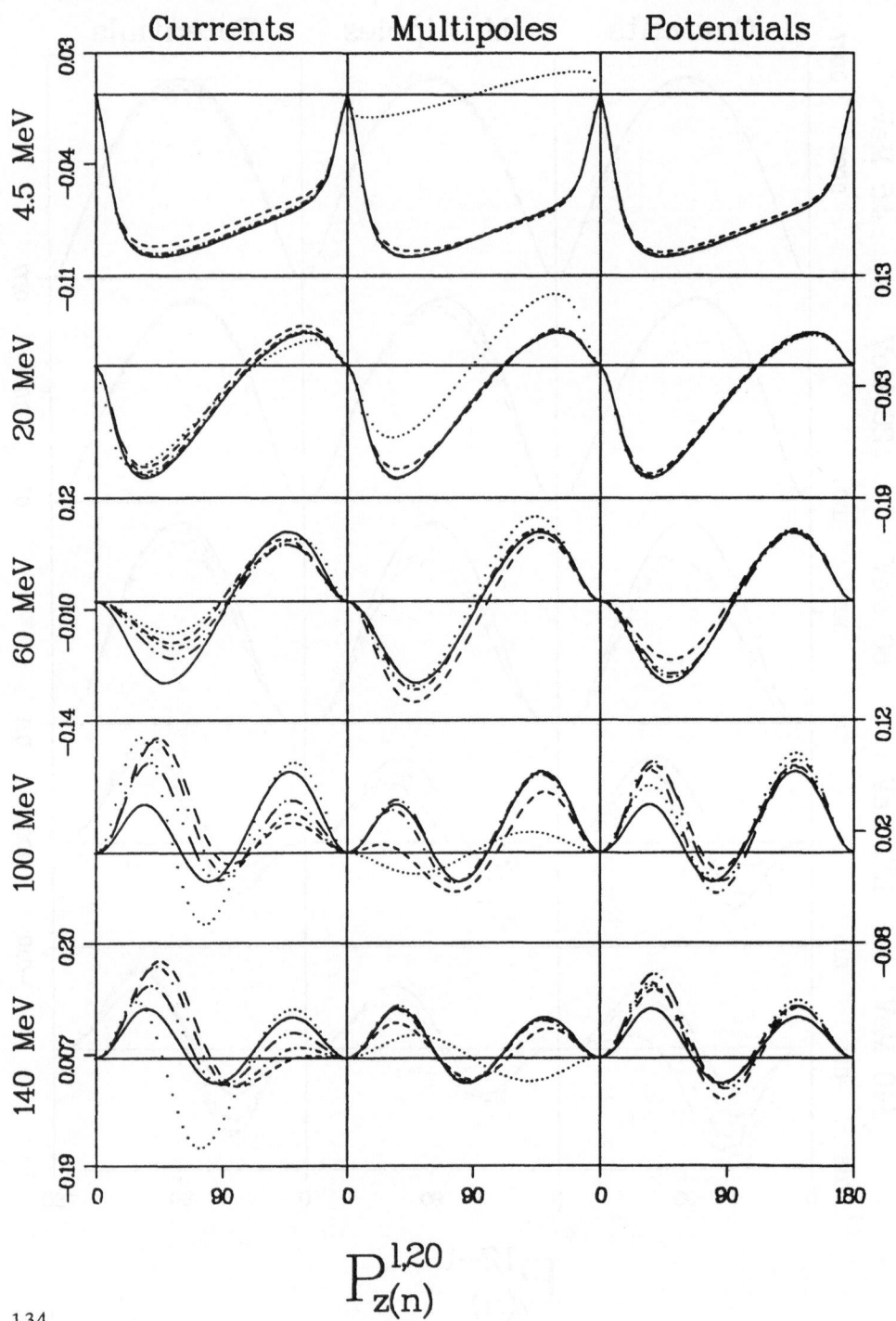

$$P_{z(n)}^{1,20}$$

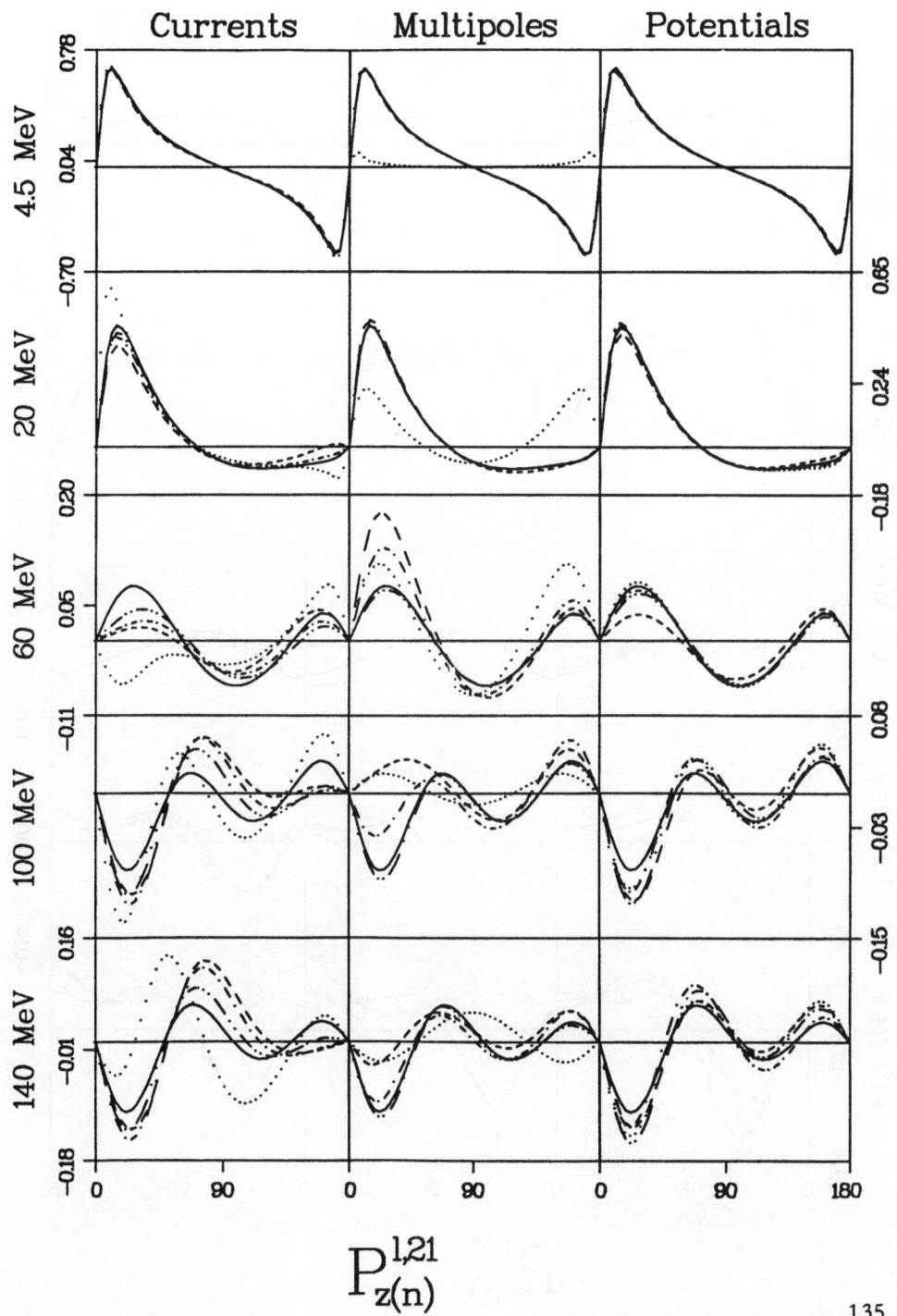

$$P_{z(n)}^{l,21}$$

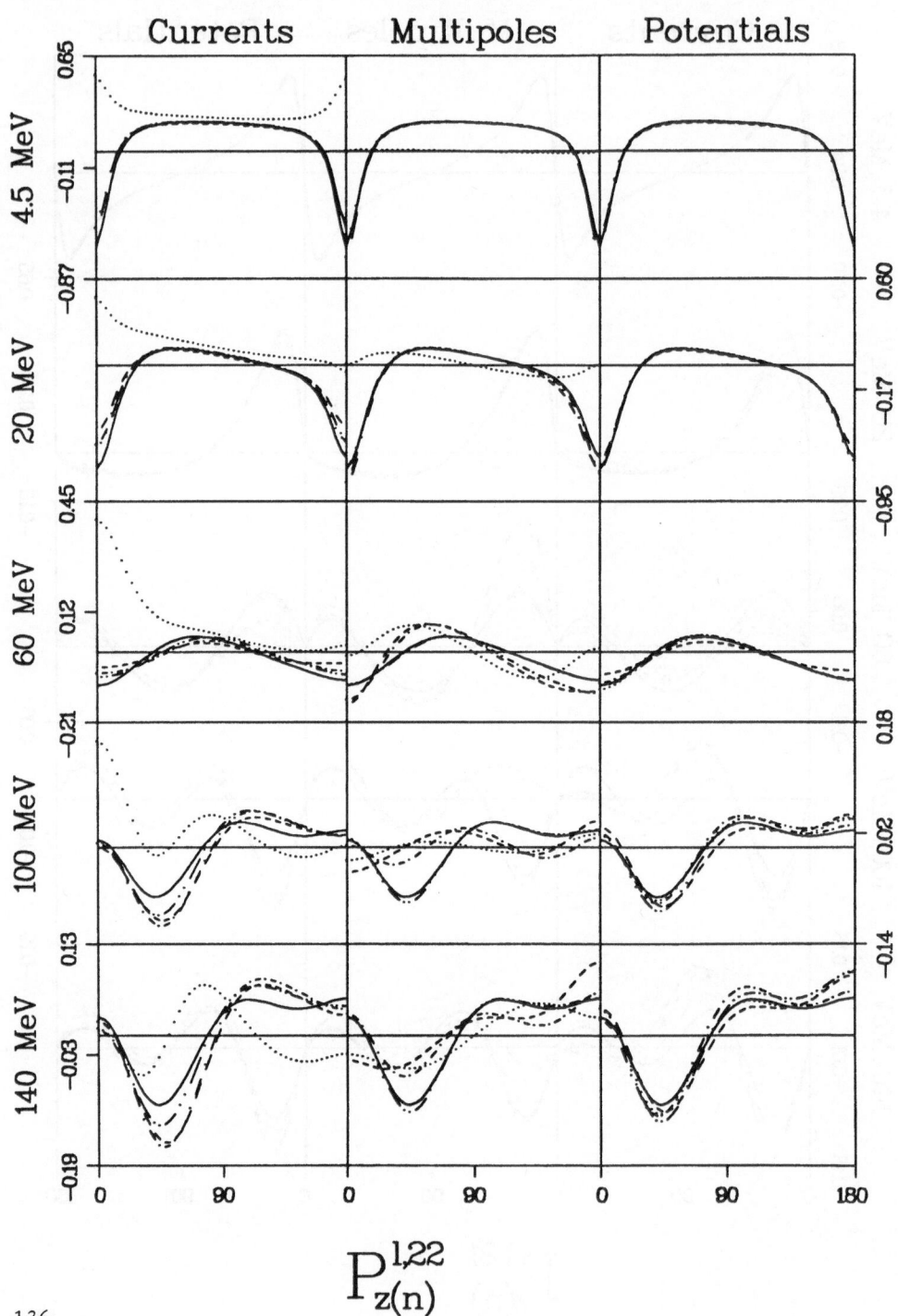

$$\mathrm{P}_{z(n)}^{l,22}$$

5.23 Proton-Neutron Spin Correlation $P_{xx}^{0,IM}$

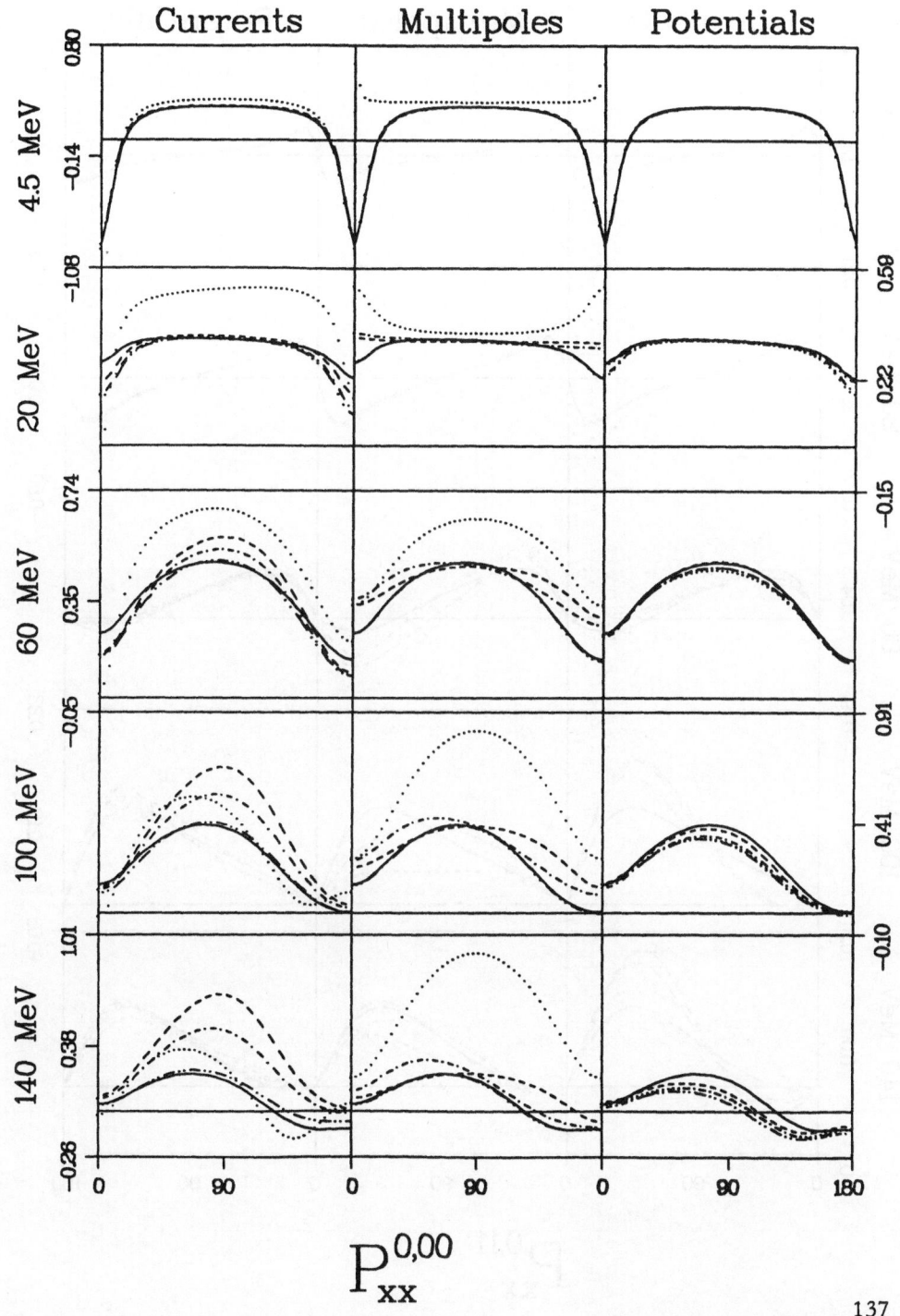

$$P_{xx}^{0,00}$$

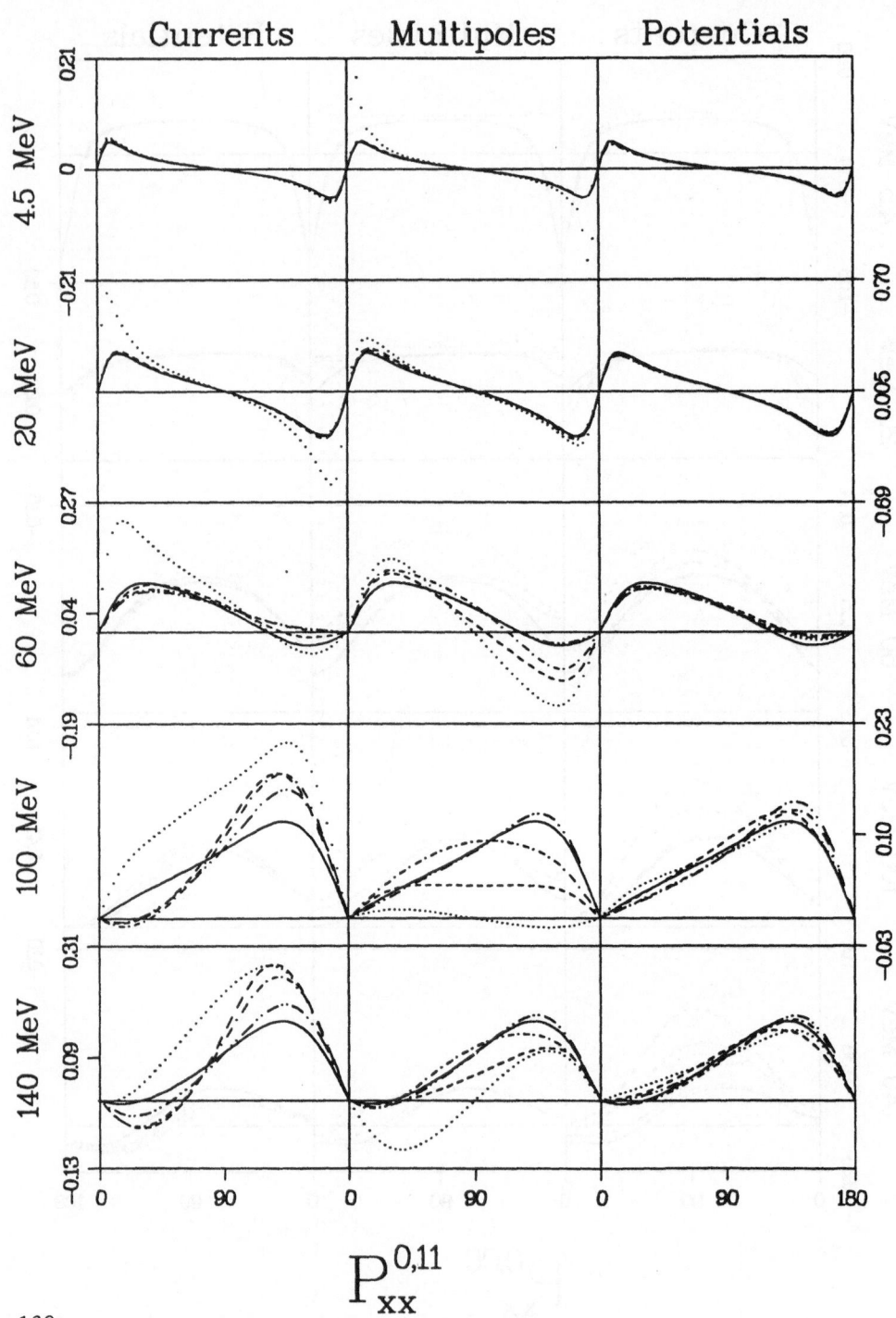

Currents   Multipoles   Potentials

$$P_{xx}^{0,11}$$

138

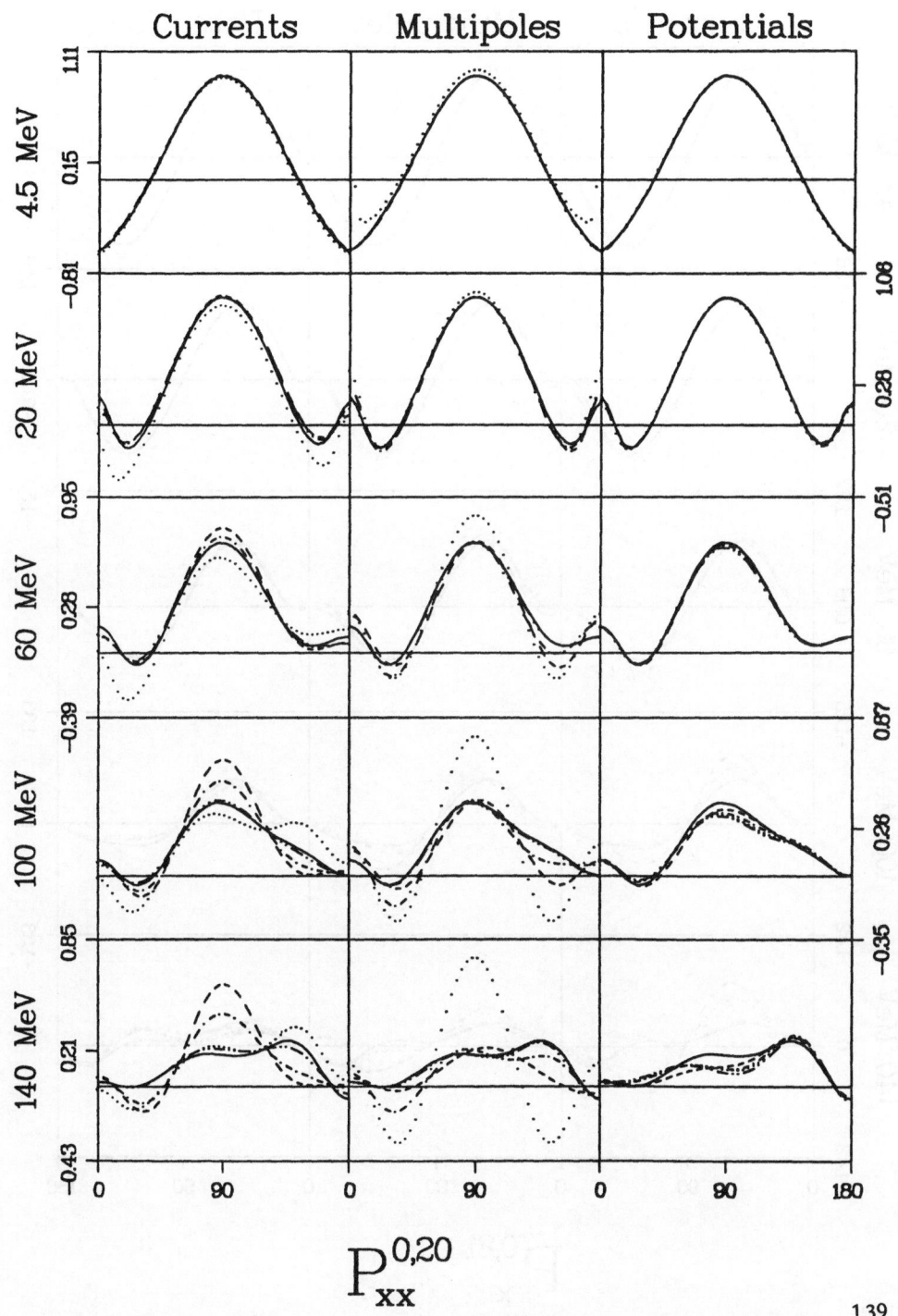

$$P_{xx}^{0,20}$$

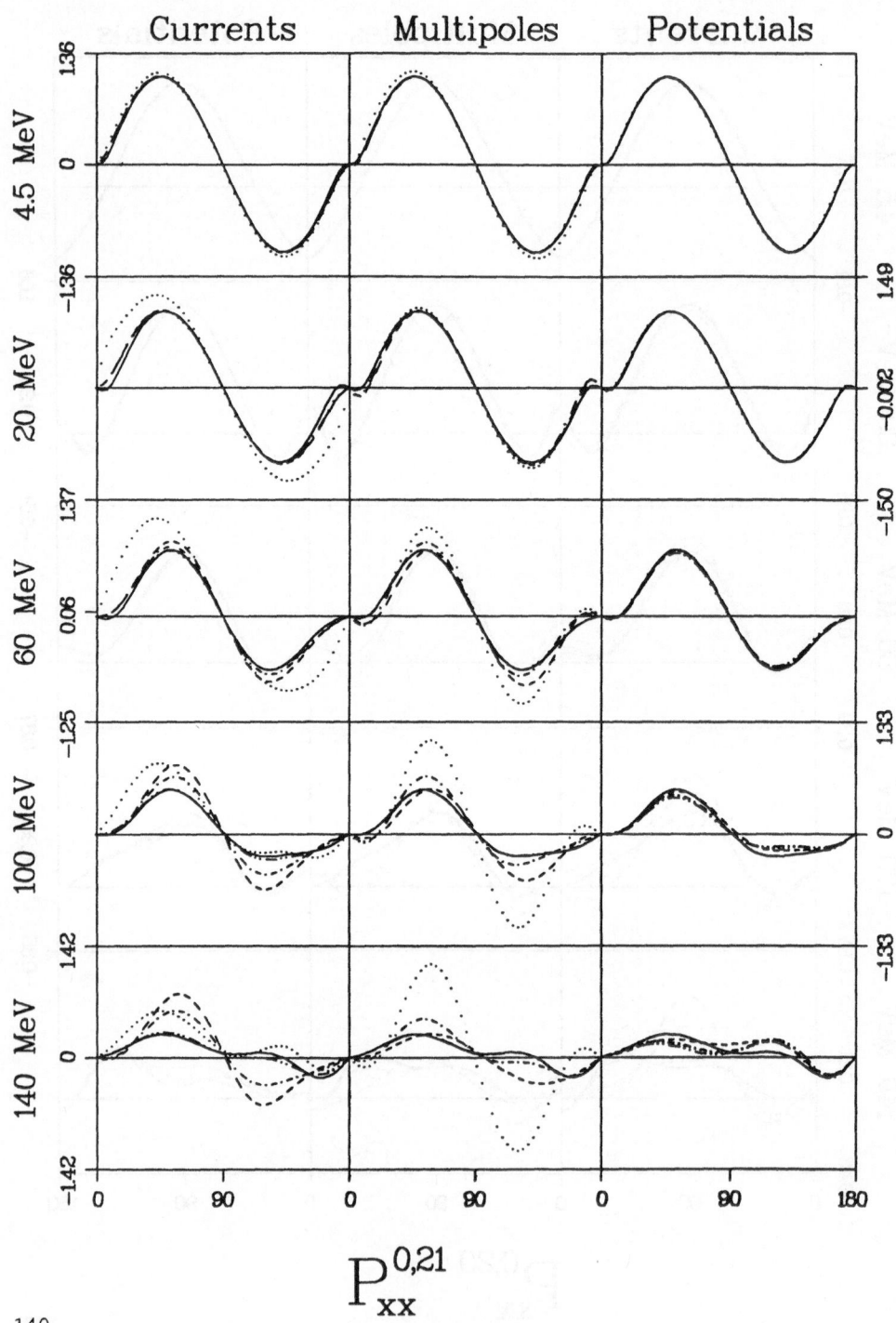

$$P_{xx}^{0,21}$$

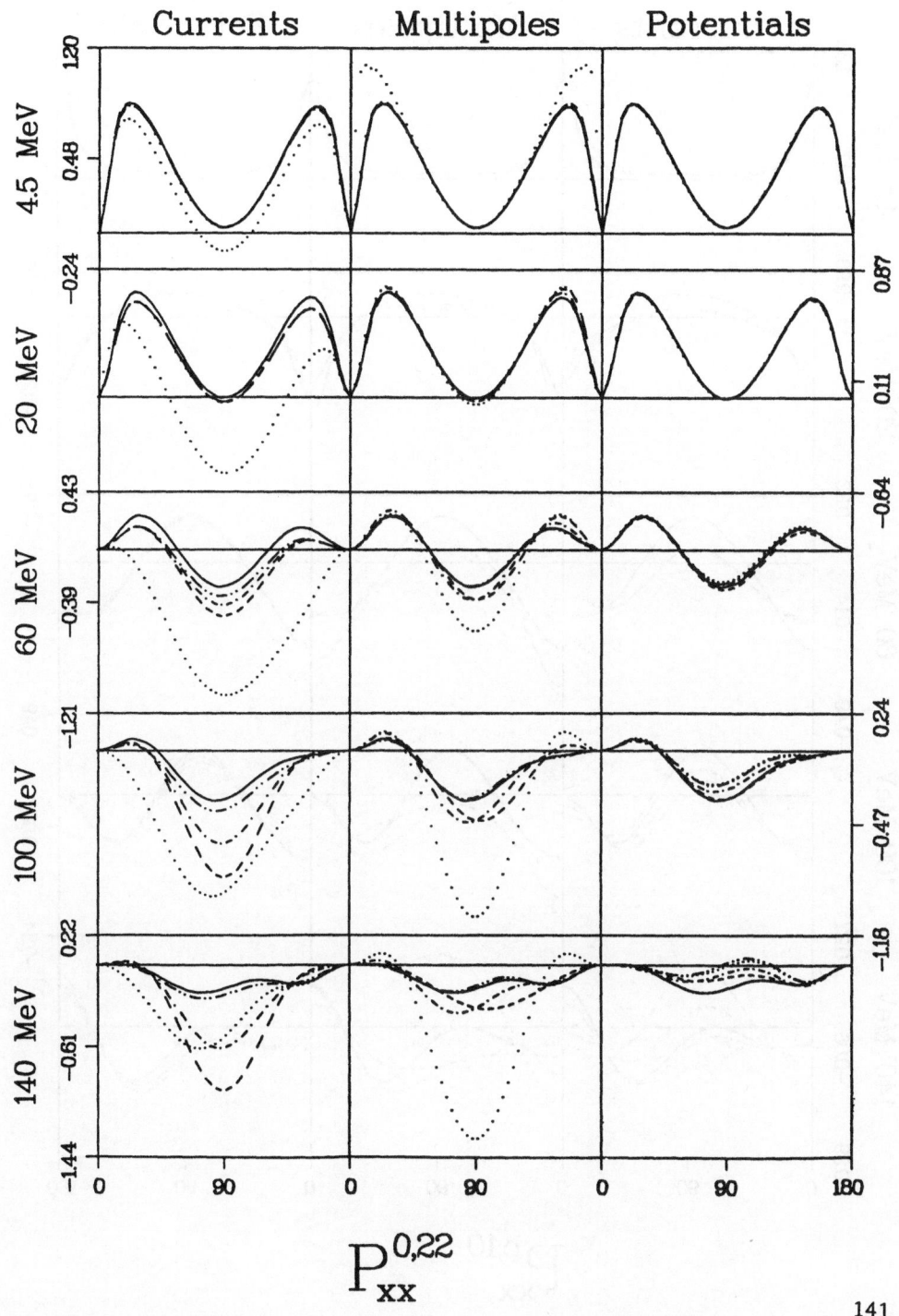

$$P_{xx}^{0,22}$$

5.24 Proton-Neutron Spin Correlation $P_{xx}^{c,IM}$

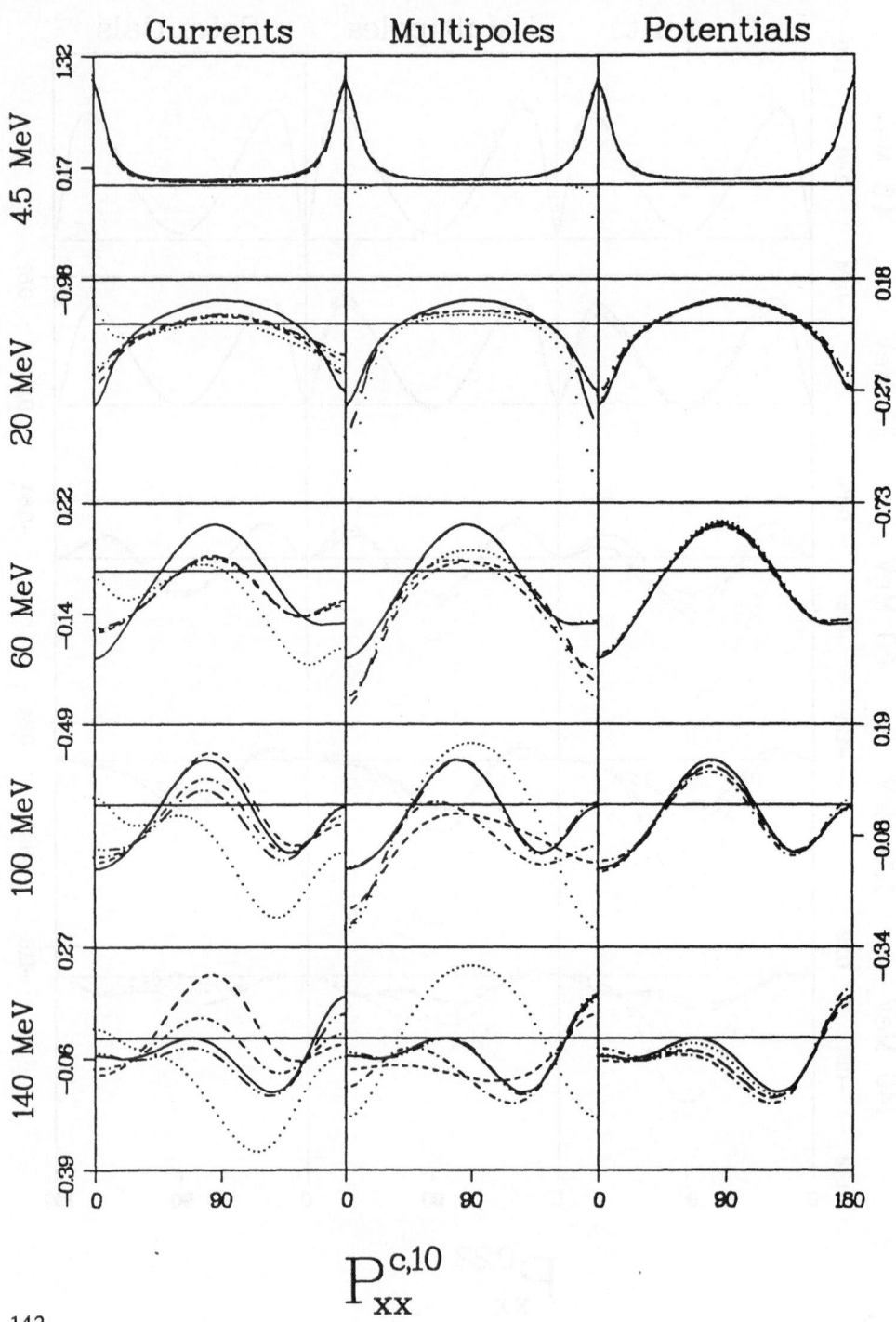

$$P_{xx}^{c,10}$$

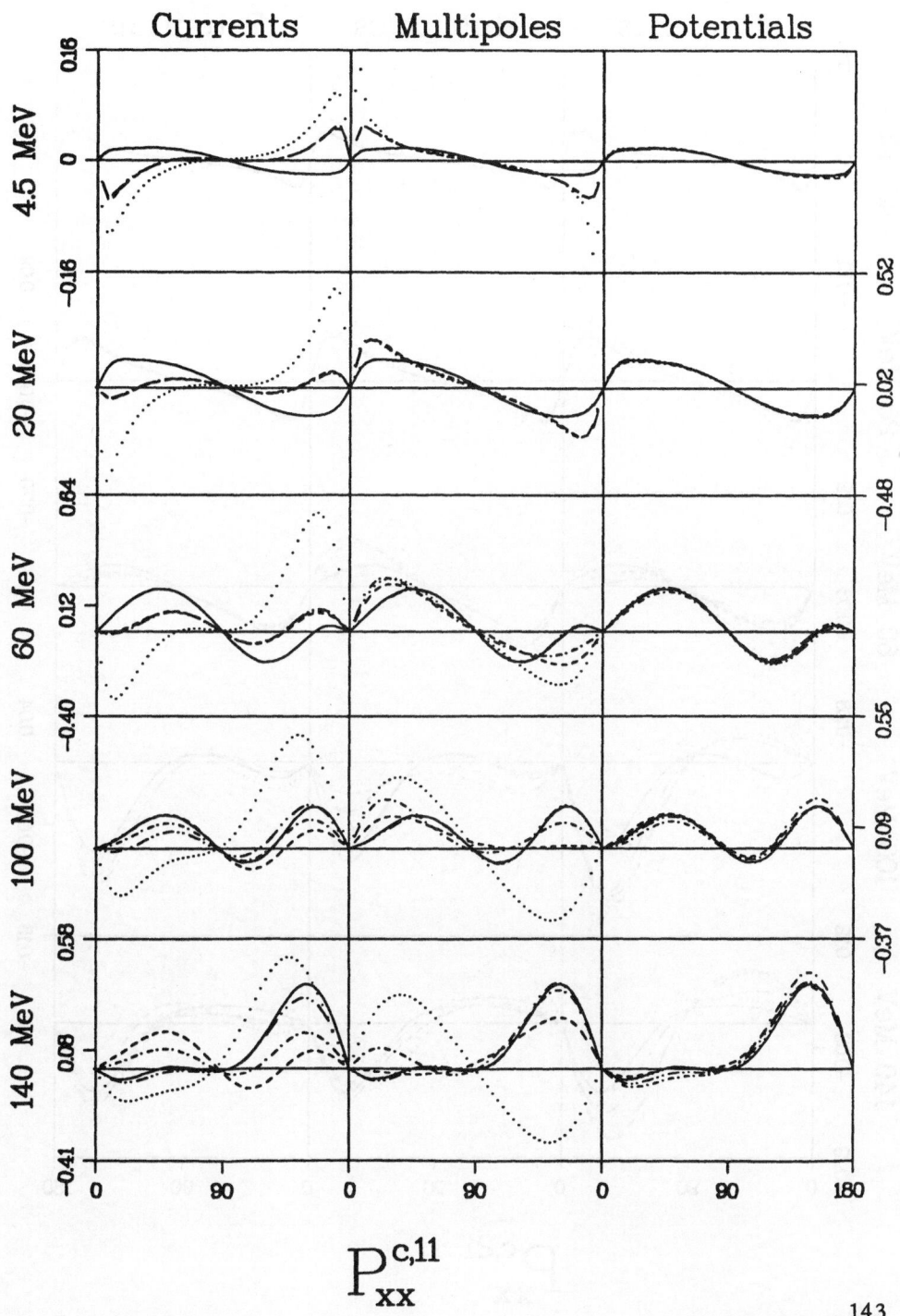

$$P_{xx}^{c,11}$$

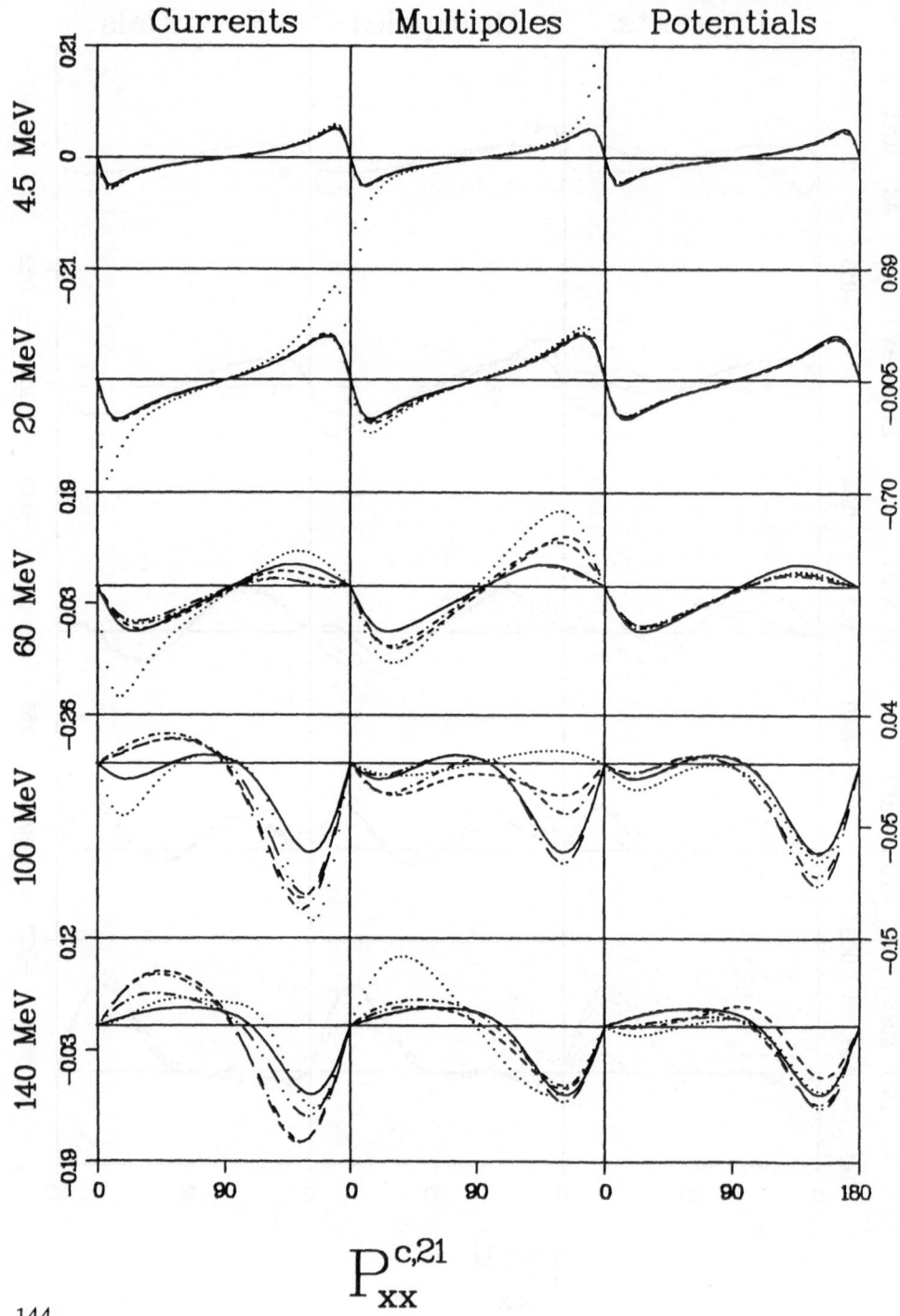

Currents   Multipoles   Potentials

$$P_{xx}^{c,21}$$

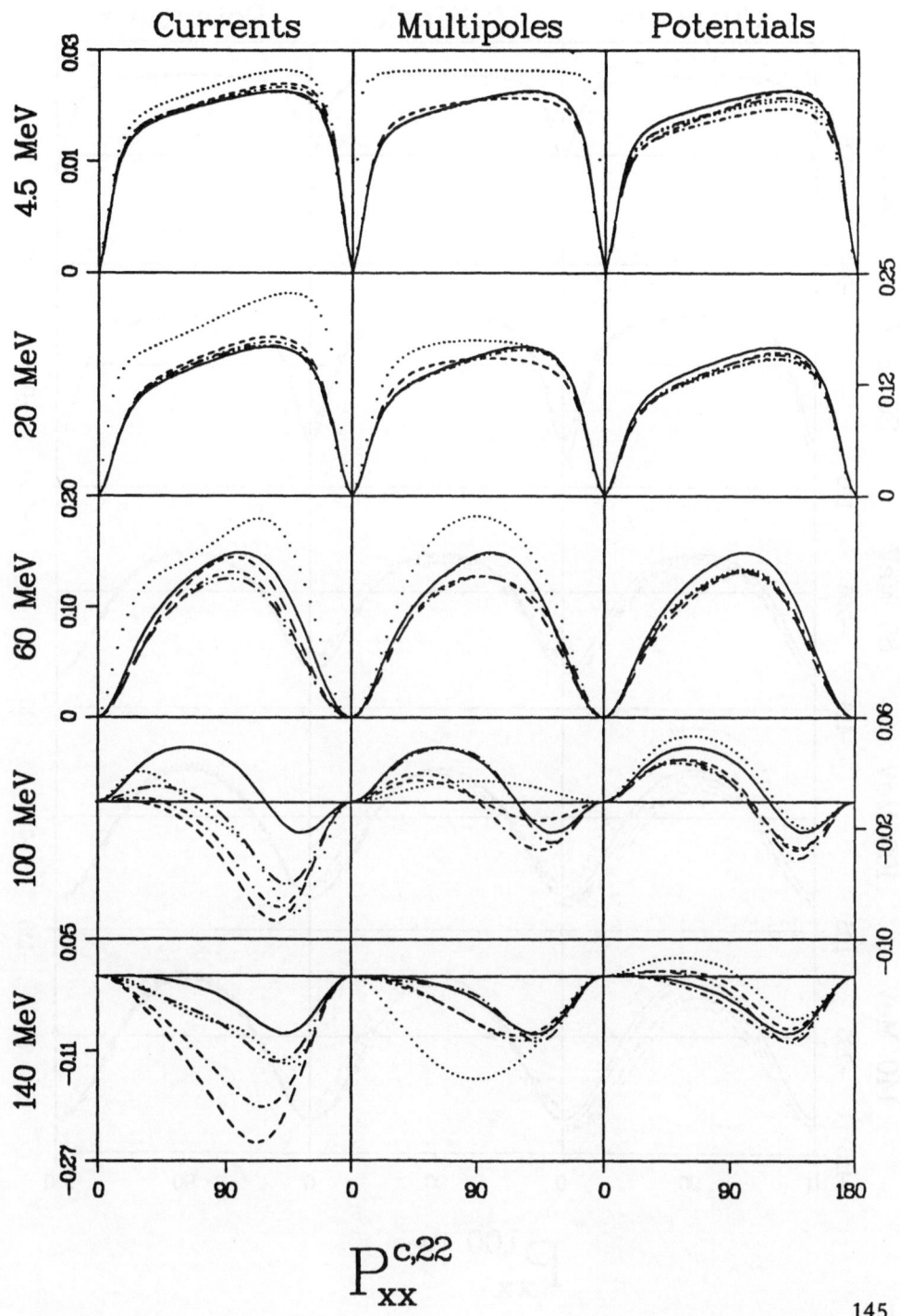

$$P_{xx}^{c,22}$$

5.25 Proton-Neutron Spin Correlation $P_{xx}^{l,IM}$

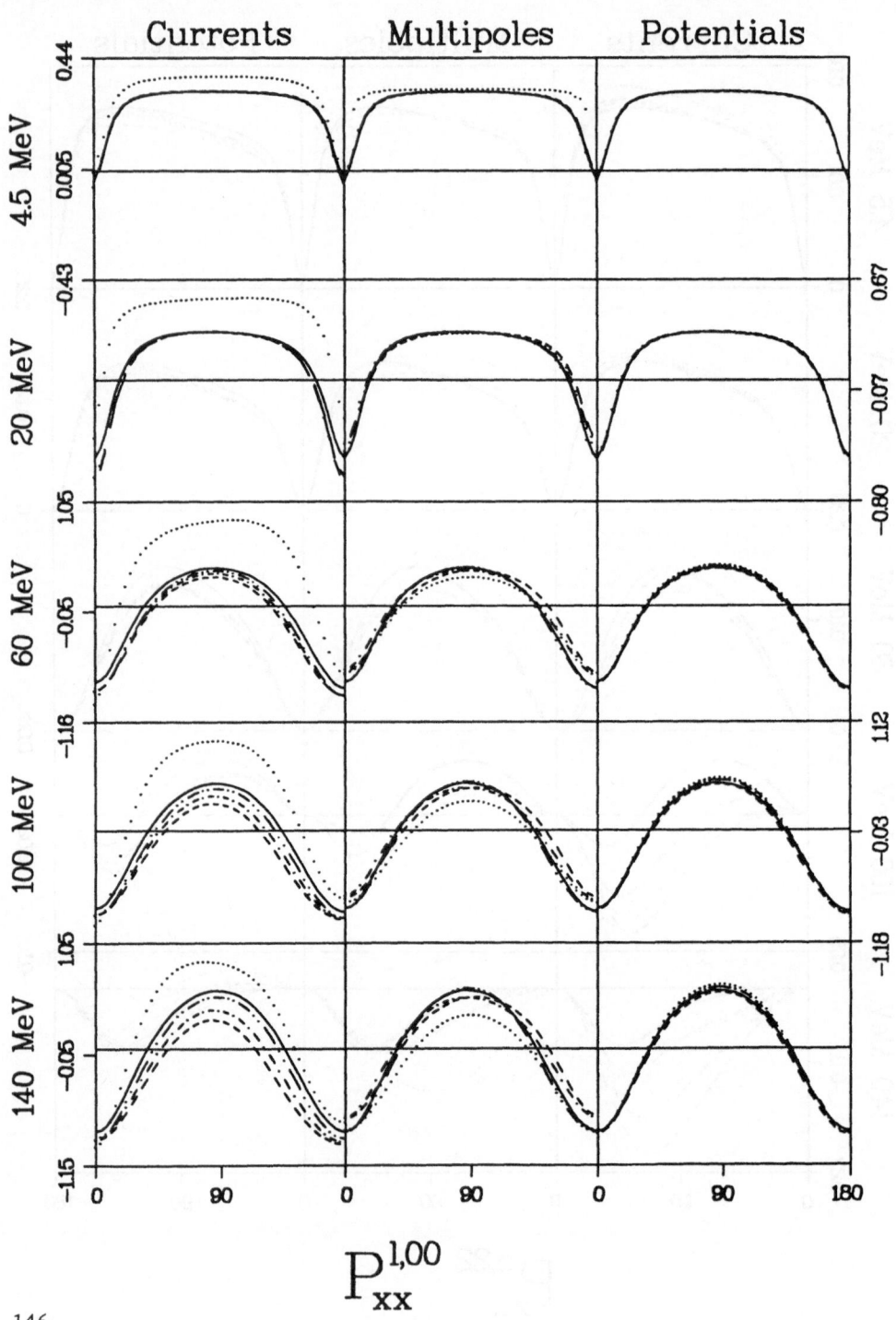

$$P_{xx}^{1,00}$$

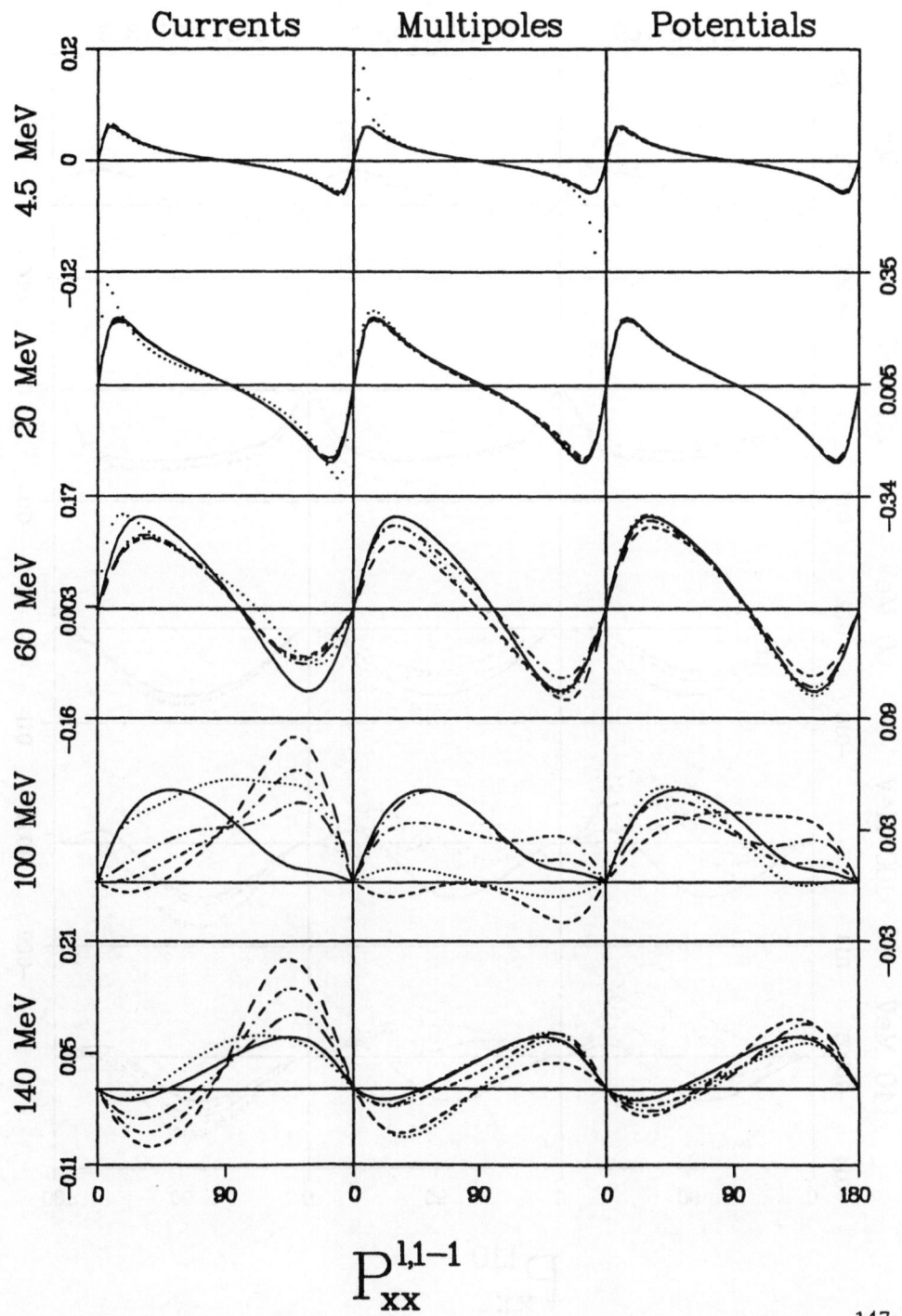

$$P_{xx}^{l,1-1}$$

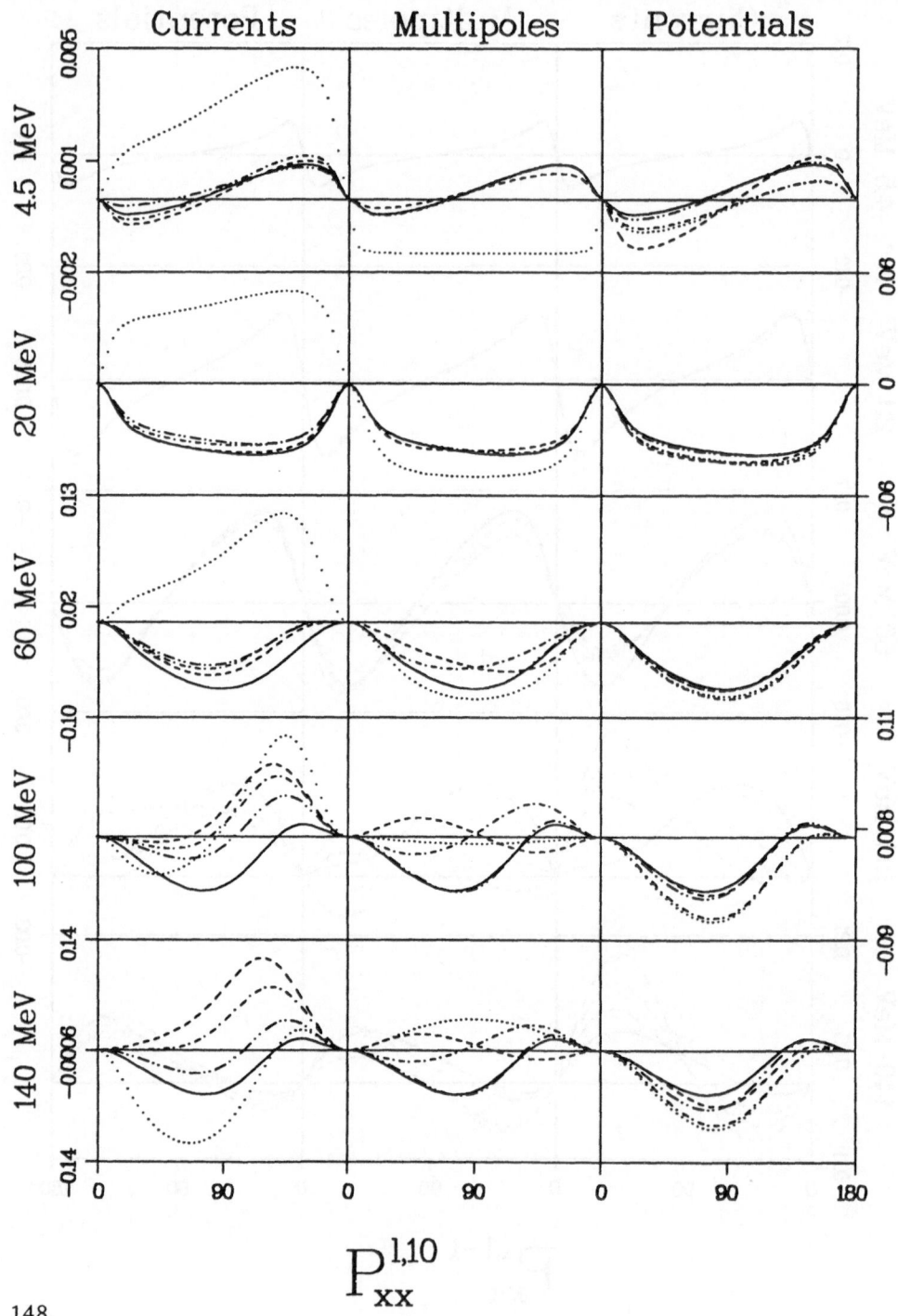

$$P_{xx}^{1,10}$$

148

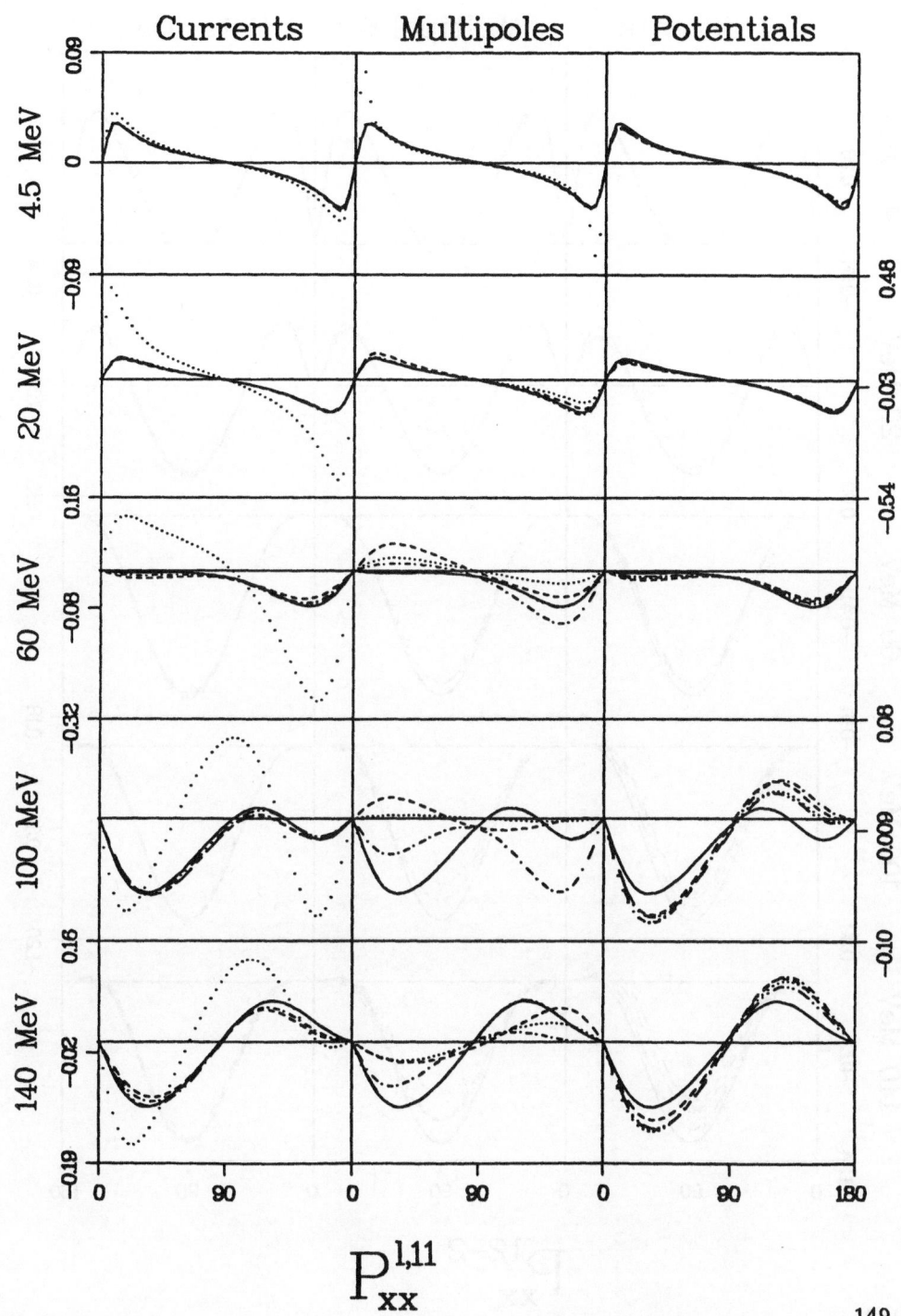

Currents   Multipoles   Potentials

4.5 MeV

20 MeV

60 MeV

100 MeV

140 MeV

$$P_{xx}^{l,11}$$

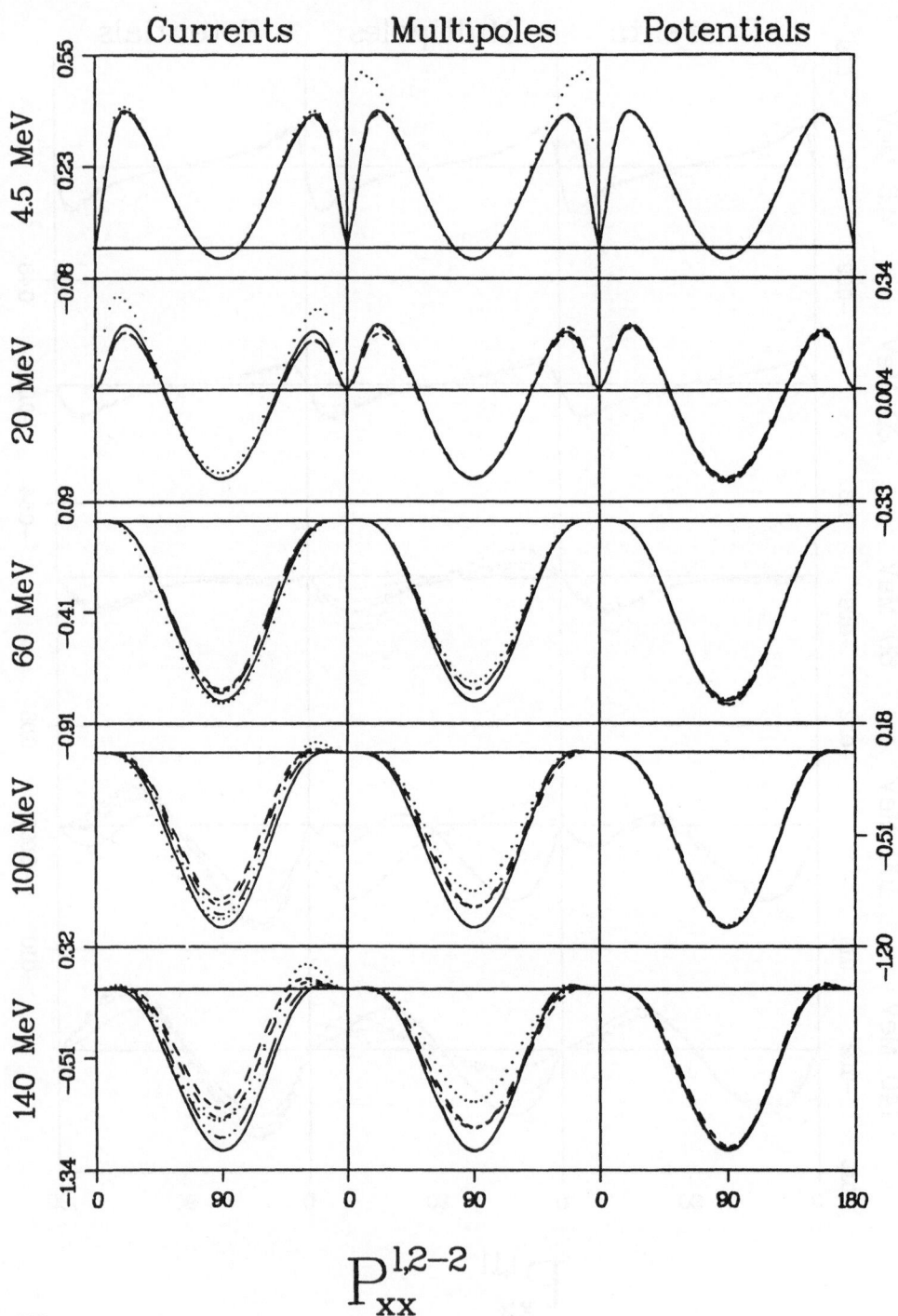

Currents Multipoles Potentials

$$P_{xx}^{1,2-2}$$

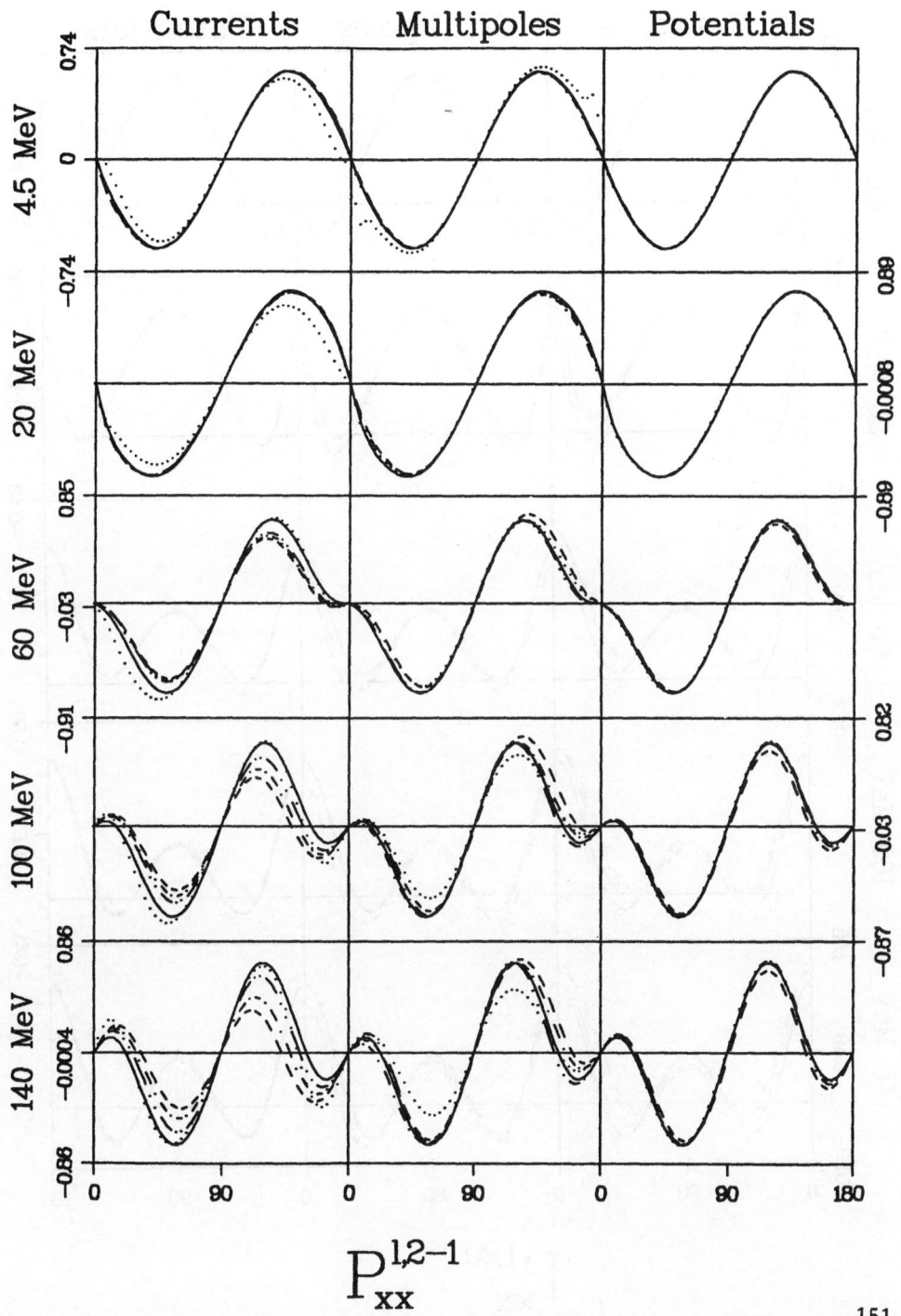

Currents   Multipoles   Potentials

4.5 MeV   20 MeV   60 MeV   100 MeV   140 MeV

$$P_{xx}^{1,2-1}$$

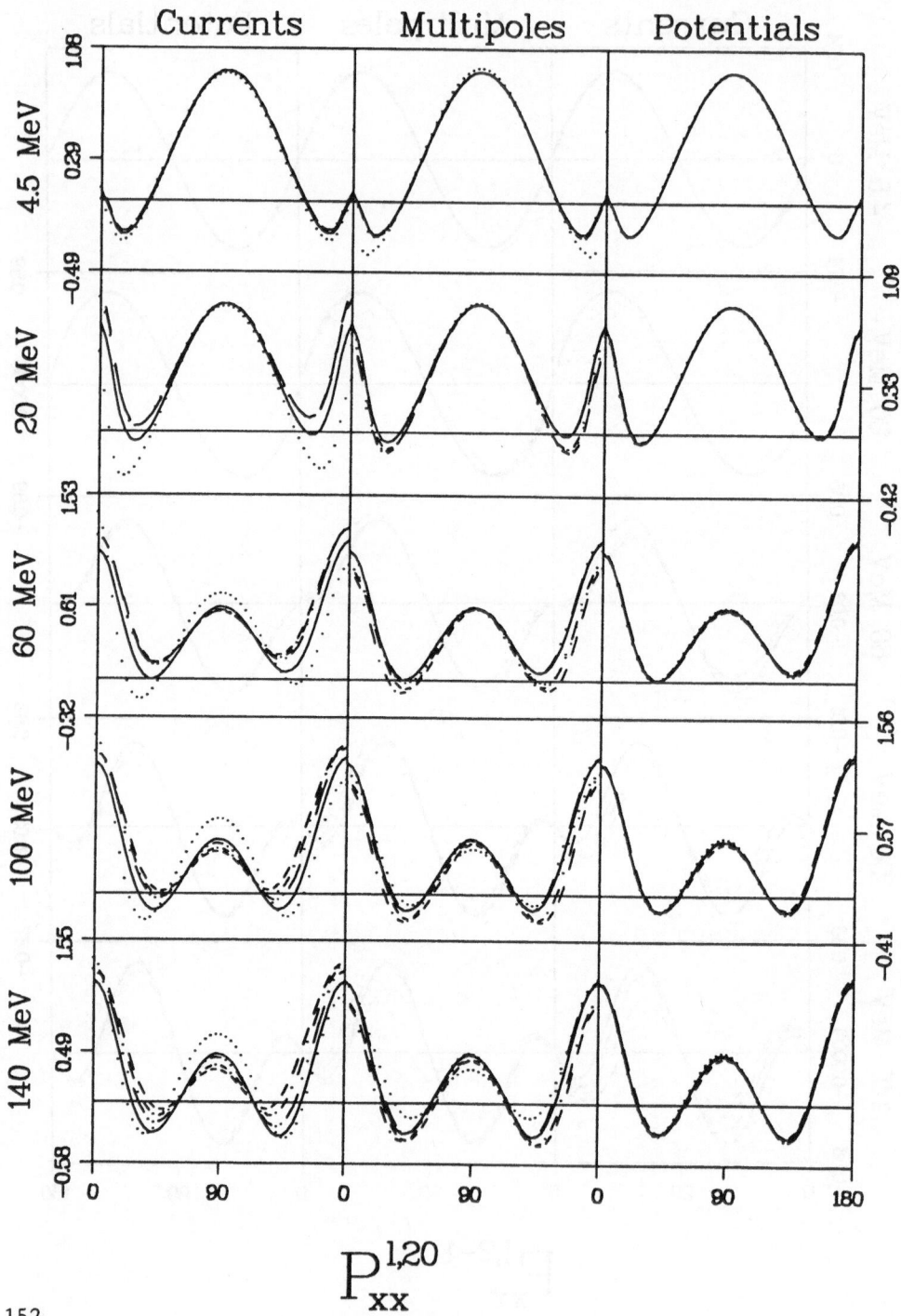

$$P_{xx}^{1,20}$$

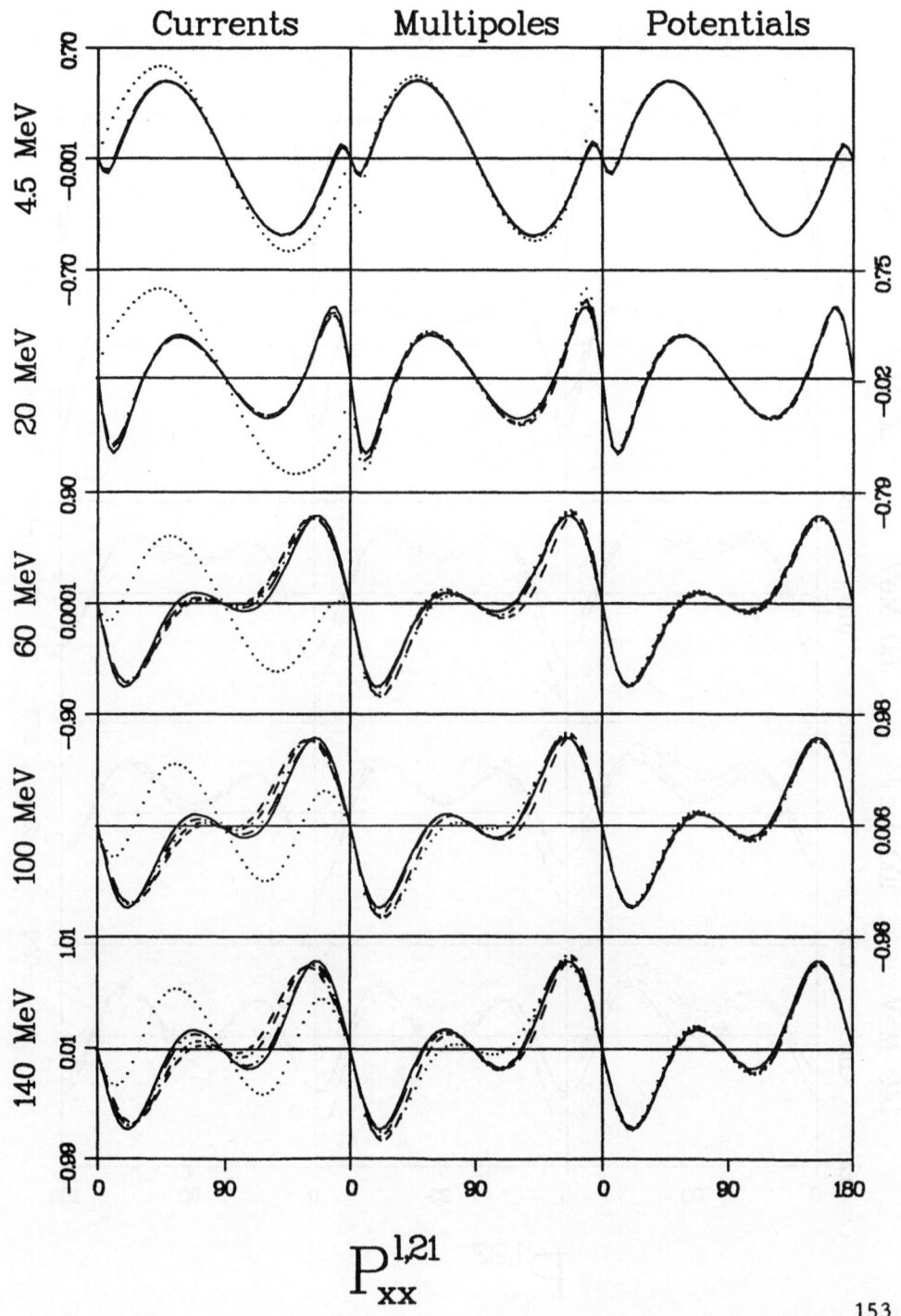

$$P_{xx}^{1,21}$$

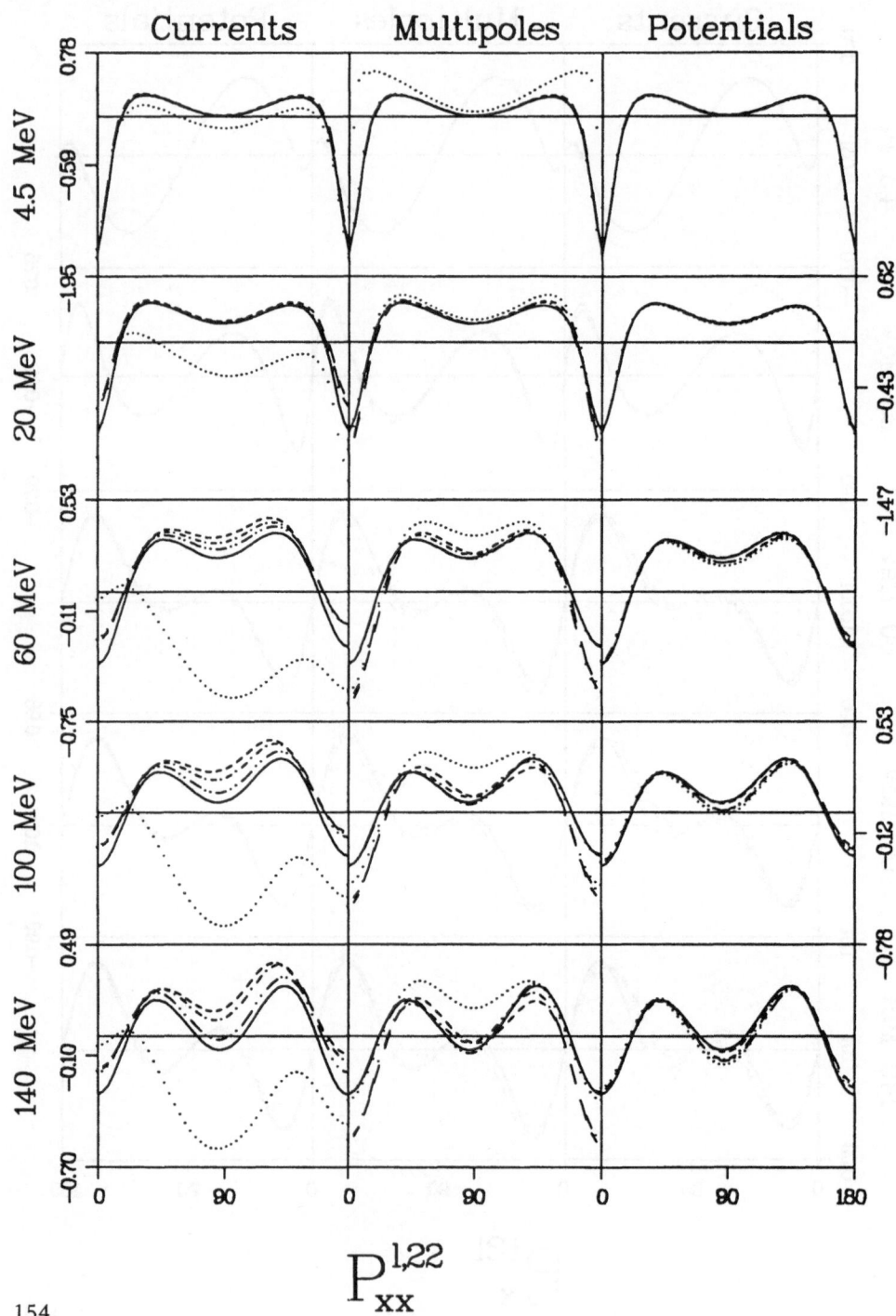

$$P_{xx}^{1,22}$$

5.26 Proton-Neutron Spin Correlation $P_{yy}^{0,IM}$

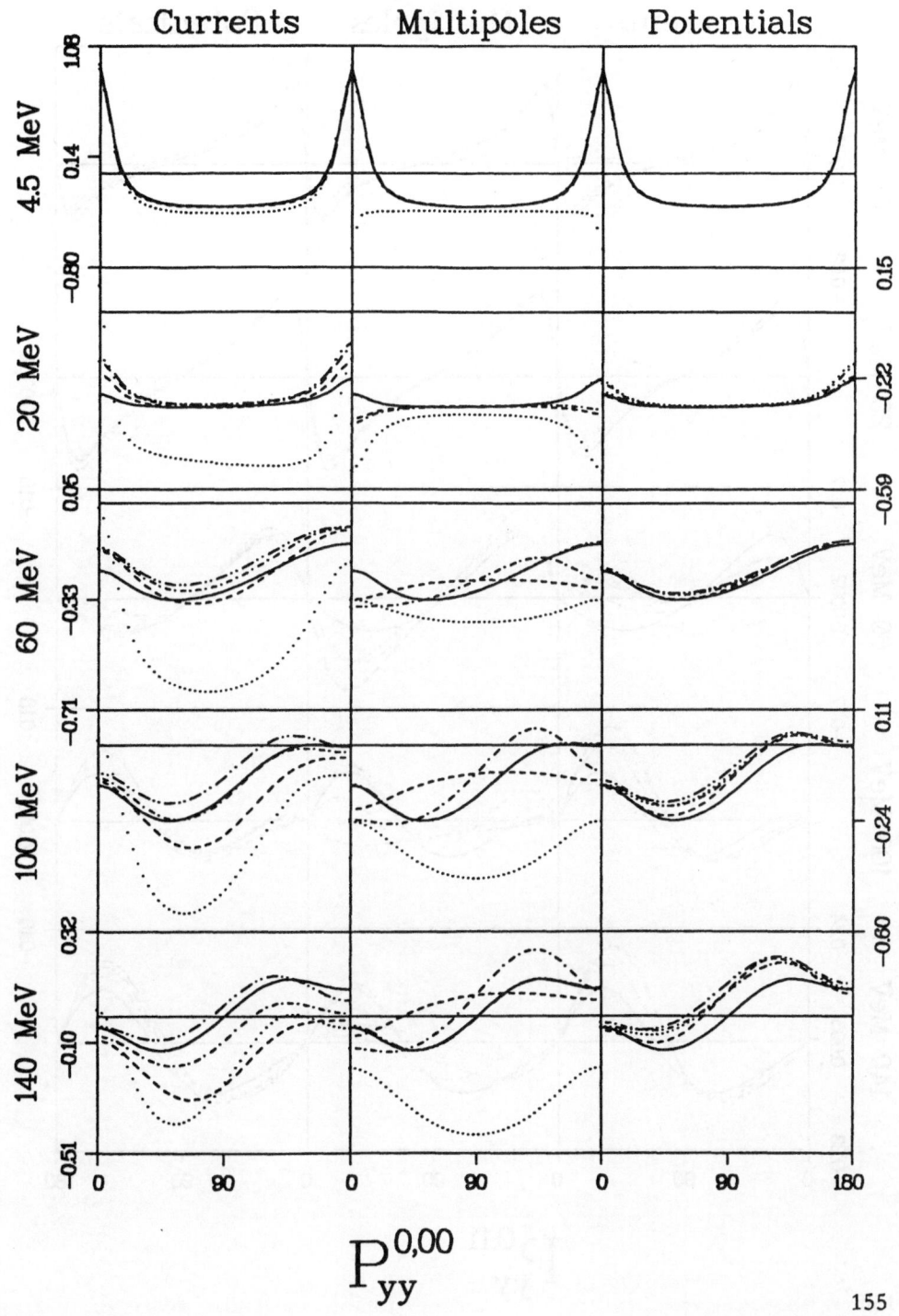

$$P_{yy}^{0,00}$$

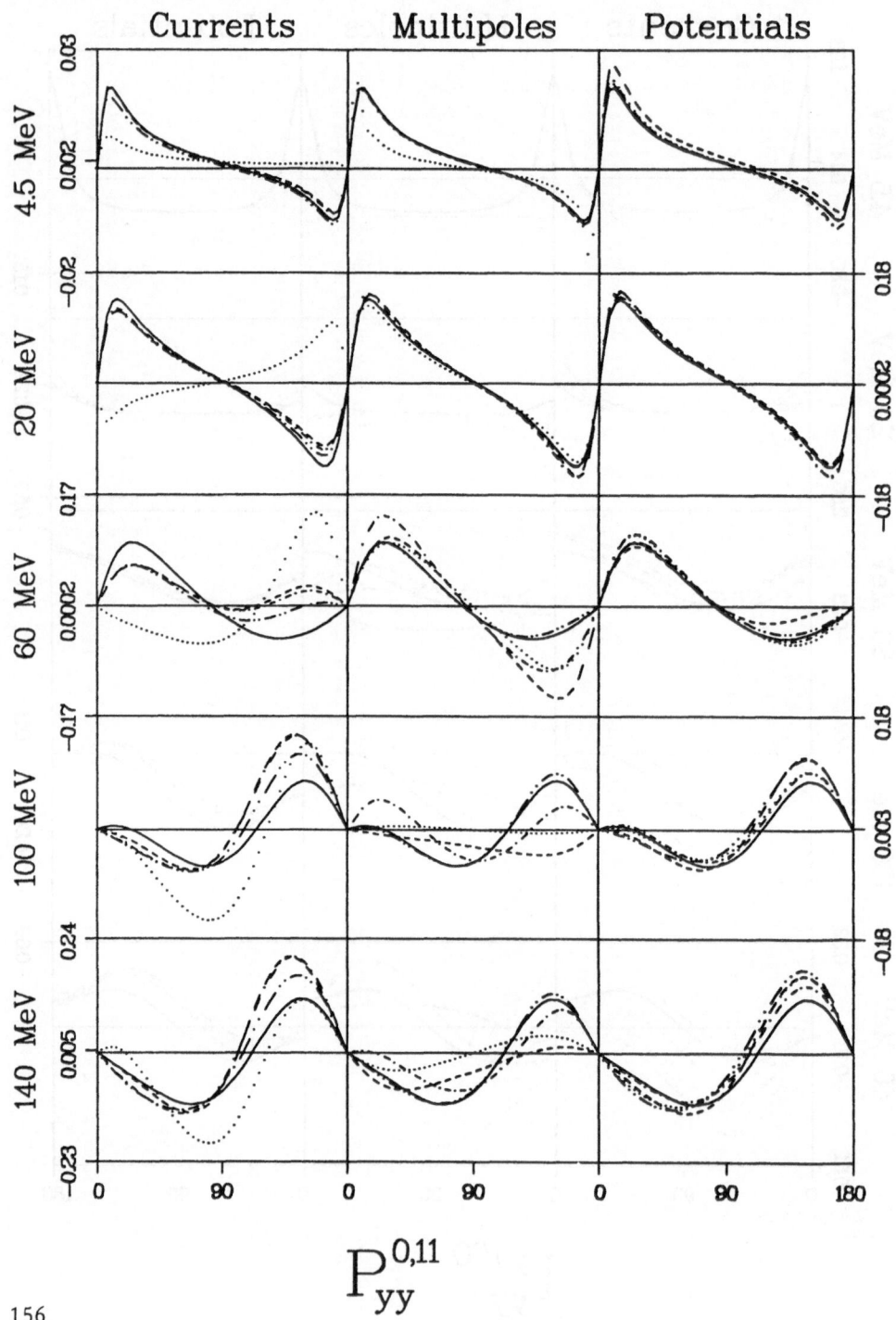

$$P_{yy}^{0,11}$$

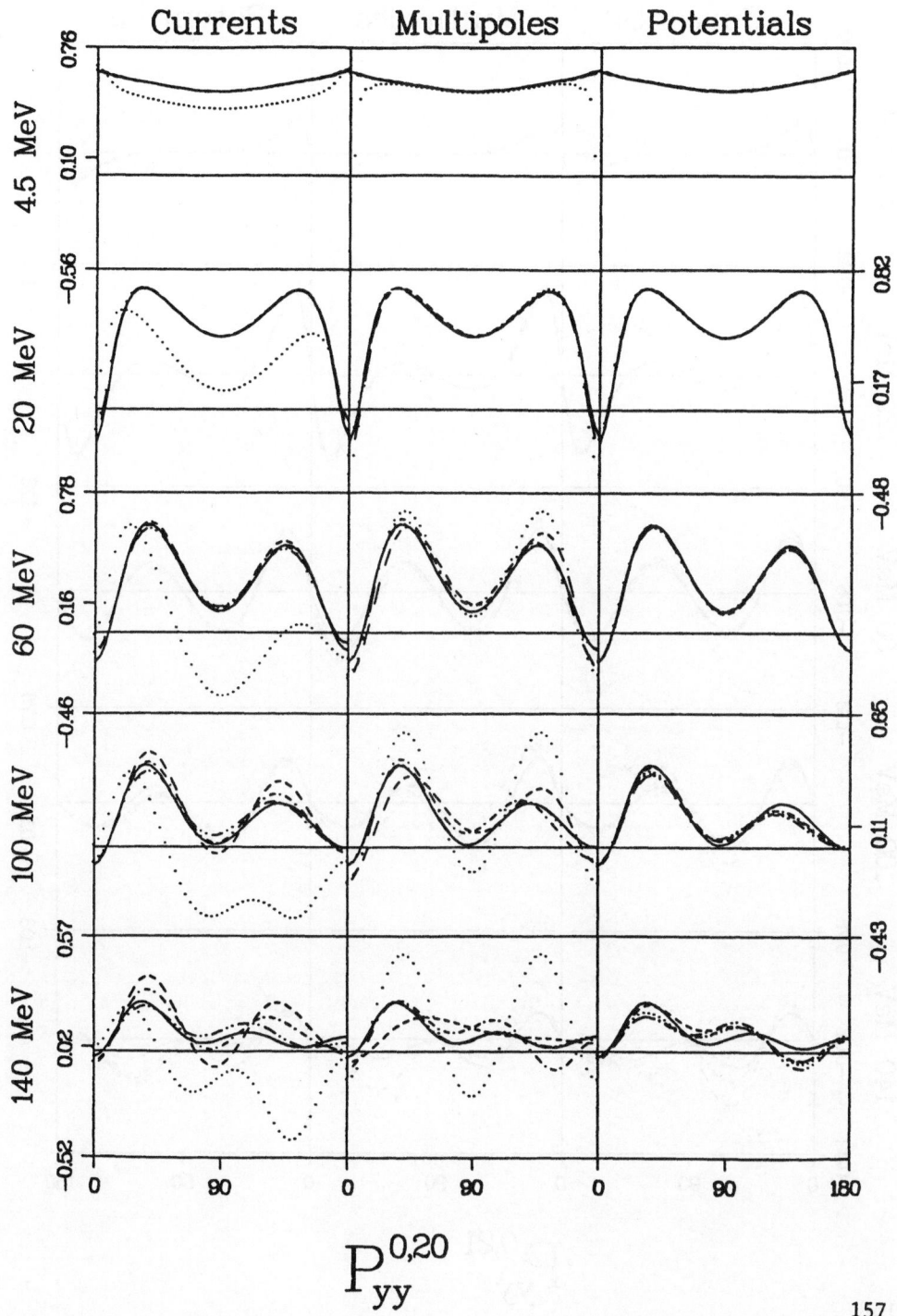

$$P_{yy}^{0,20}$$

157

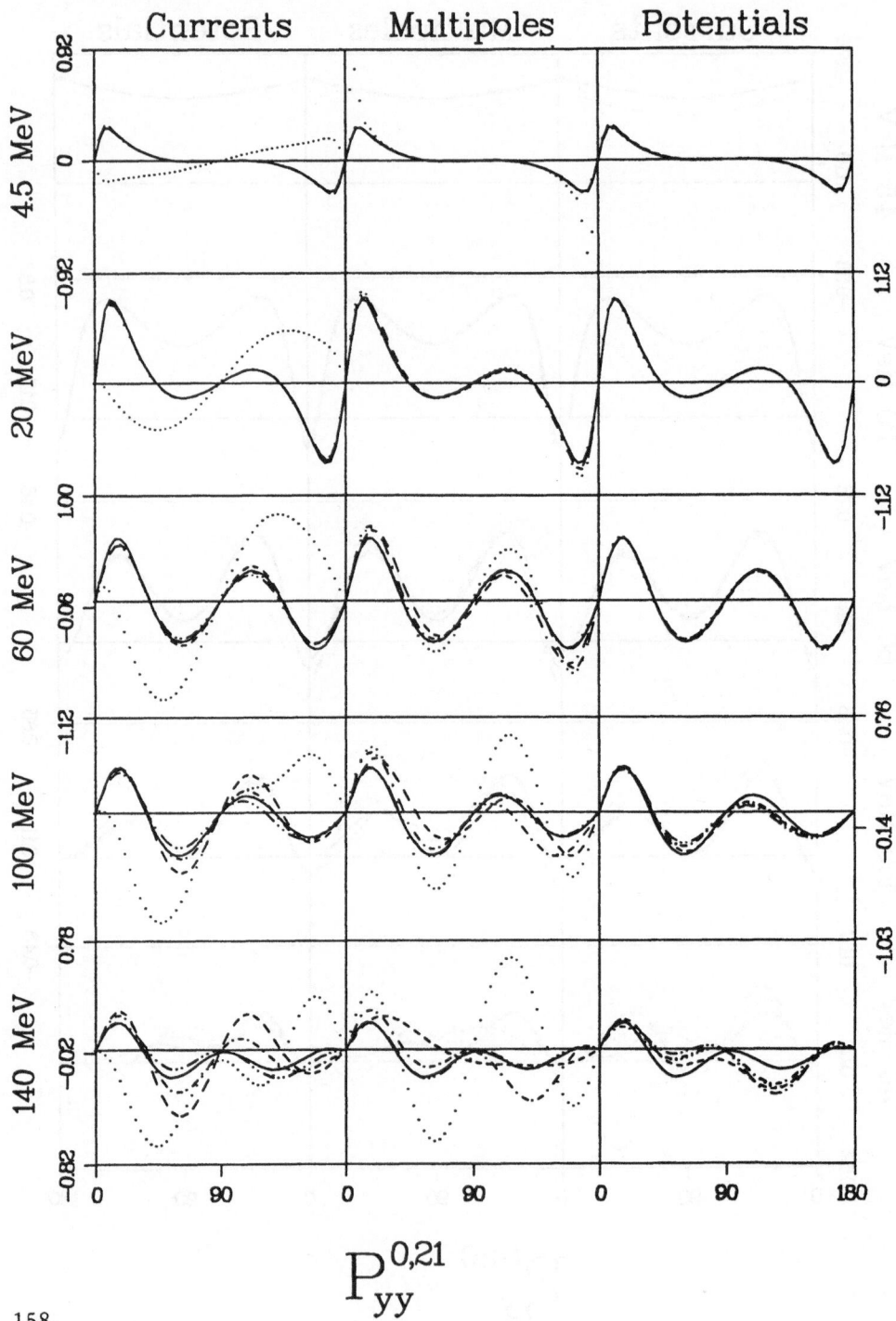

Currents Multipoles Potentials

4.5 MeV 20 MeV 60 MeV 100 MeV 140 MeV

$$P_{yy}^{0,21}$$

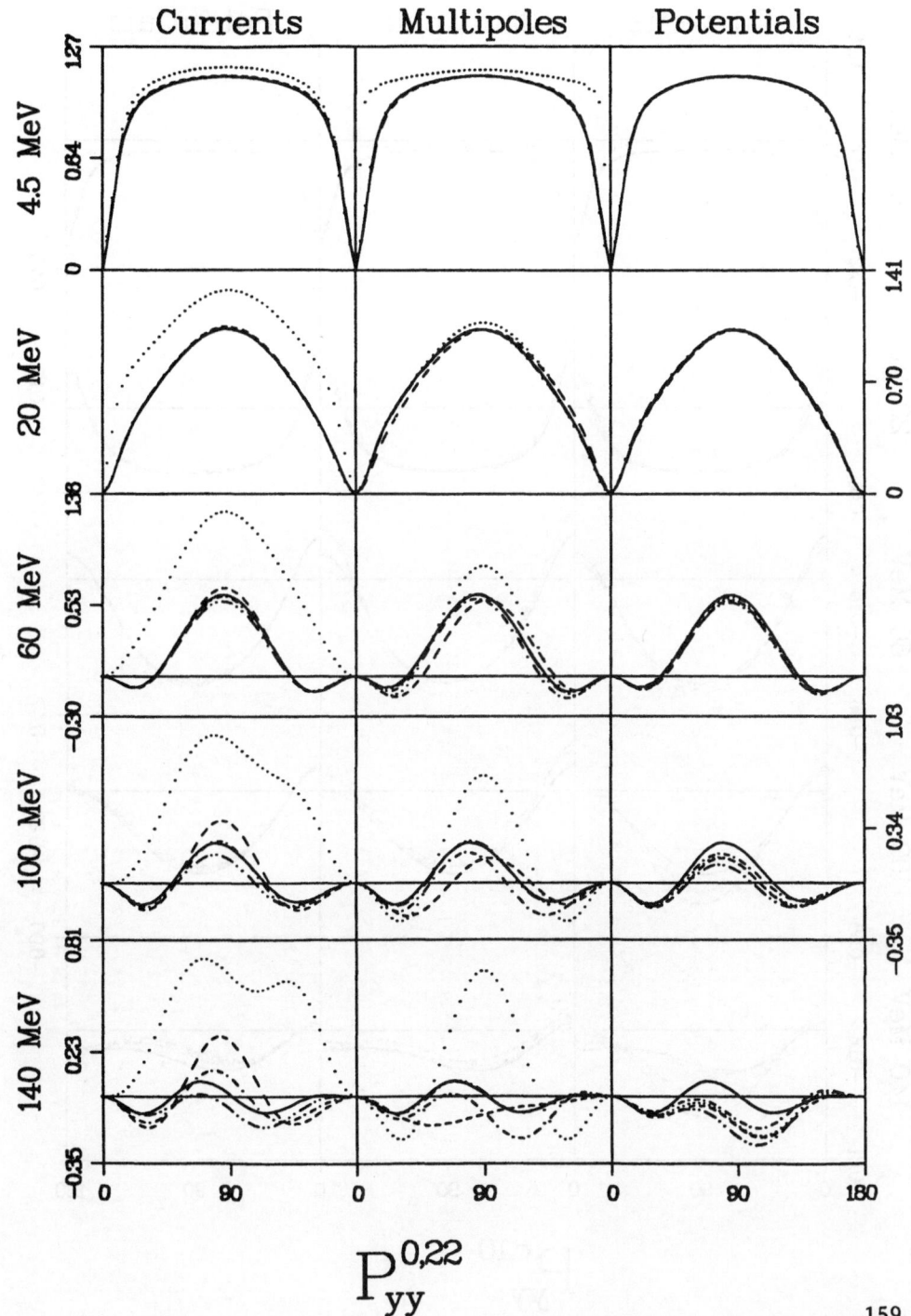

$$P_{yy}^{0,22}$$

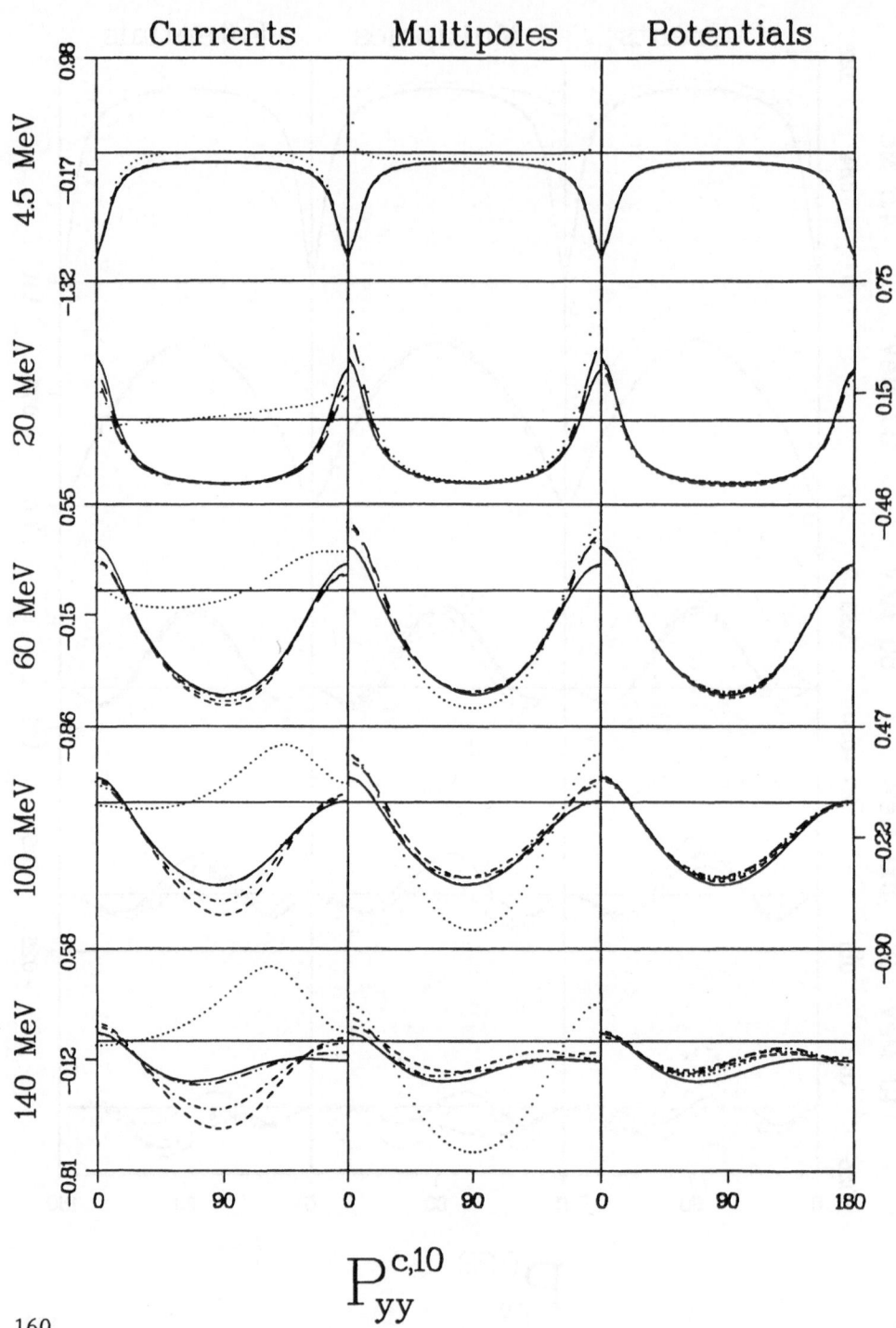

$$P_{yy}^{c,10}$$

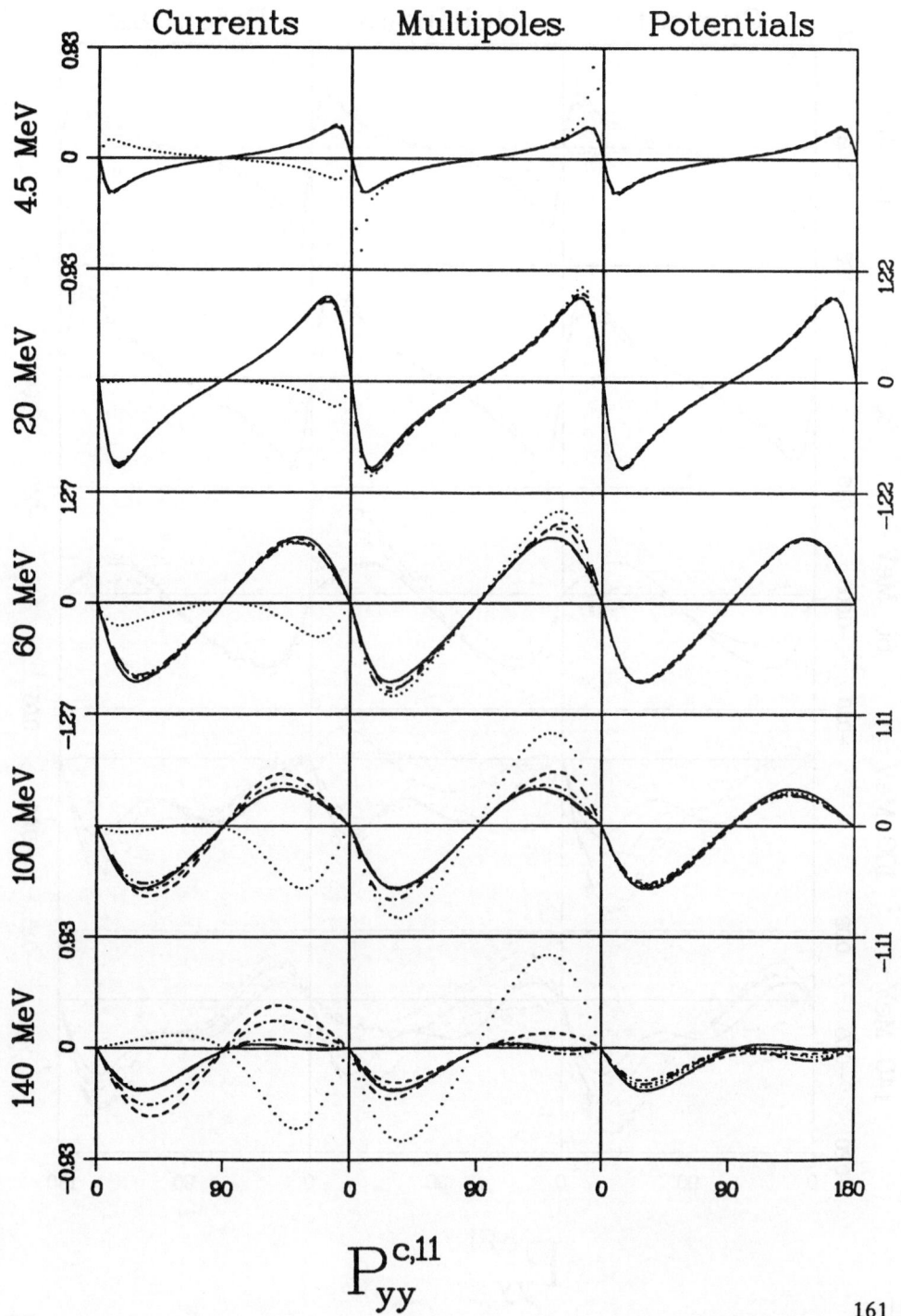

Currents   Multipoles   Potentials

4.5 MeV   20 MeV   60 MeV   100 MeV   140 MeV

$$P_{yy}^{c,11}$$

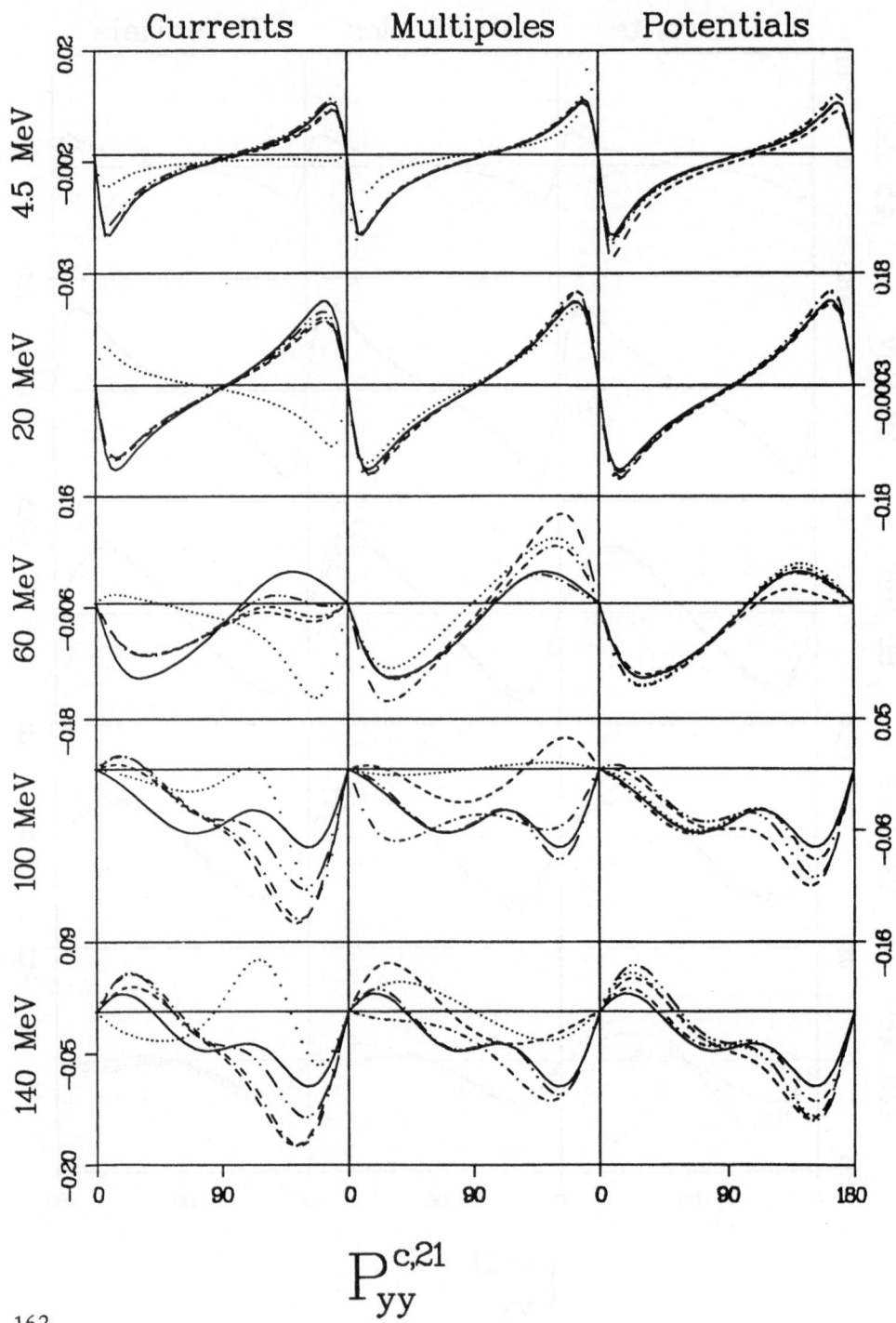

Currents   Multipoles   Potentials

4.5 MeV

20 MeV

60 MeV

100 MeV

140 MeV

$$P_{yy}^{c,21}$$

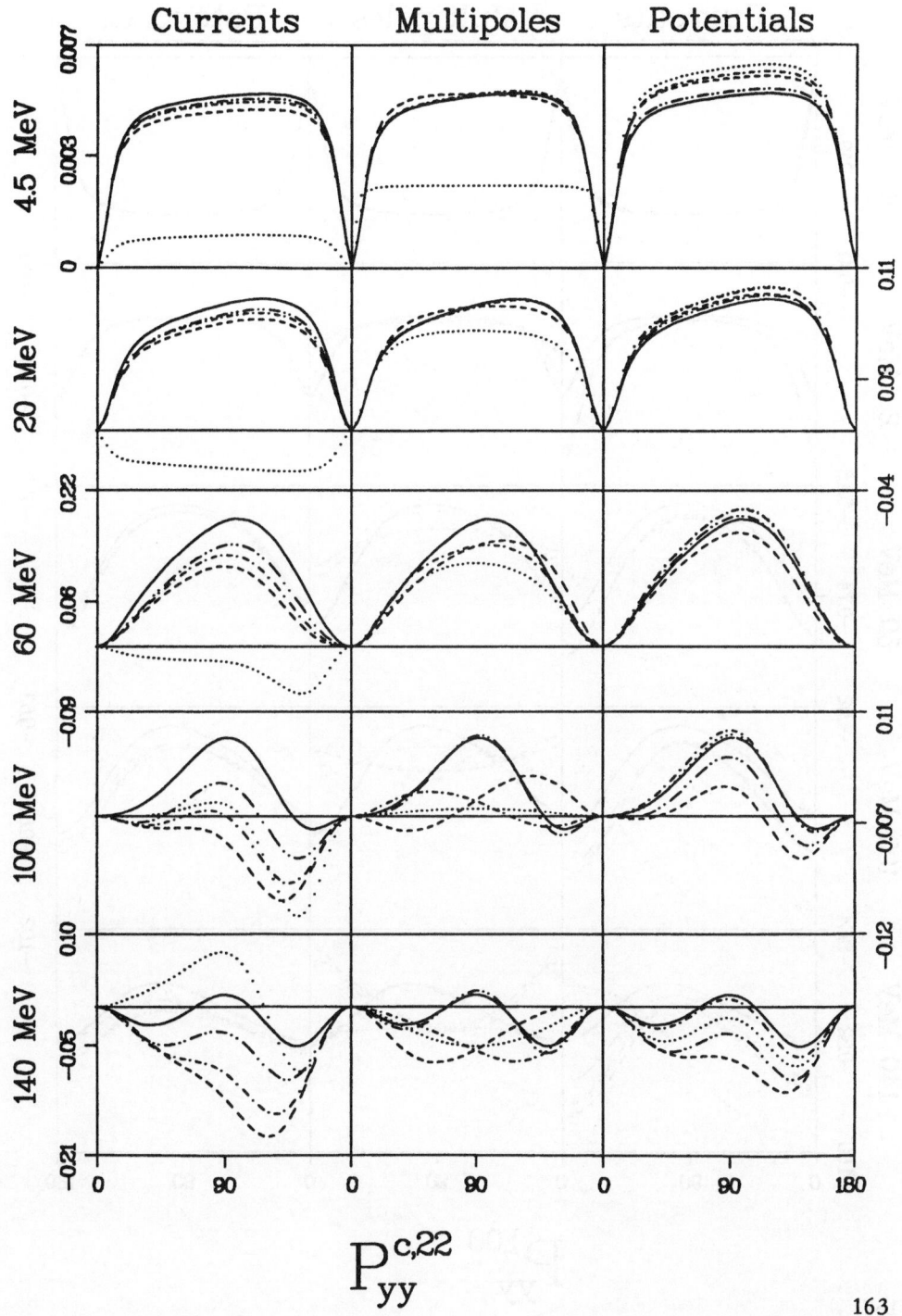

$$P_{yy}^{c,22}$$

5.28 Proton-Neutron Spin Correlation $P^{l,IM}_{yy}$

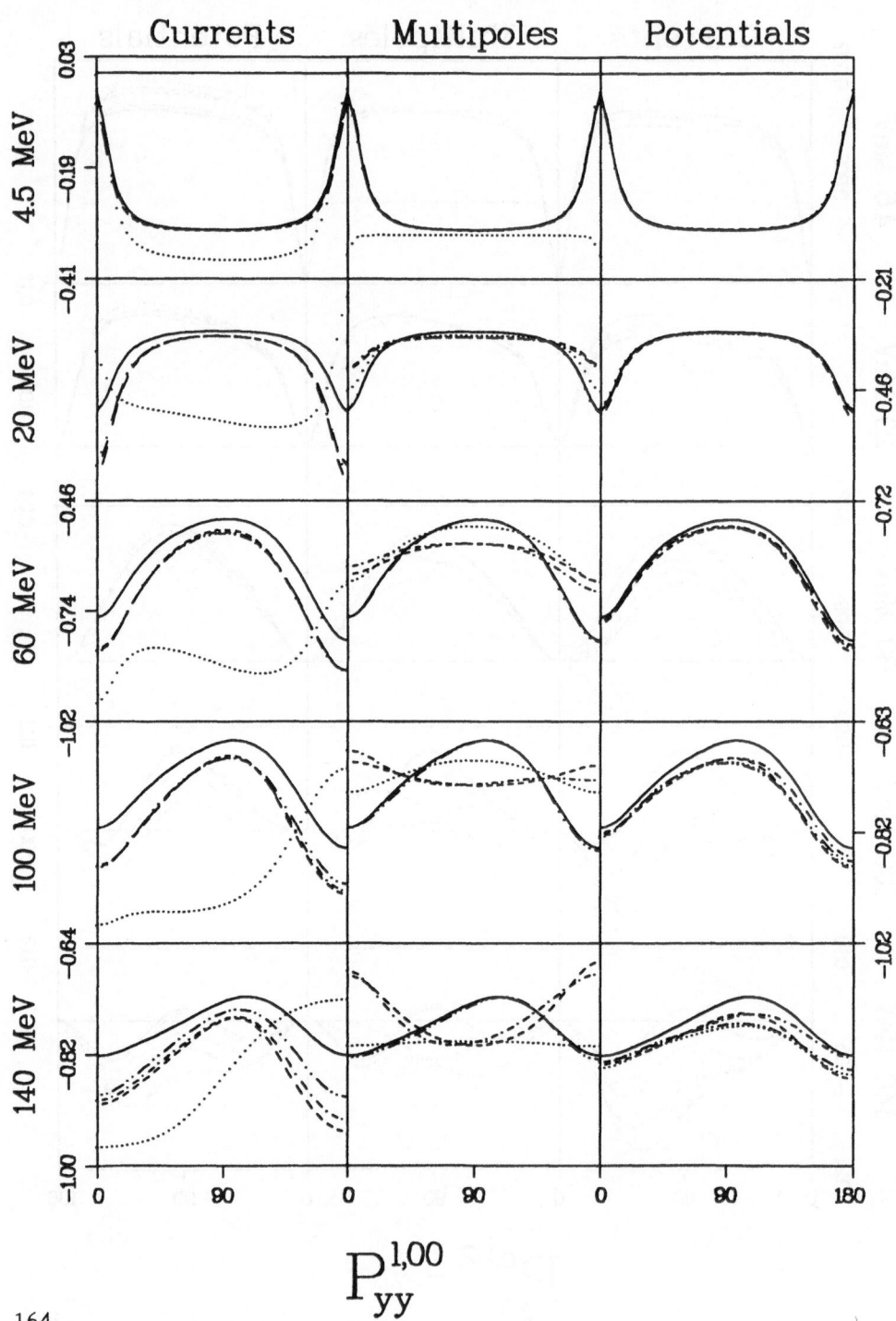

$$P^{1,00}_{yy}$$

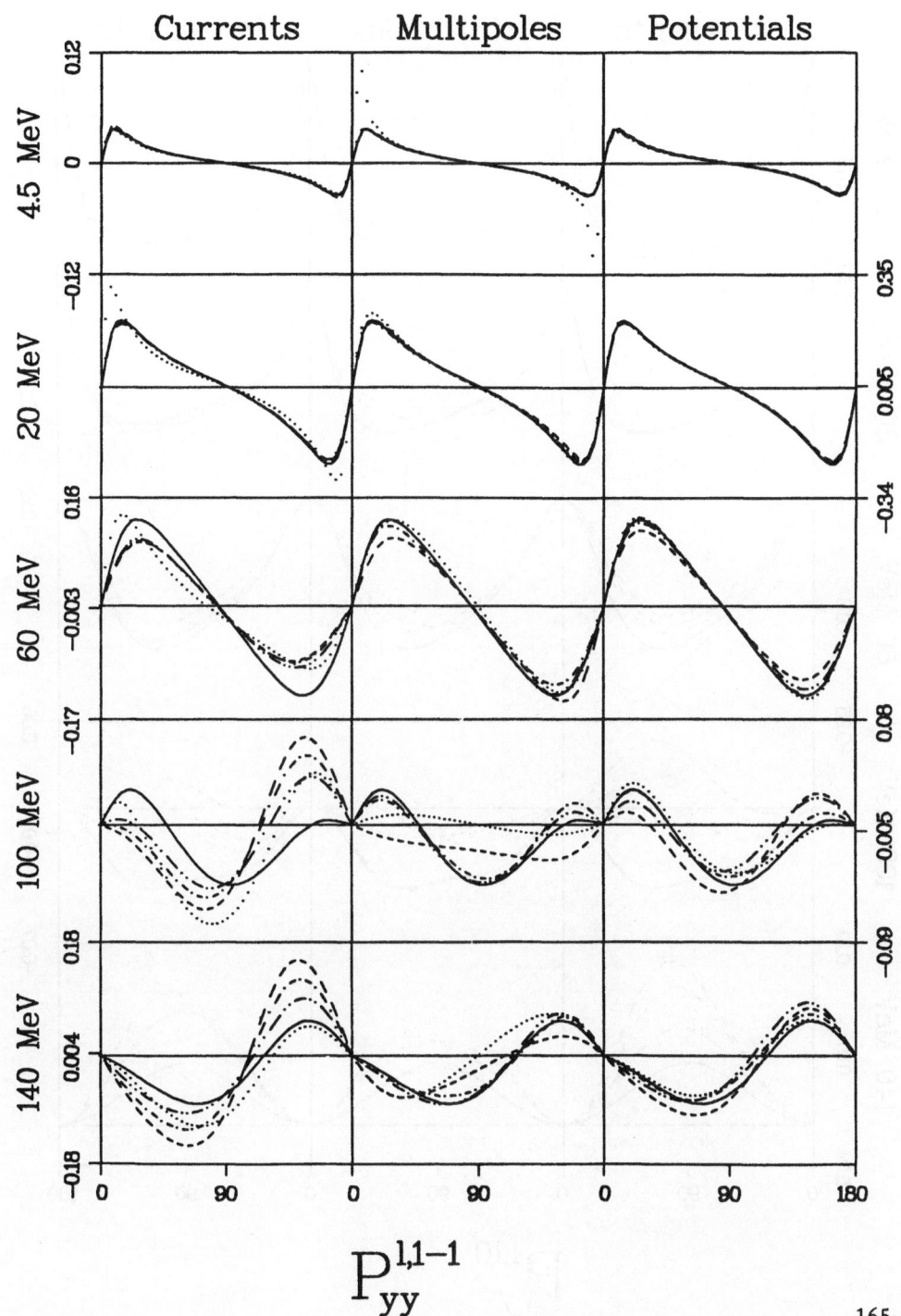

$$P_{yy}^{l,1-1}$$

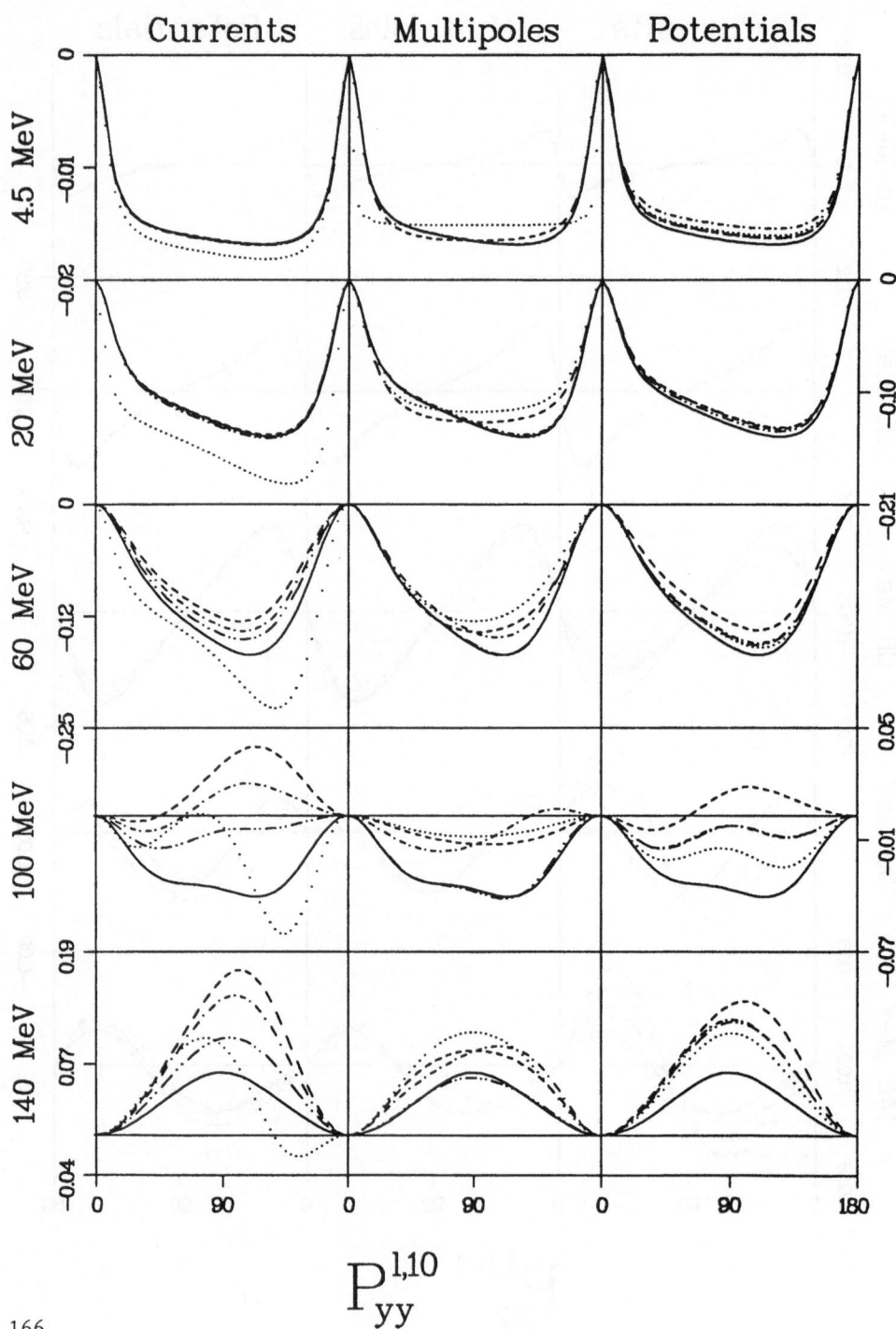

$$P_{yy}^{1,10}$$

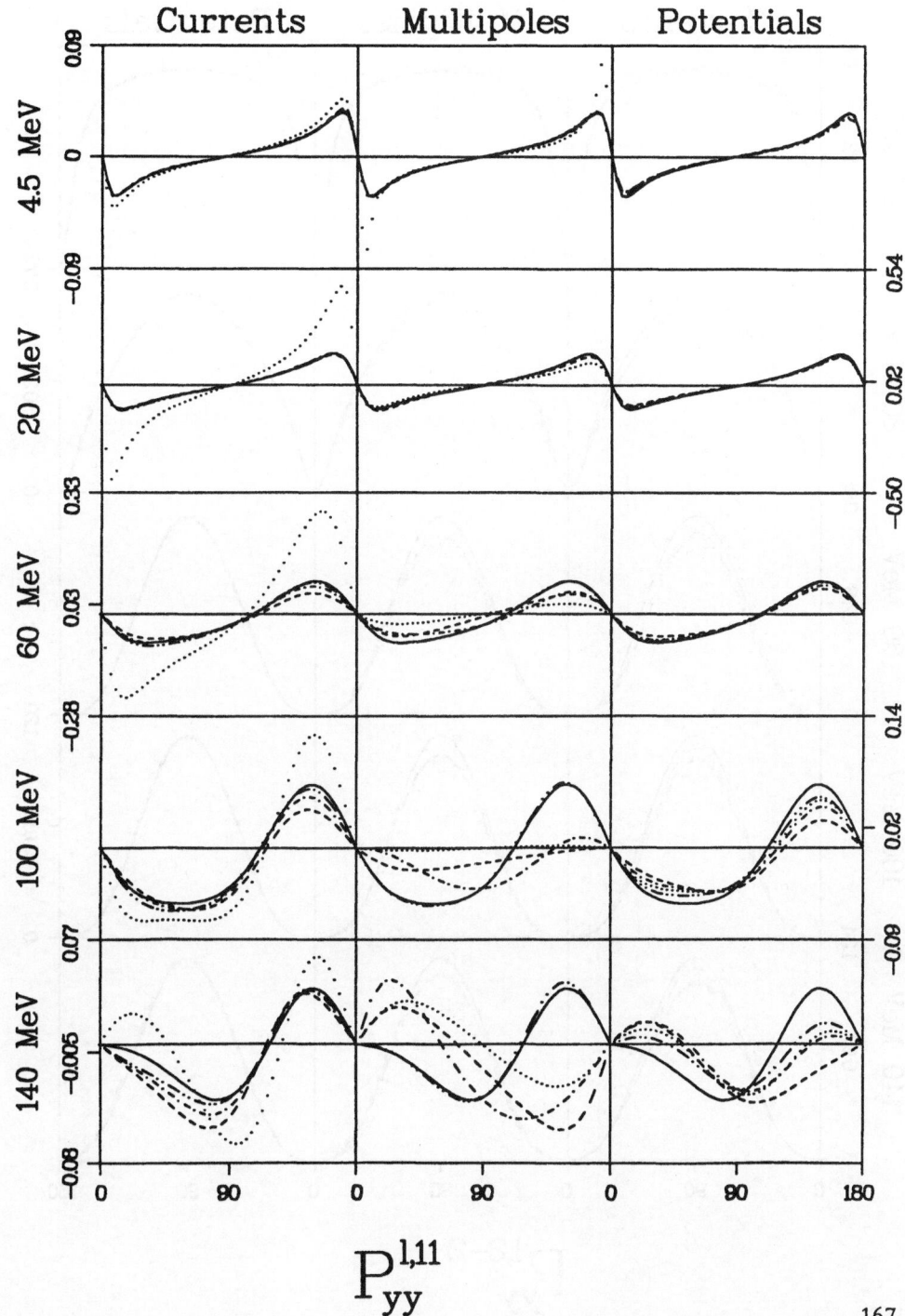

$$P_{yy}^{l,11}$$

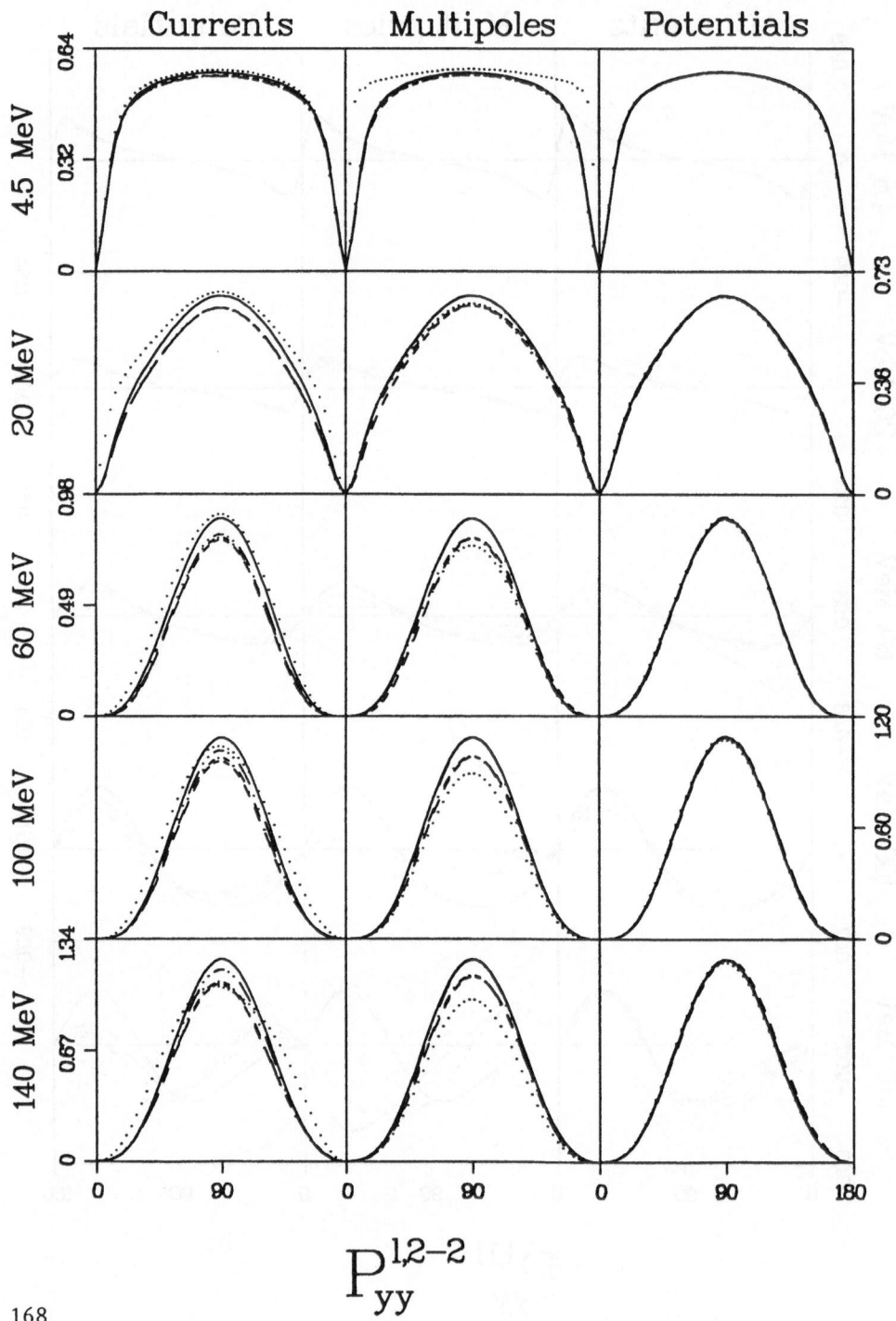

$$P_{yy}^{1,2-2}$$

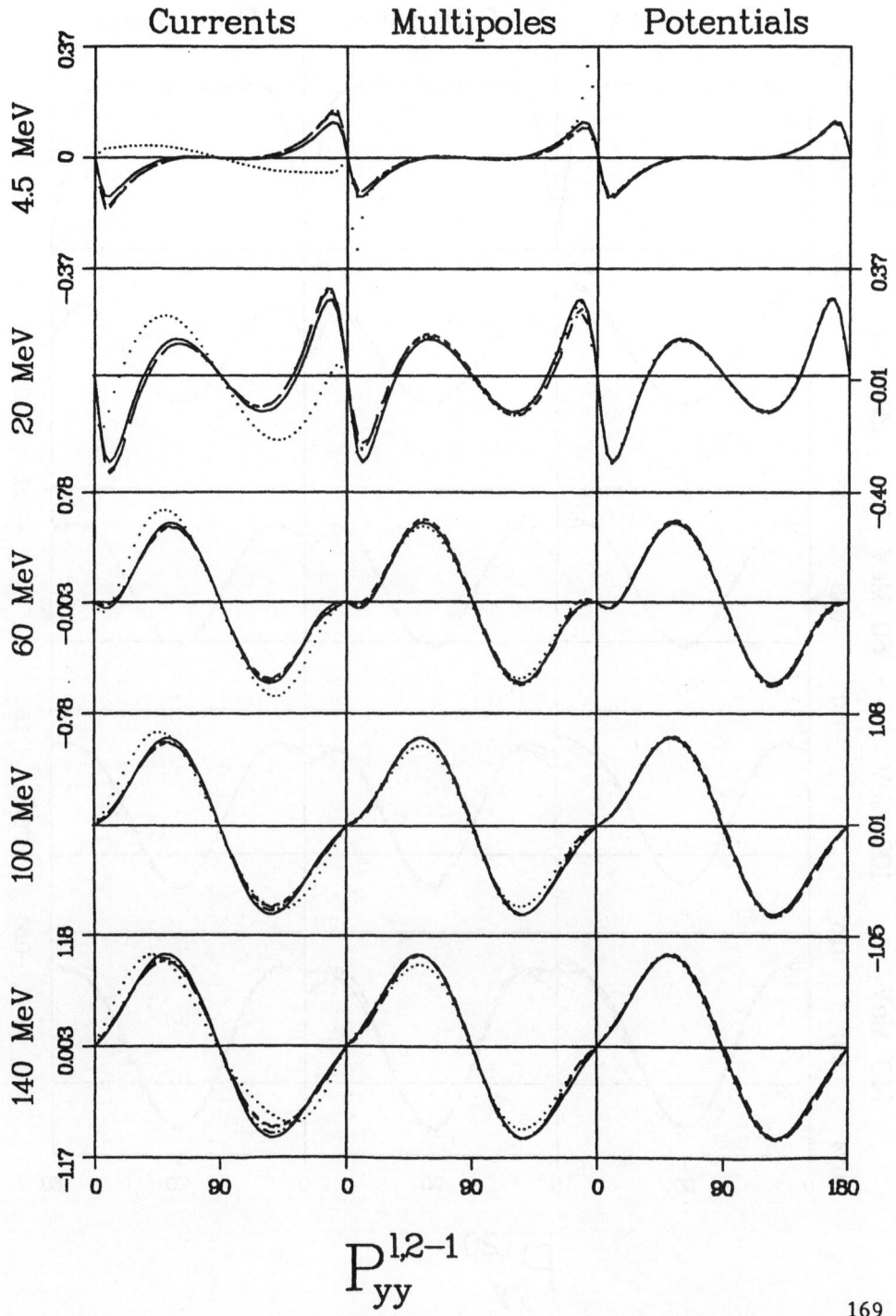

$$P_{yy}^{1,2-1}$$

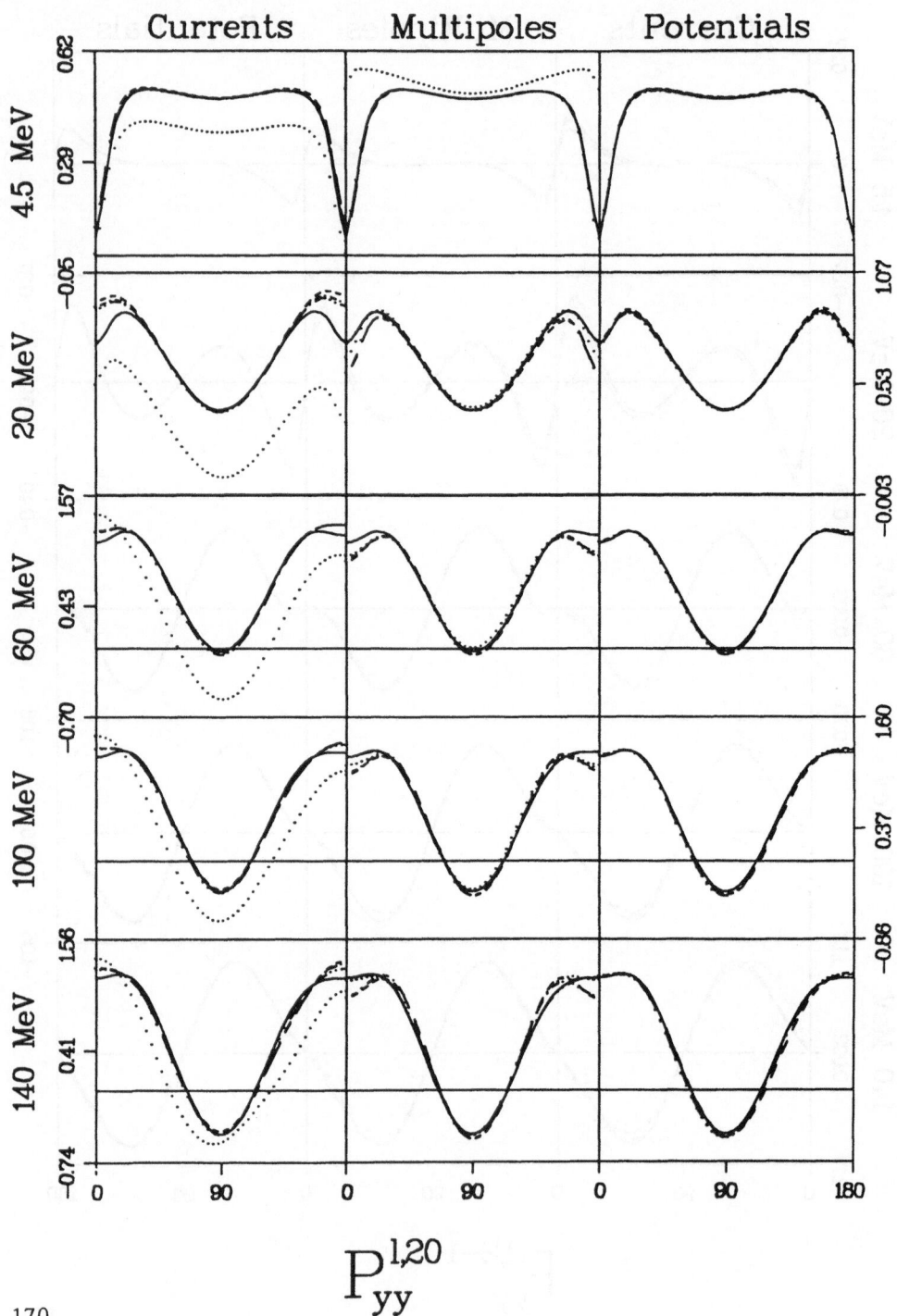

$$P_{yy}^{1,20}$$

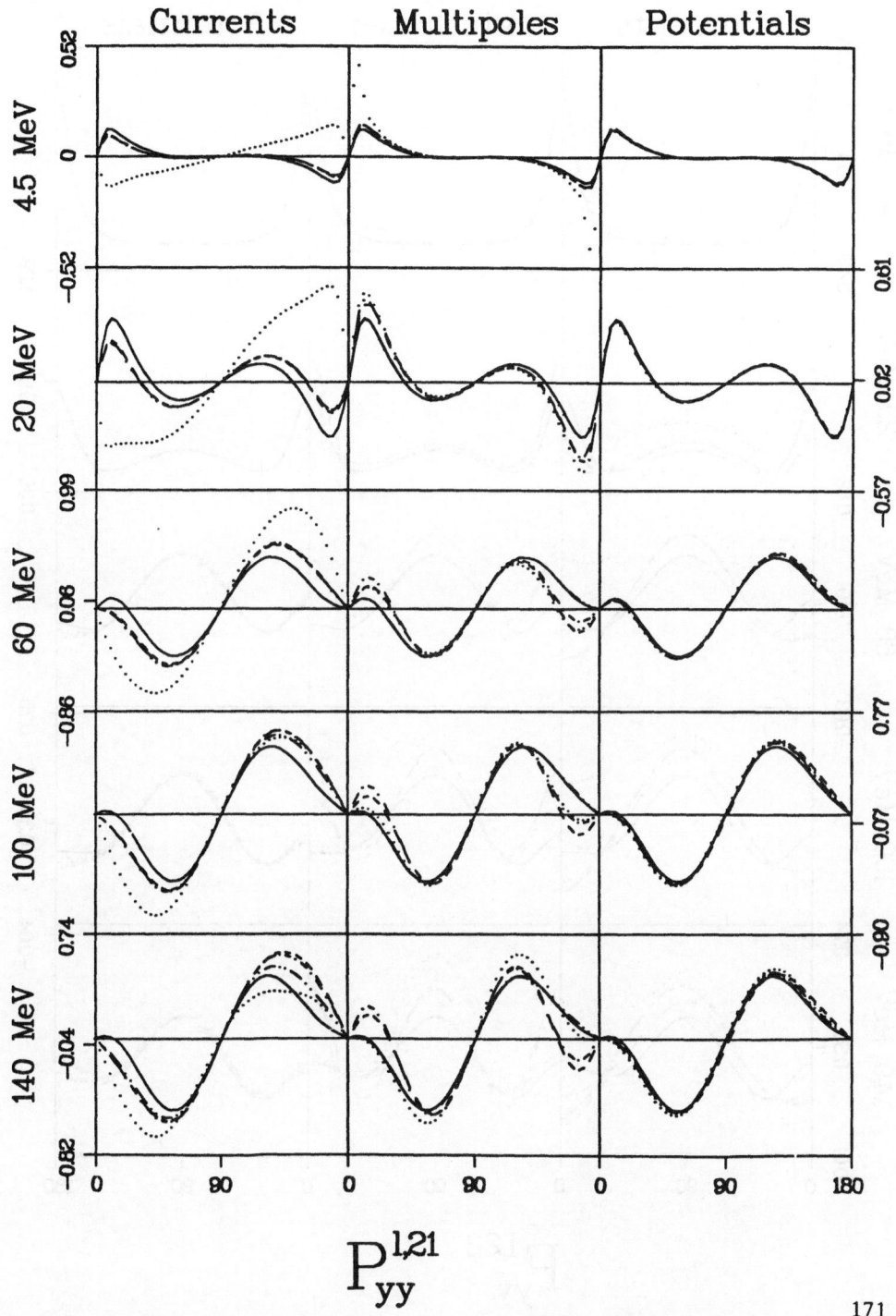

$$P_{yy}^{1,21}$$

Proton-Neutron Spin Correlation $P_{yy}^{l,IM}$

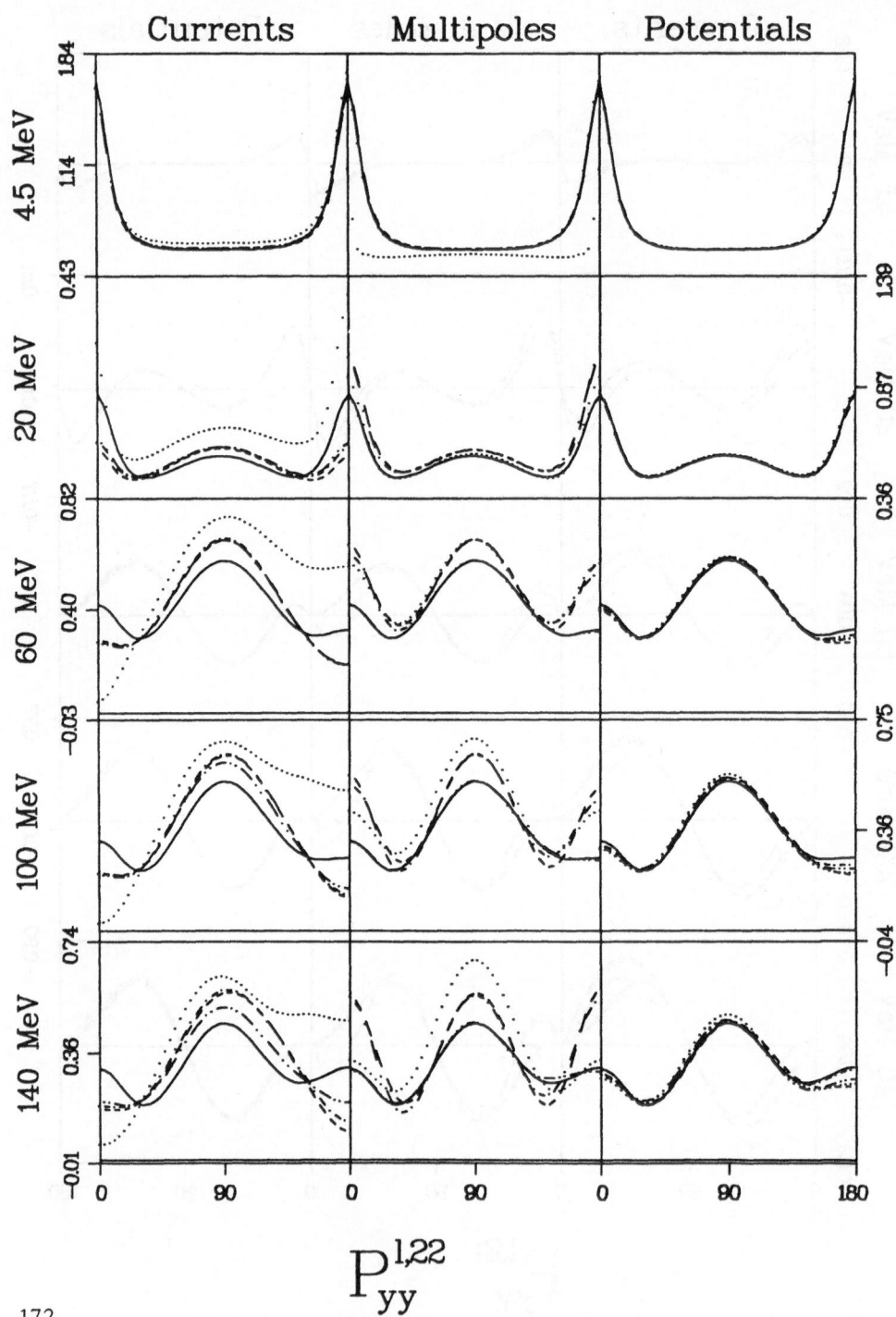

$$P_{yy}^{l,22}$$

5.29 Proton-Neutron Spin Correlation $P_{zz}^{0,IM}$

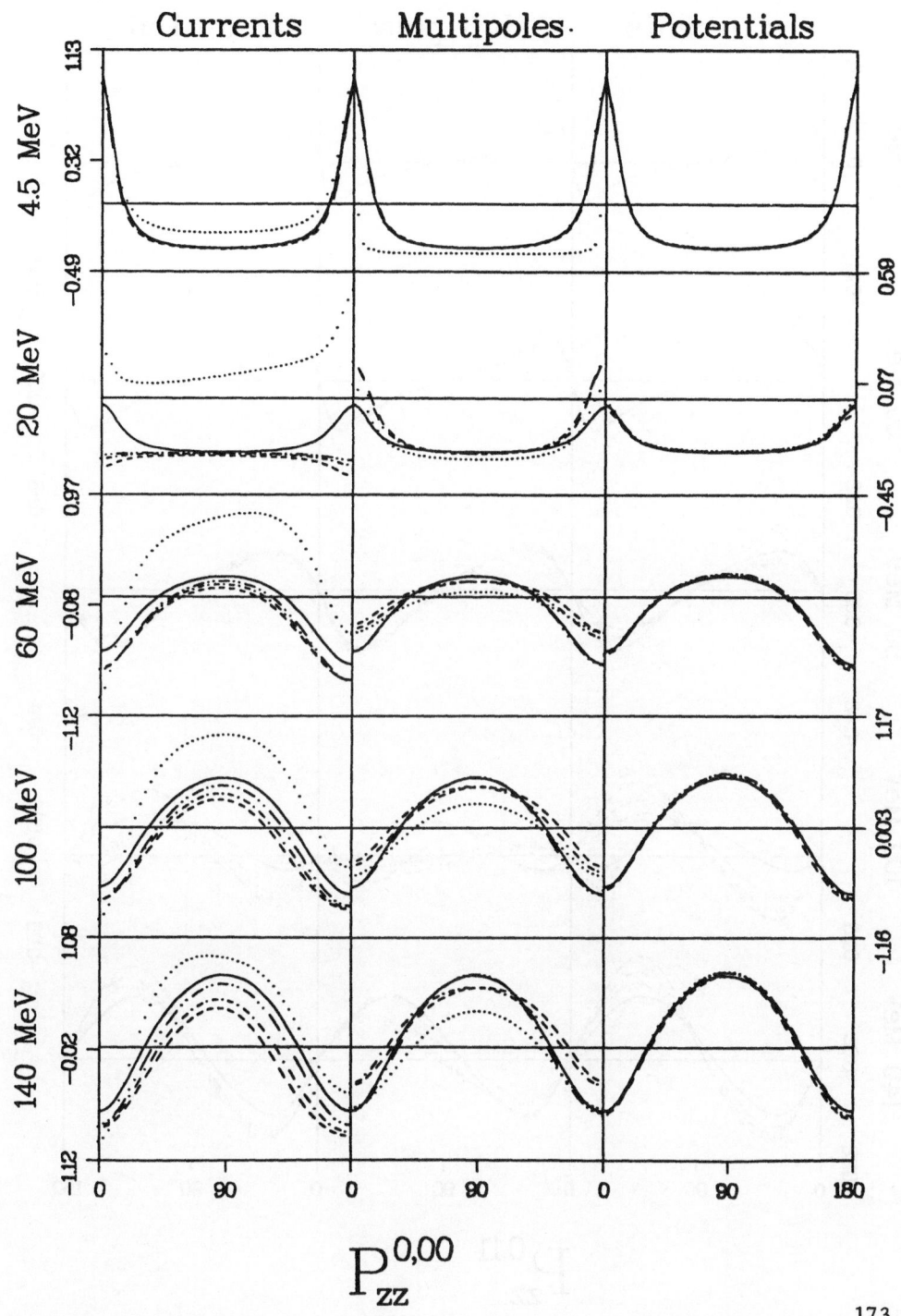

$$P_{zz}^{0,00}$$

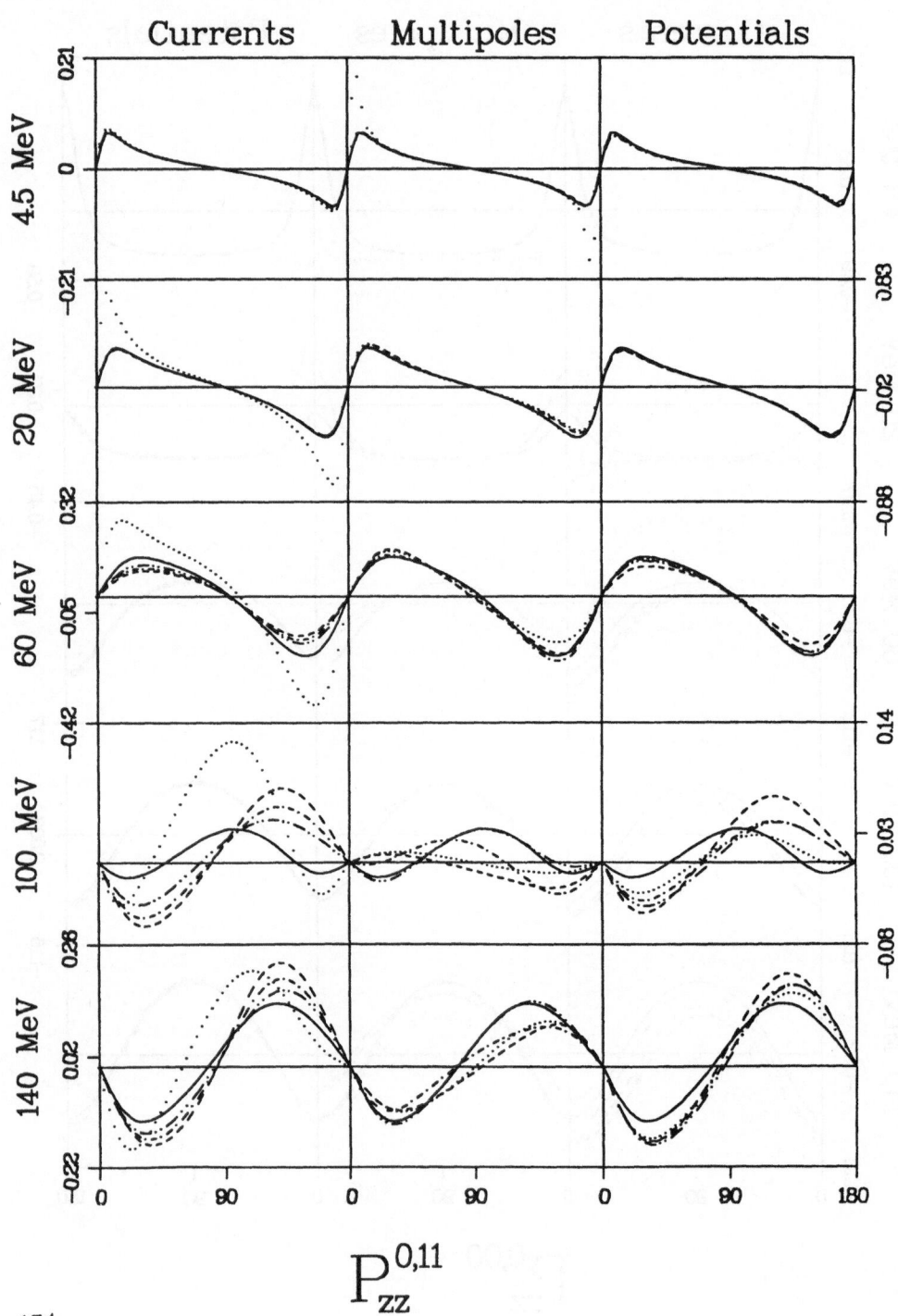

$$P_{zz}^{0,11}$$

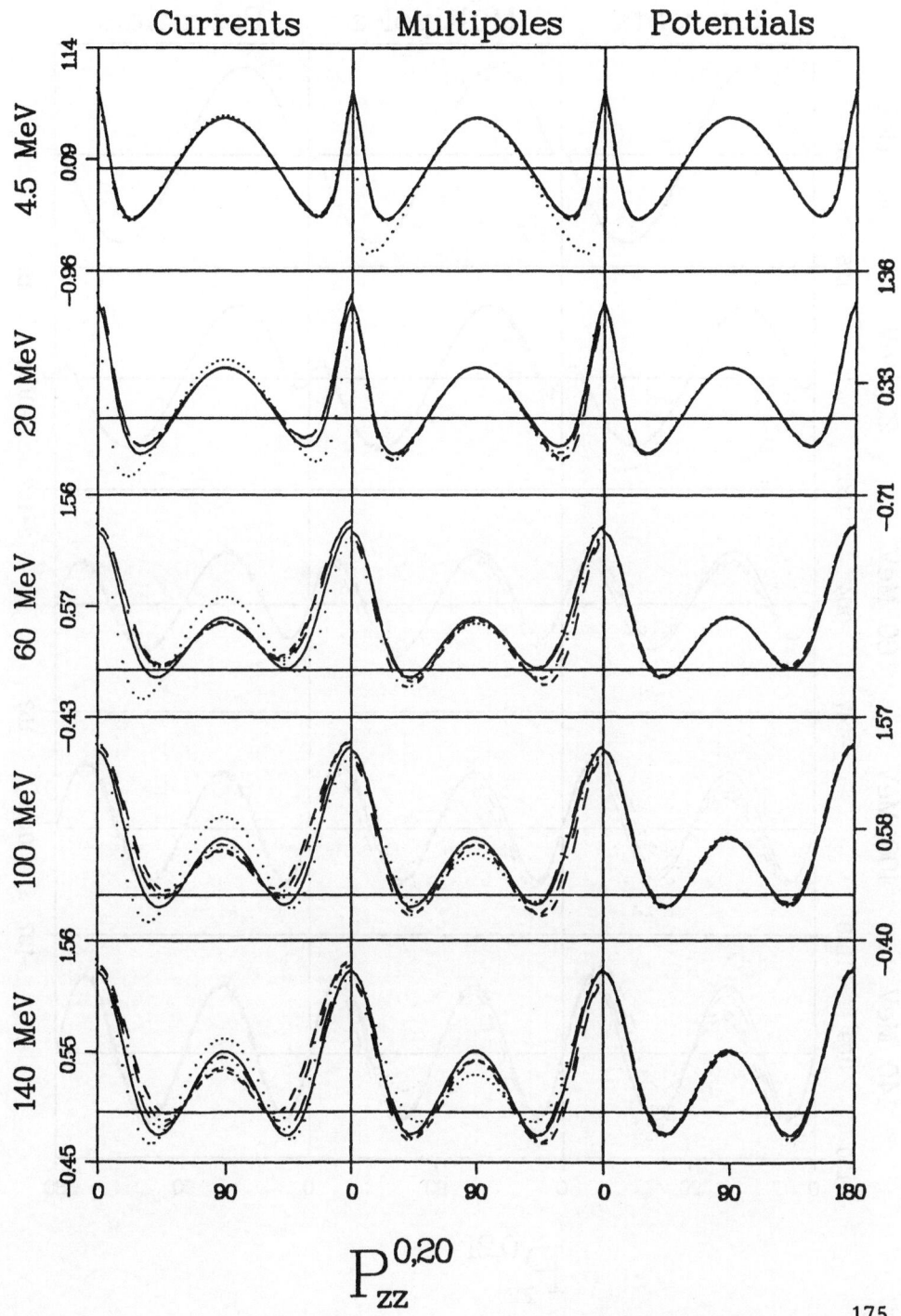

$$P_{zz}^{0,20}$$

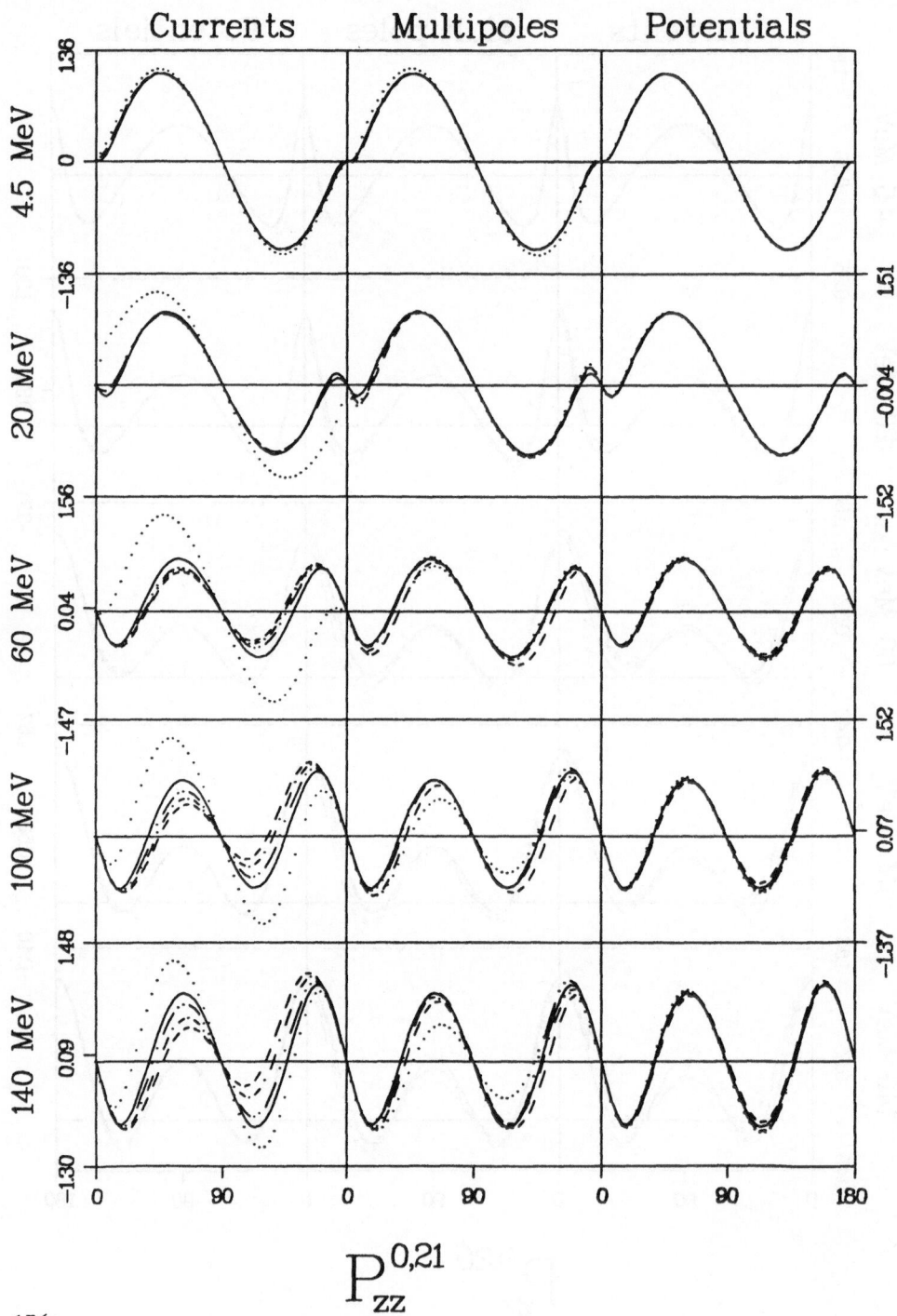

$$P_{zz}^{0,21}$$

Proton-Neutron Spin Correlation $P_{zz}^{0,IM}$

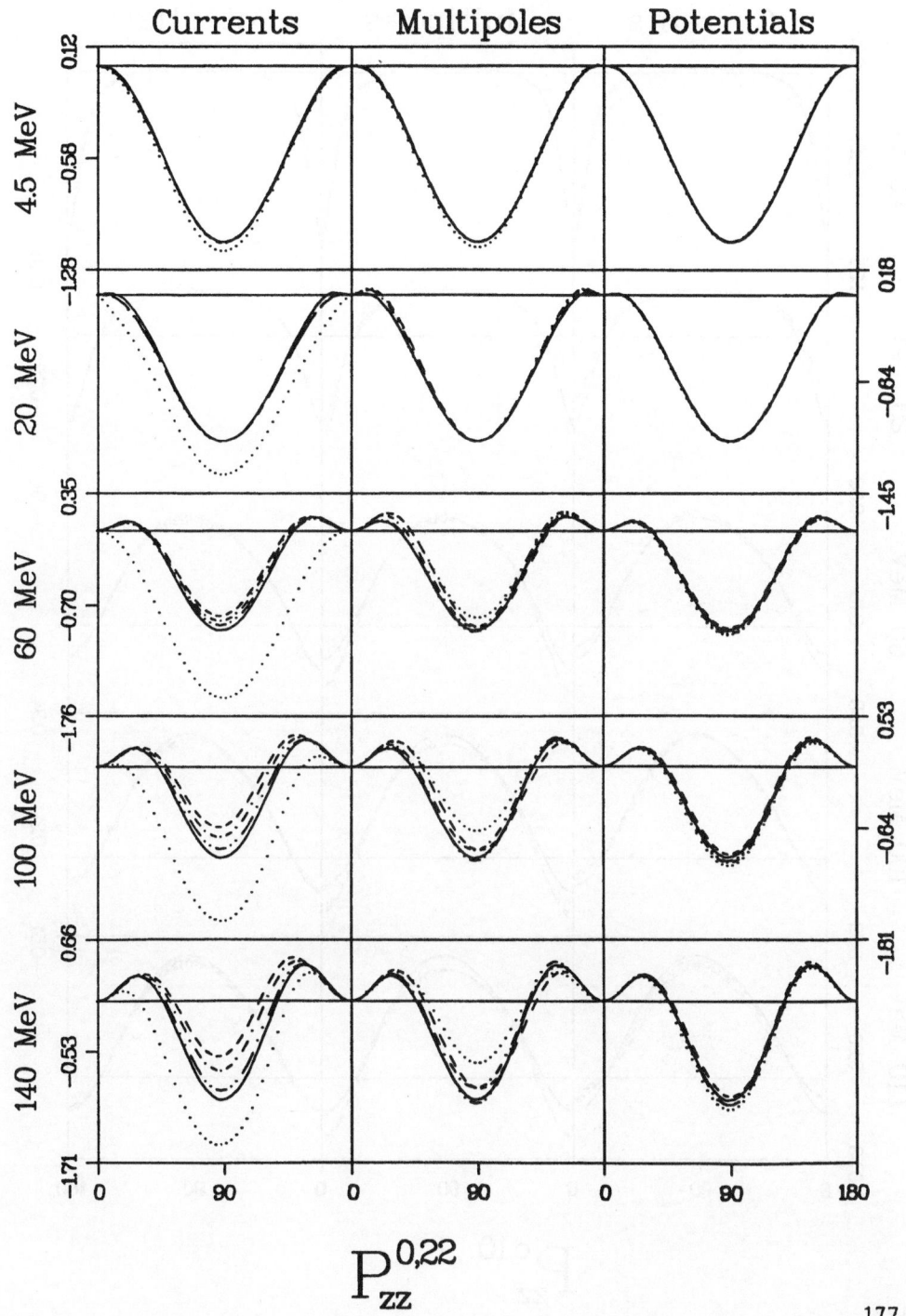

$$P_{zz}^{0,22}$$

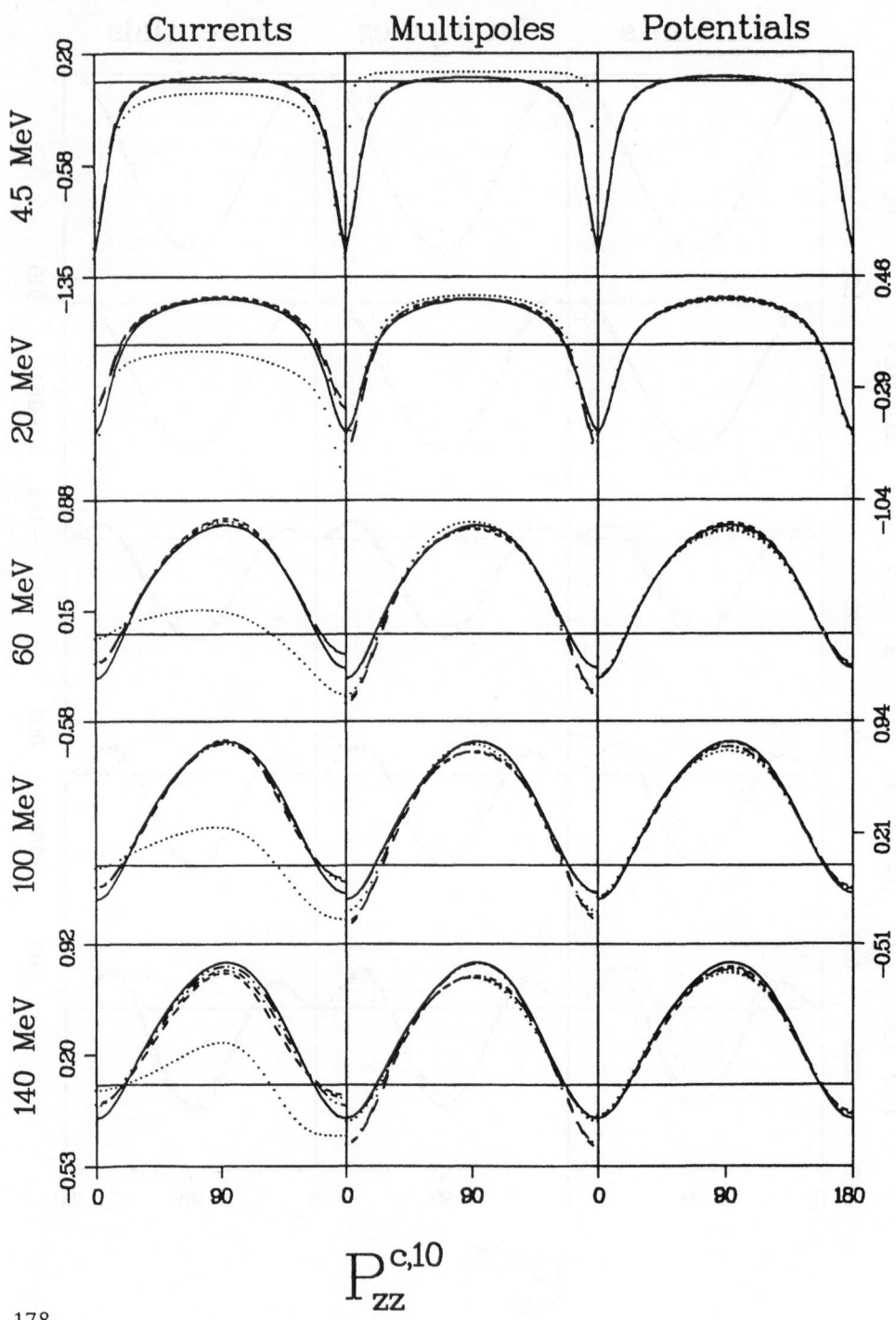

$$P_{zz}^{c,10}$$

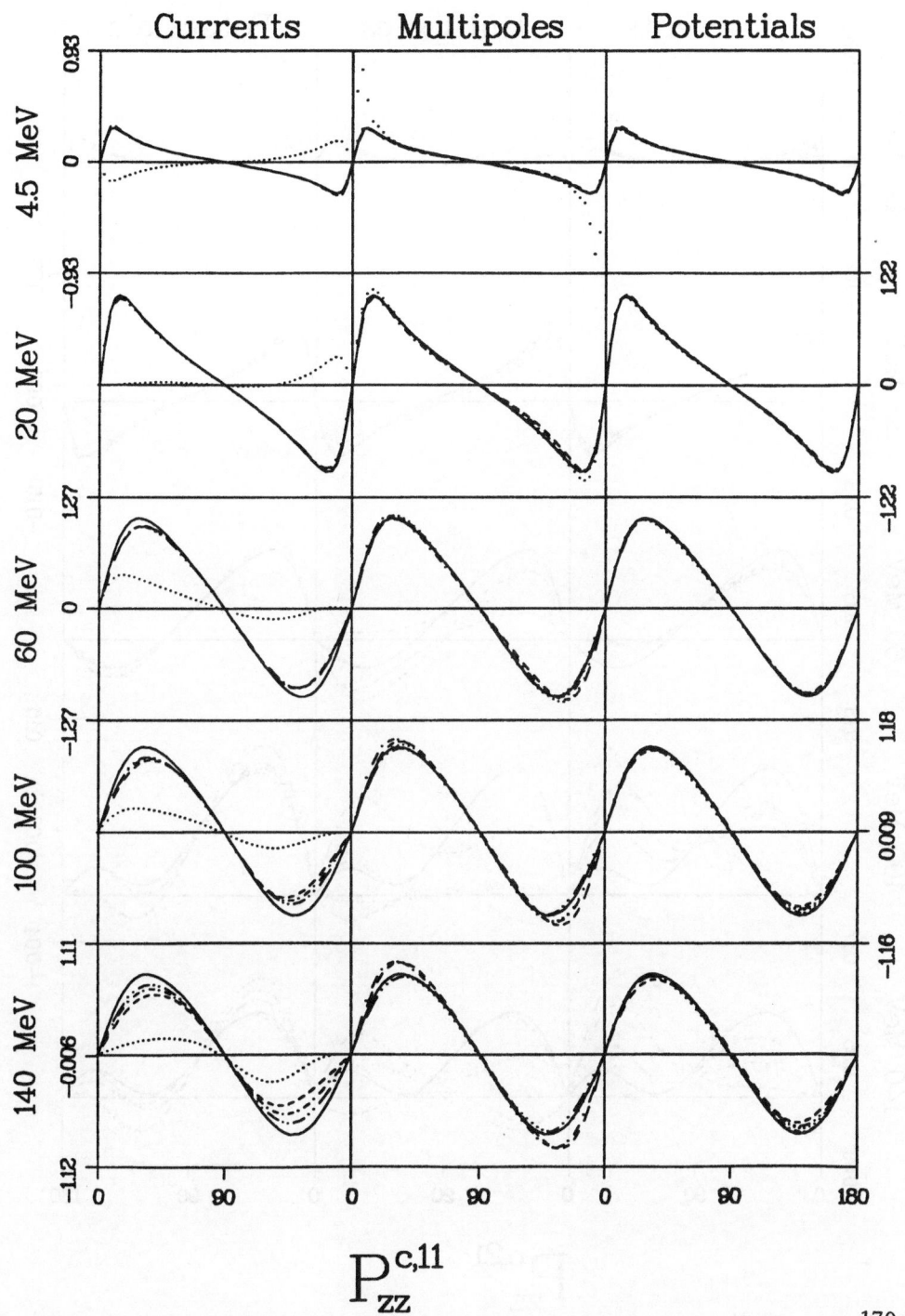

$$P_{zz}^{c,11}$$

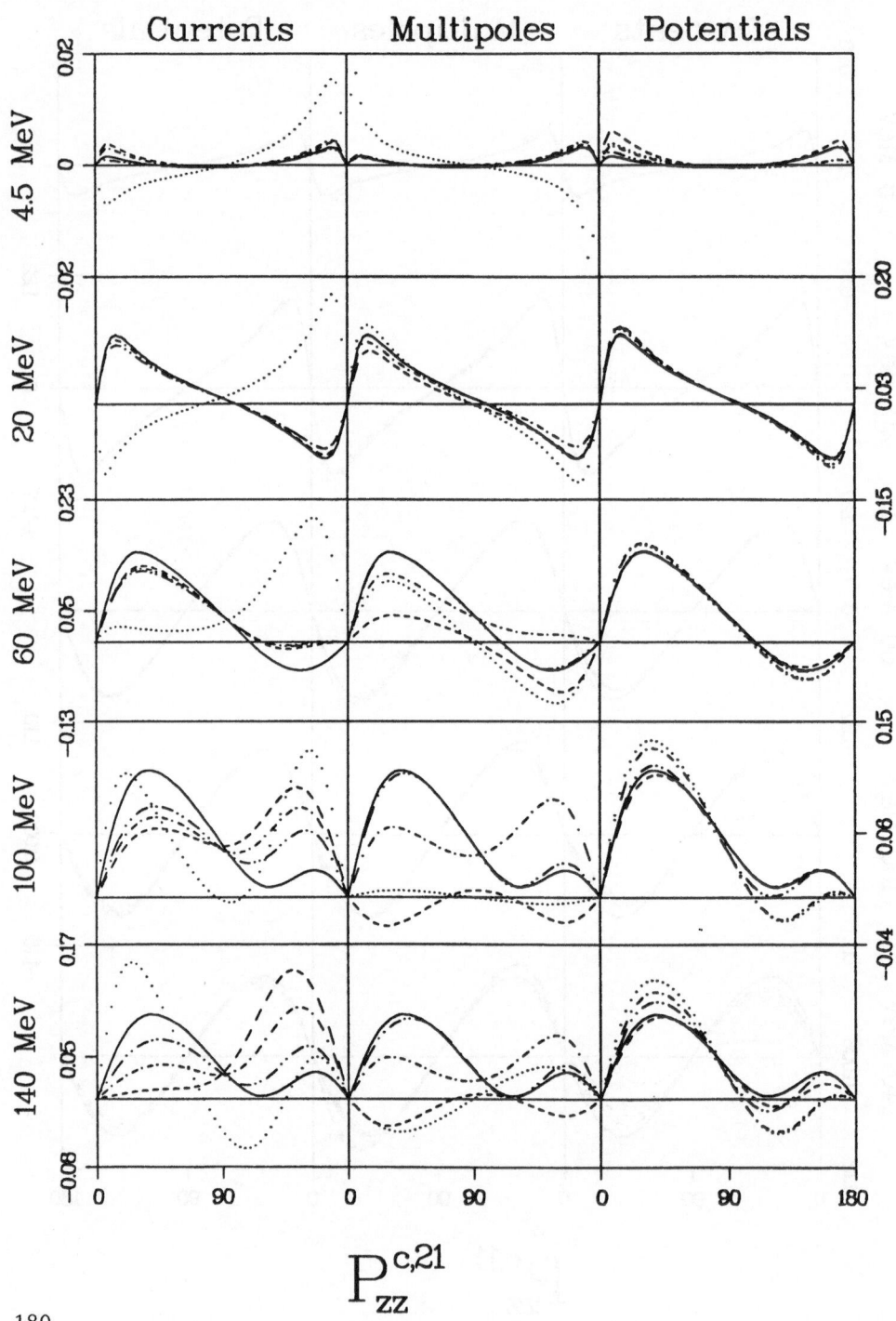

Currents   Multipoles   Potentials

4.5 MeV

20 MeV

60 MeV

100 MeV

140 MeV

$$P_{zz}^{c,21}$$

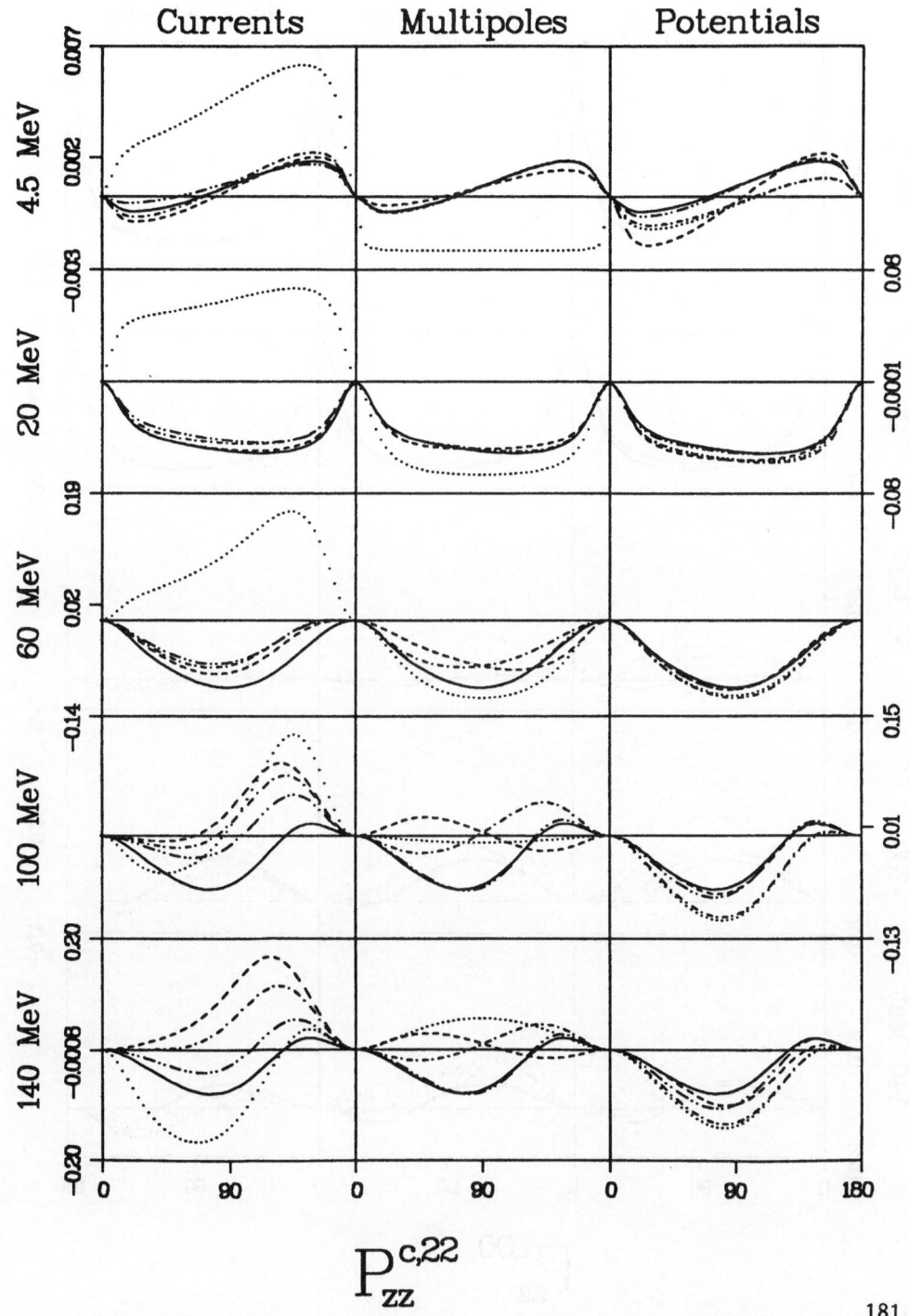

$$P_{zz}^{c,22}$$

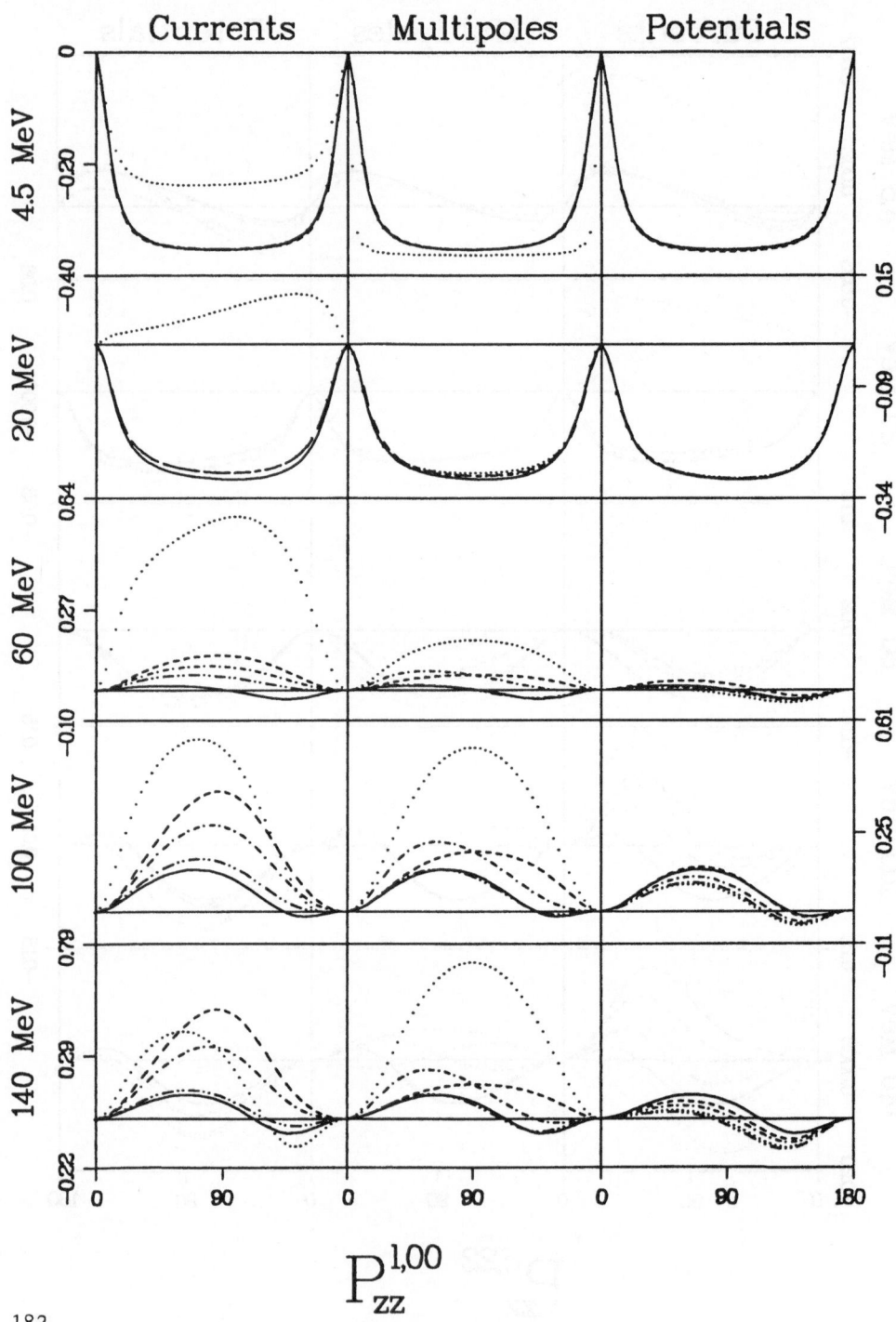

$$P_{zz}^{1,00}$$

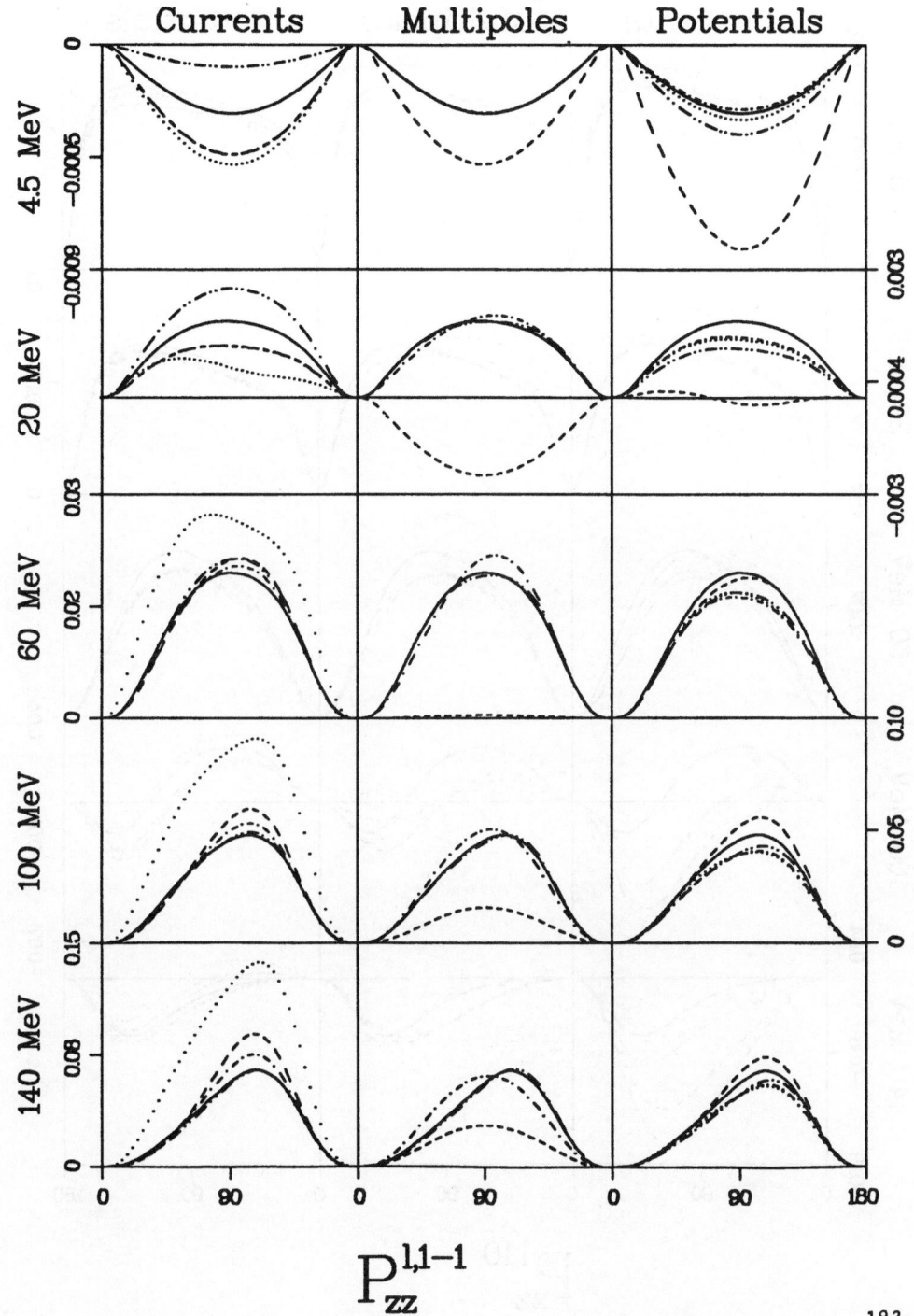

Currents   Multipoles   Potentials

$$P_{zz}^{l,1-1}$$

Currents   Multipoles   Potentials

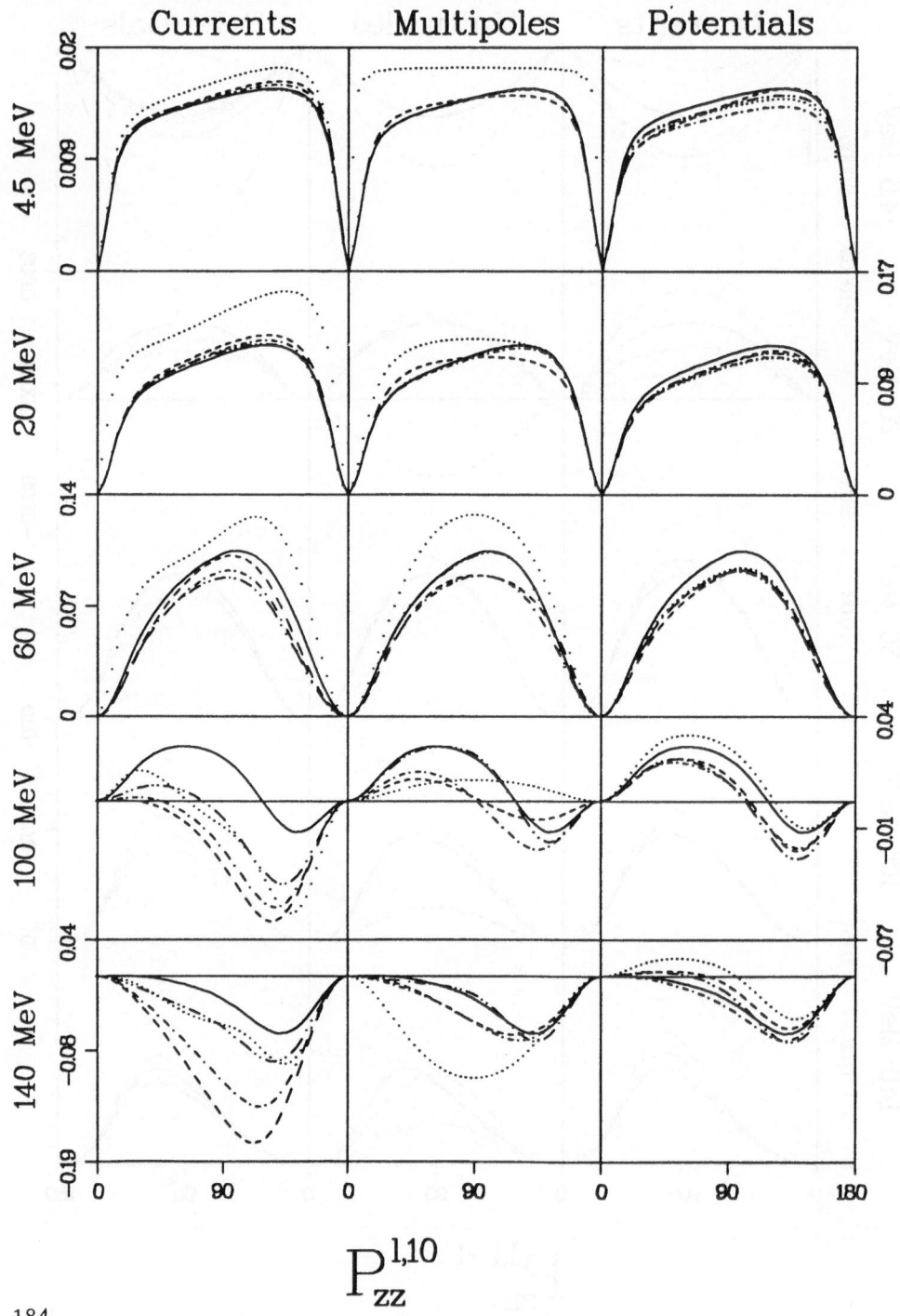

$$P_{zz}^{1,10}$$

184

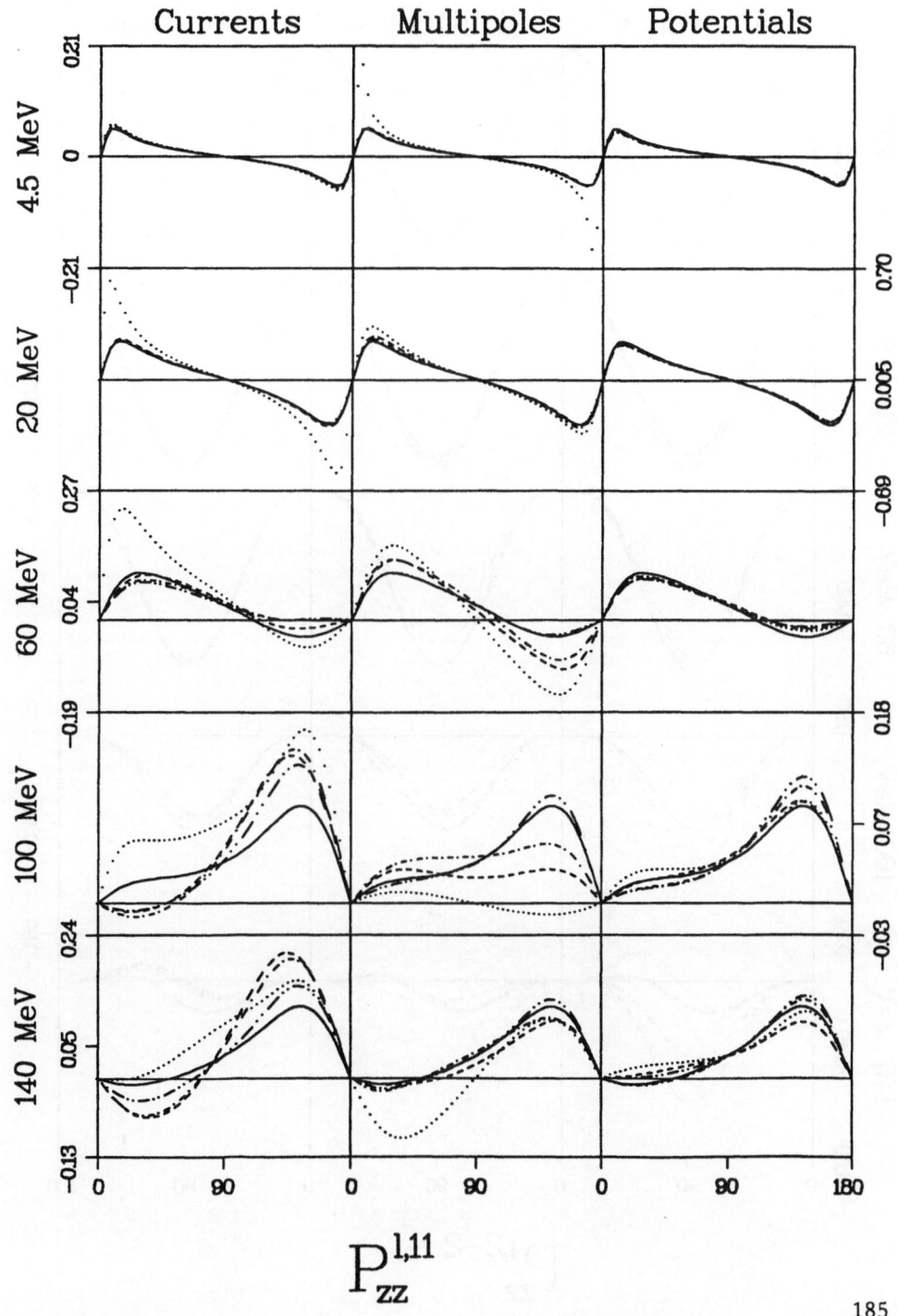

Currents   Multipoles   Potentials

4.5 MeV   20 MeV   60 MeV   100 MeV   140 MeV

$$P_{zz}^{1,11}$$

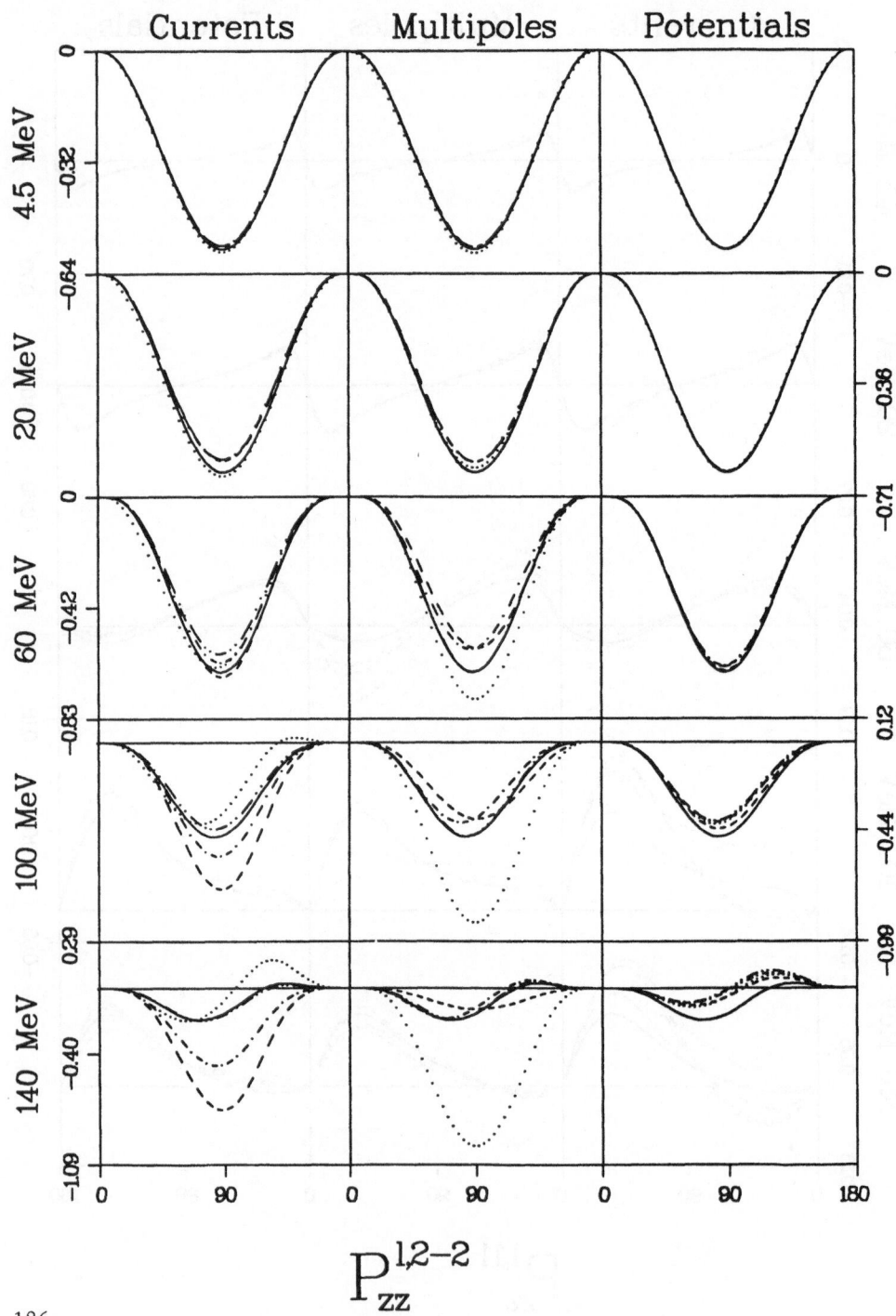

$$P_{zz}^{1,2-2}$$

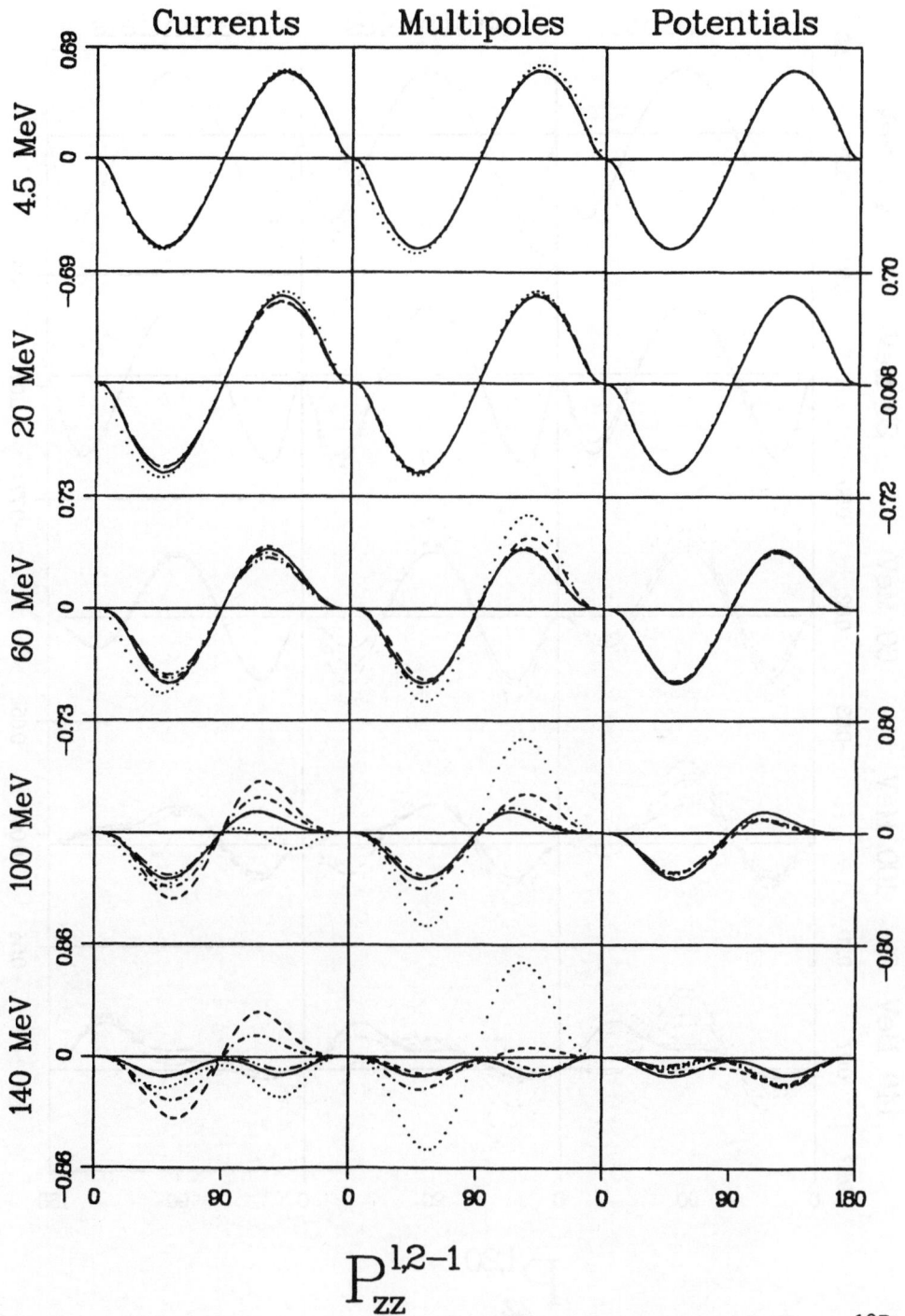

$$P_{ZZ}^{1,2-1}$$

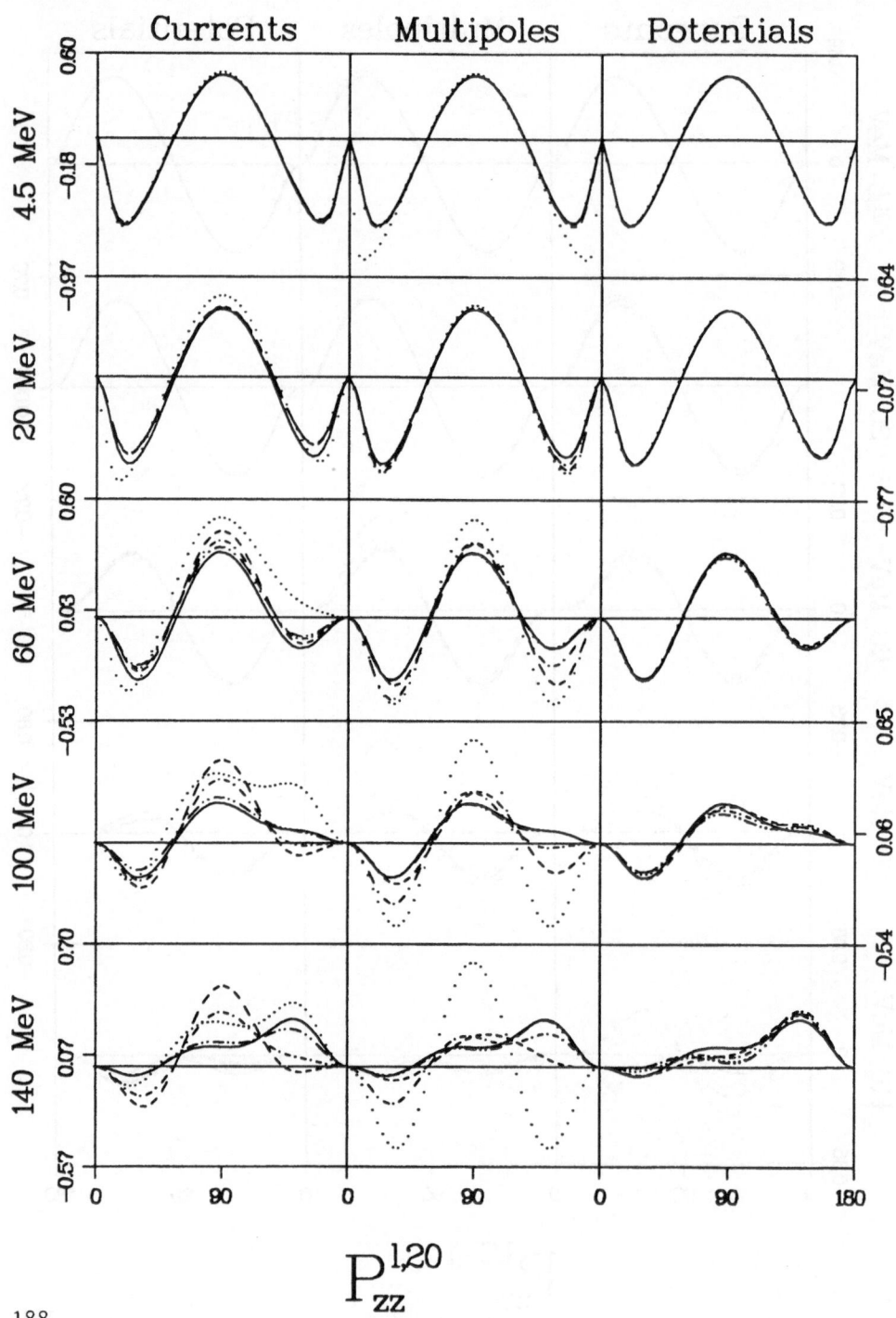

$$P_{zz}^{l,20}$$

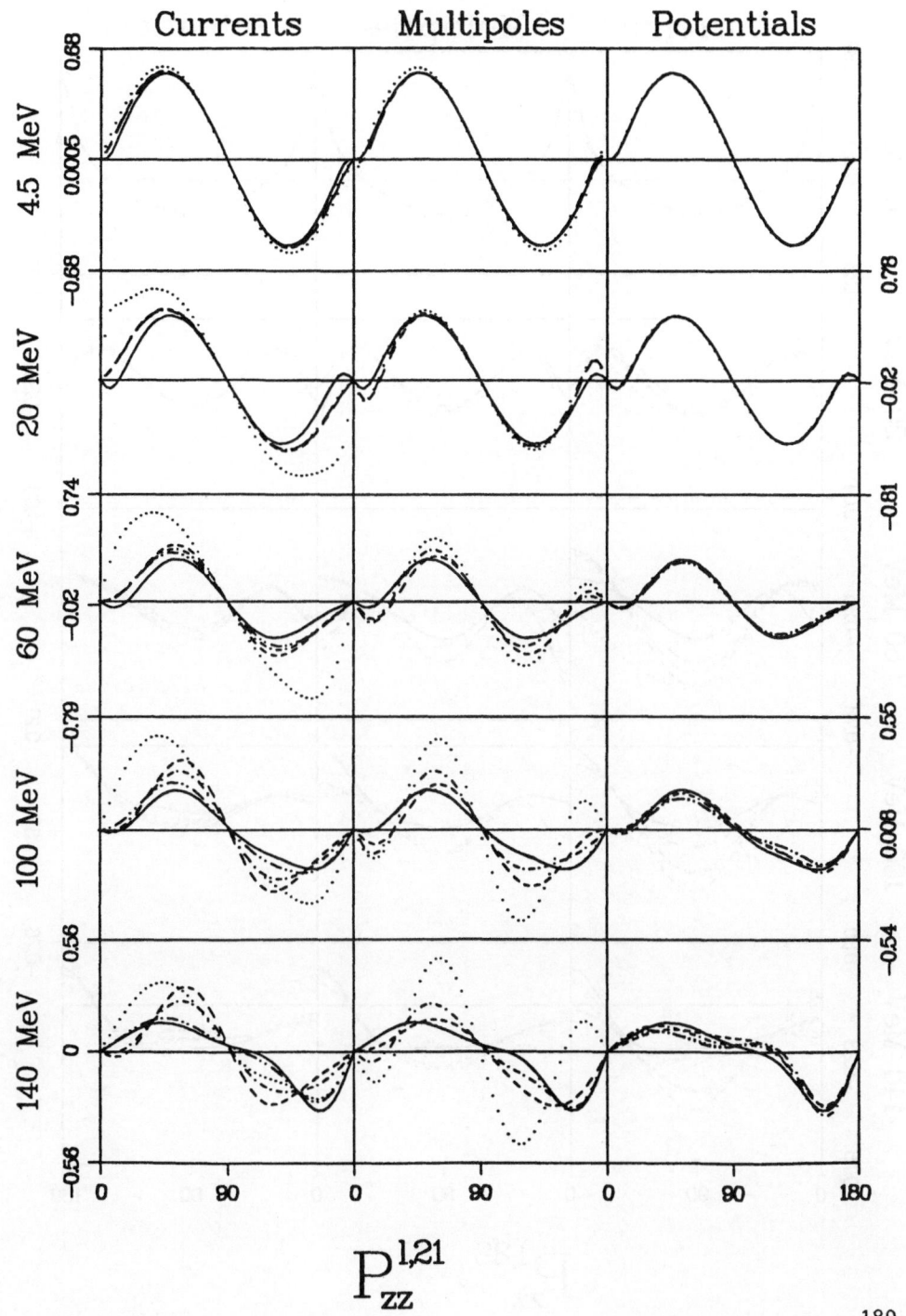

$$P_{zz}^{1,21}$$

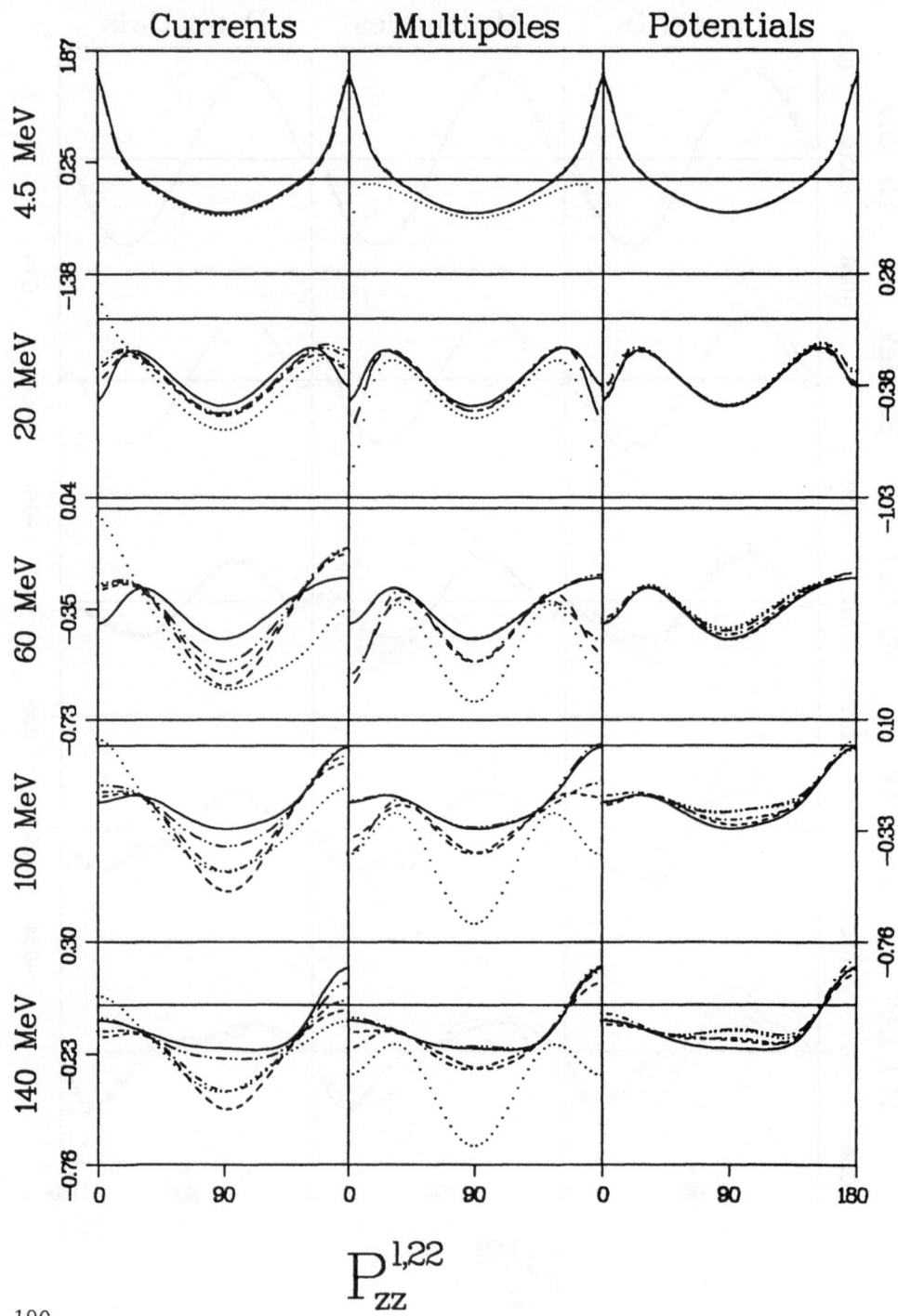

$$P_{zz}^{l,22}$$

5.32 Proton-Neutron Spin Correlation $P_{xz}^{0,IM}$

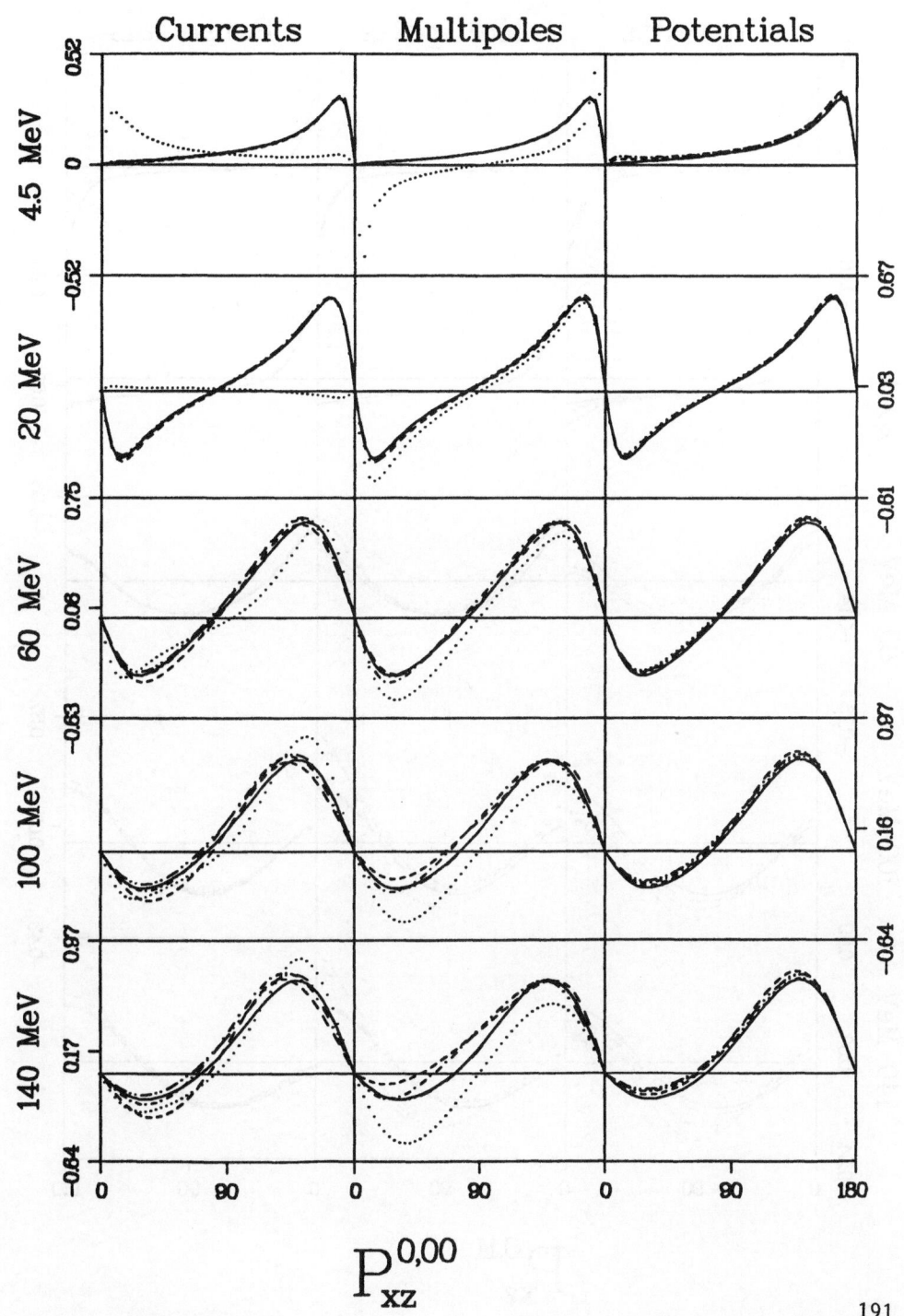

$$P_{xz}^{0,00}$$

Currents   Multipoles   Potentials

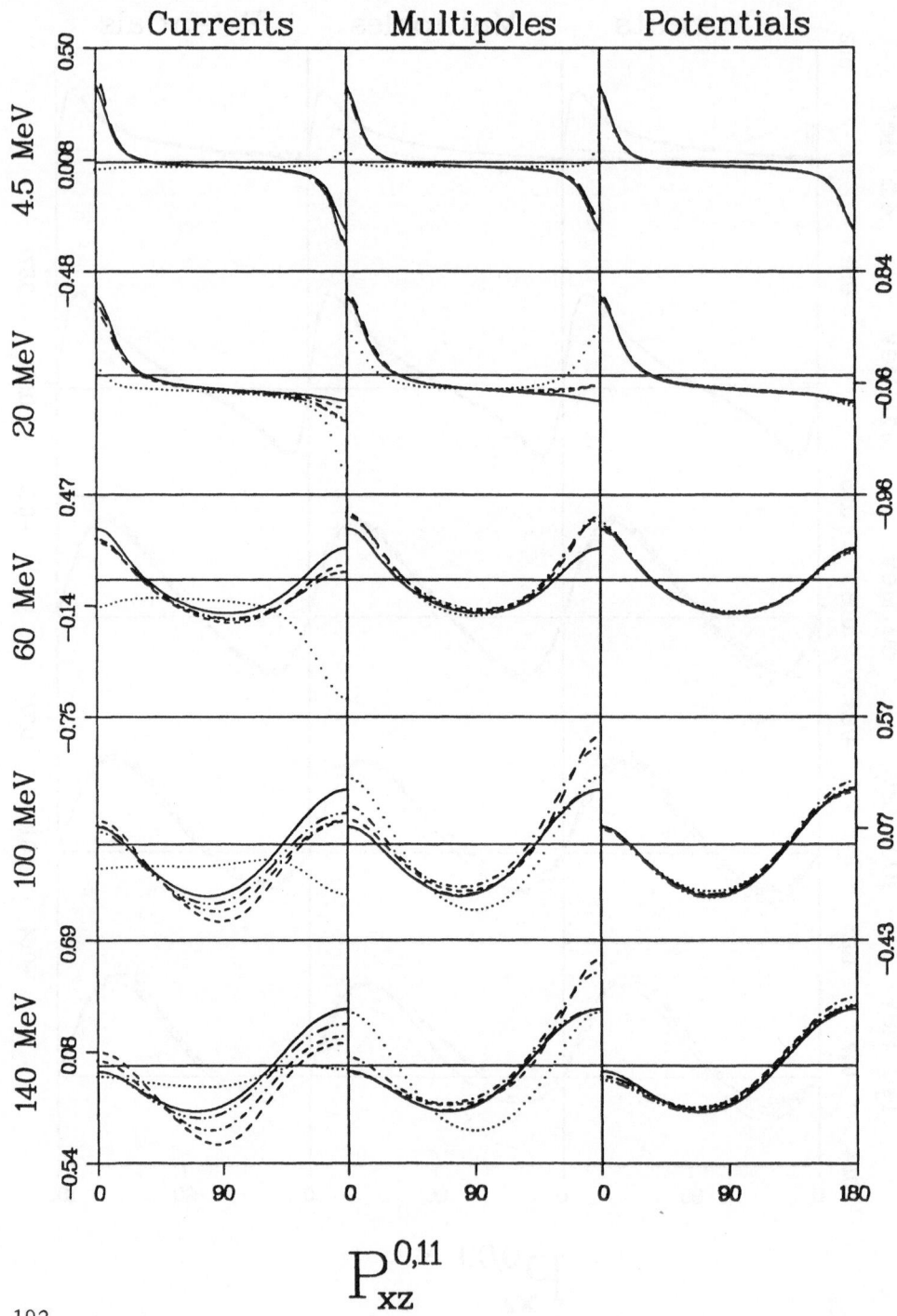

$$P_{xz}^{0,11}$$

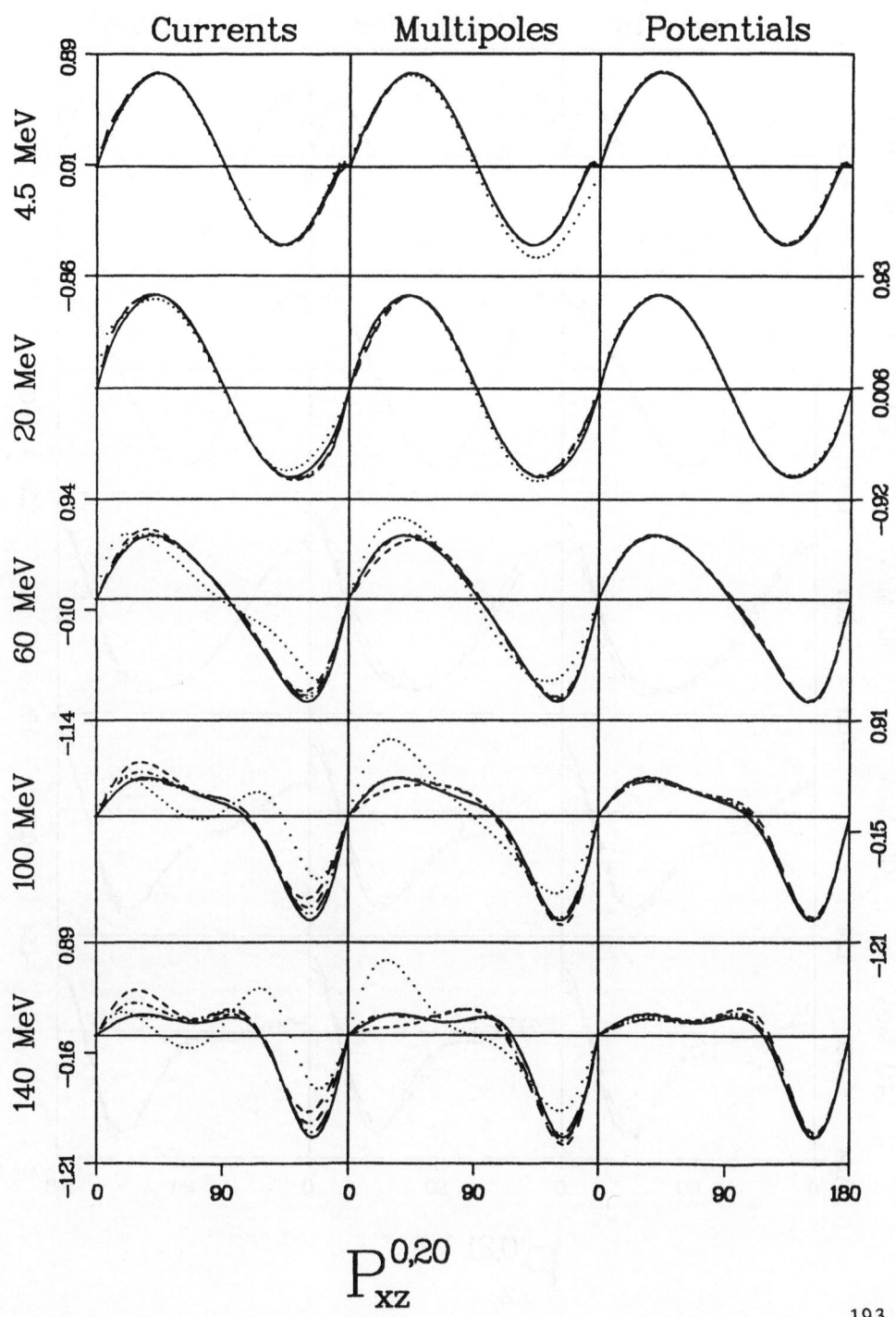

$$P_{xz}^{0,20}$$

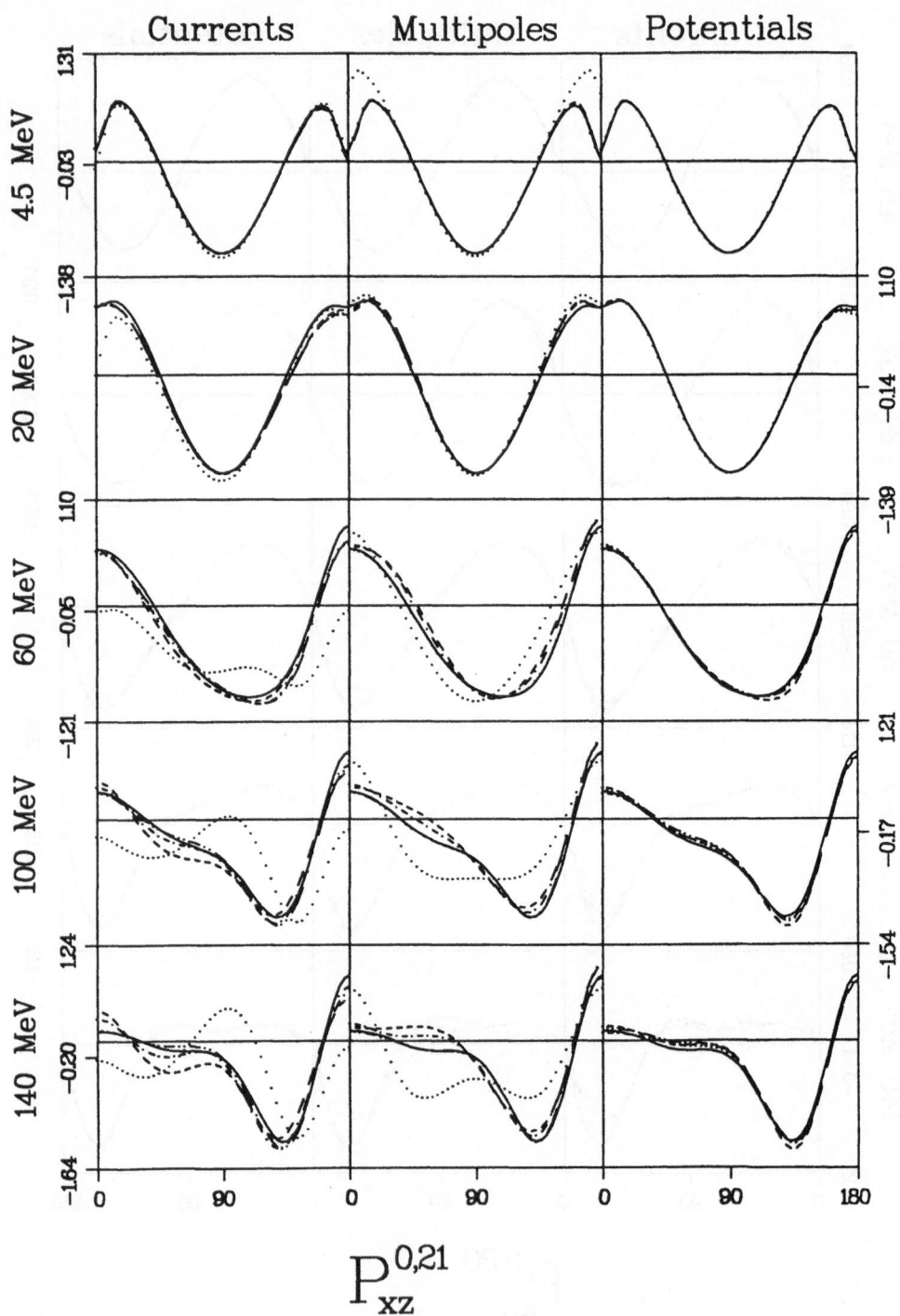

$$P_{xz}^{0,21}$$

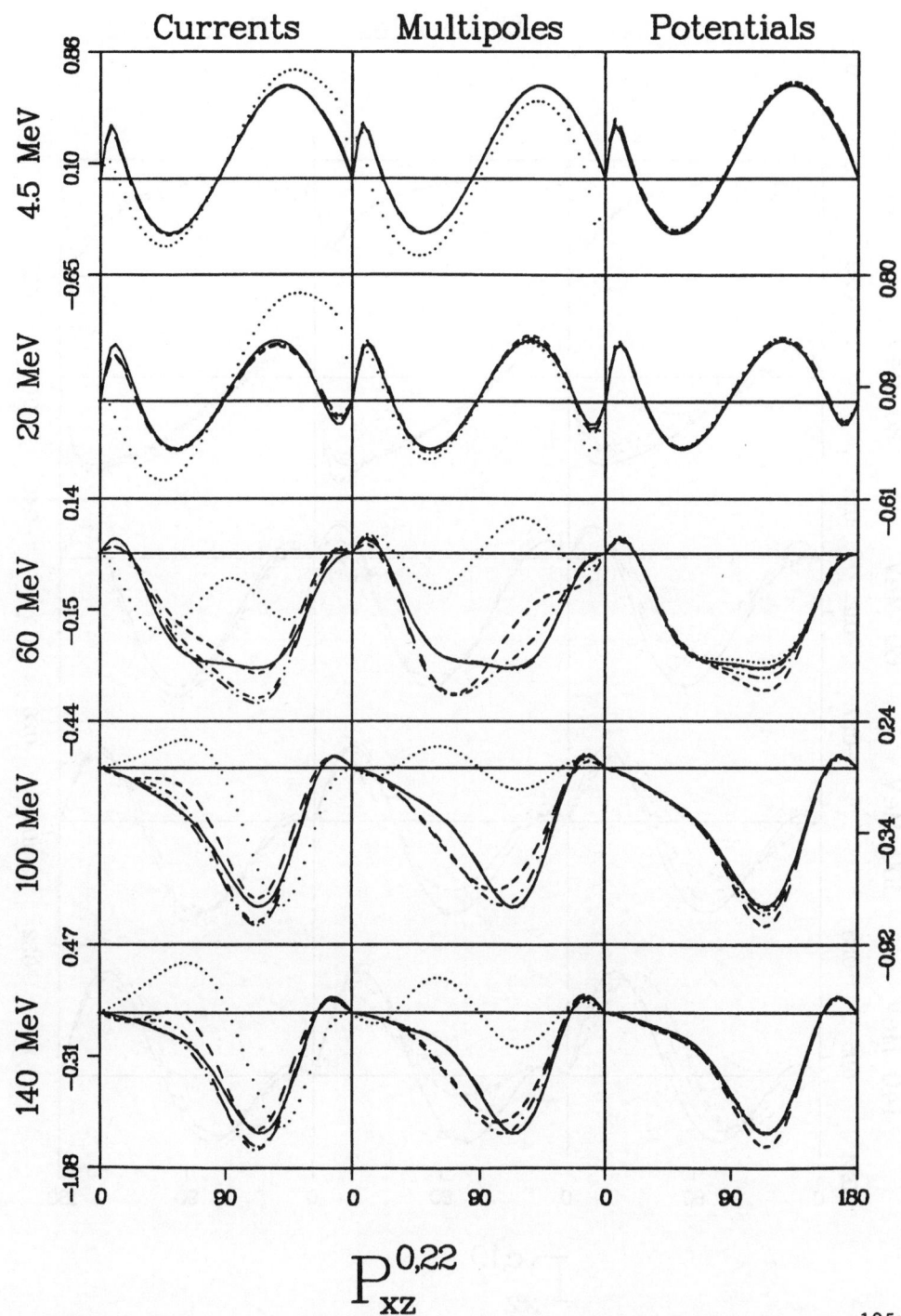

$$P_{xz}^{0,22}$$

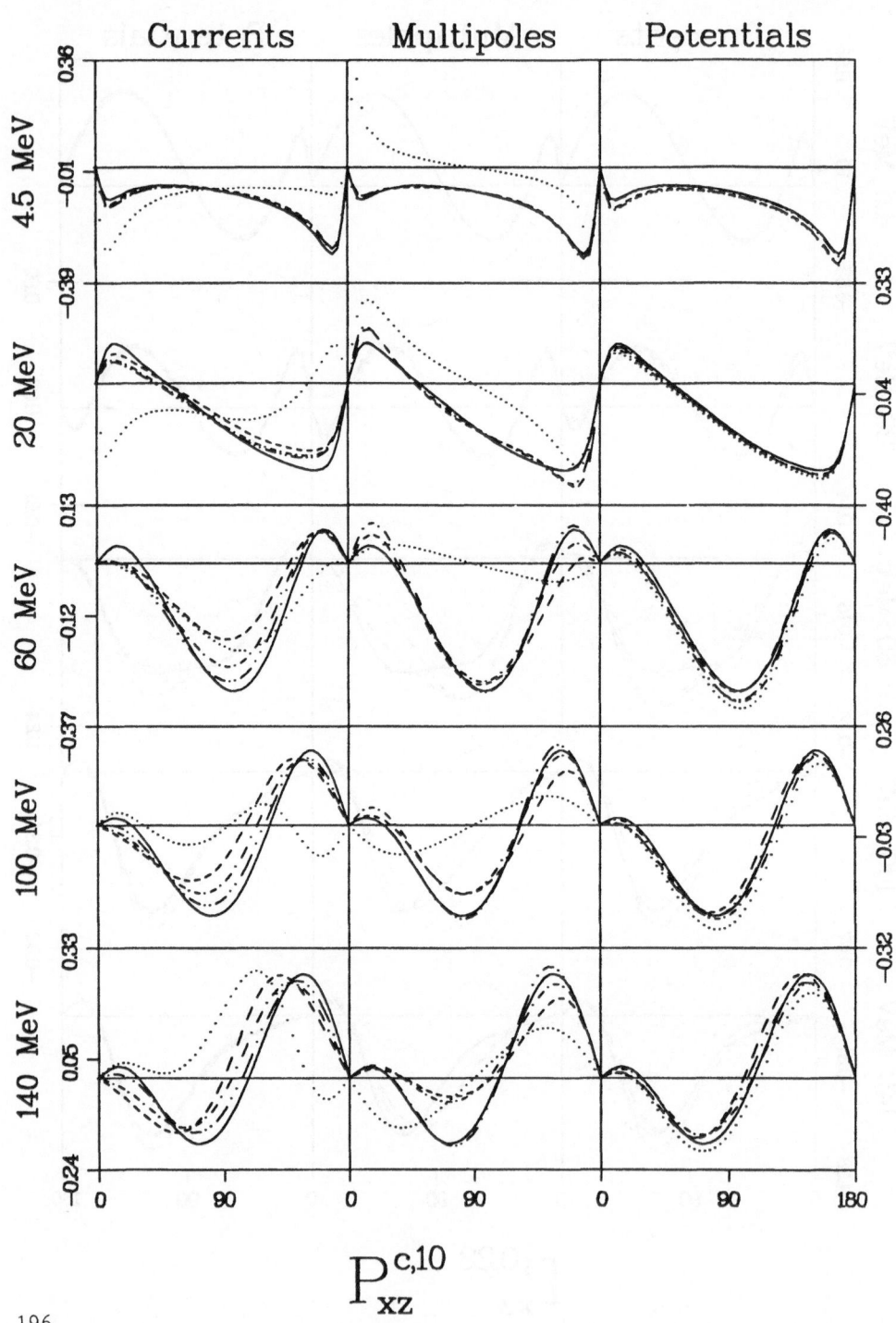

$$P_{xz}^{c,10}$$

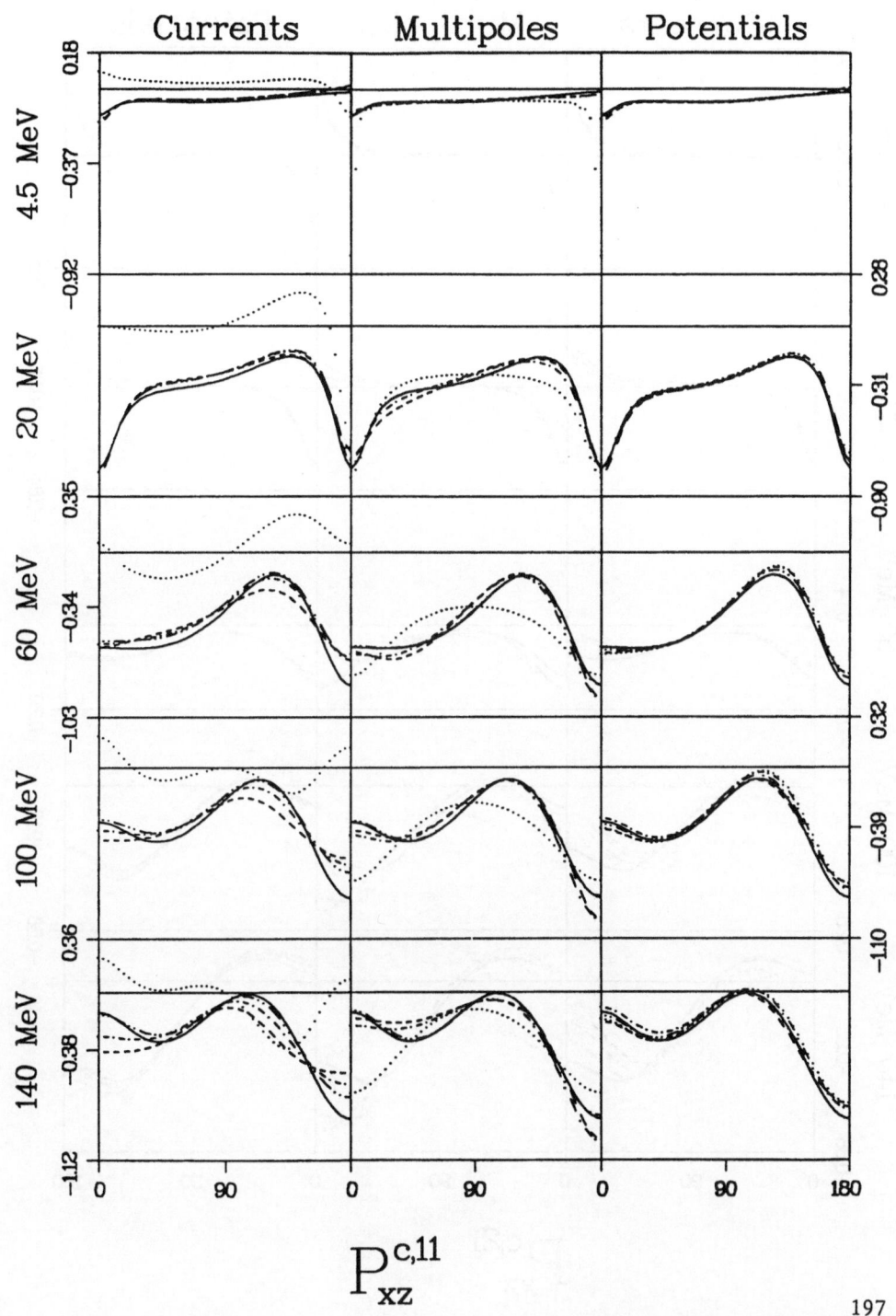

$$P_{xz}^{c,11}$$

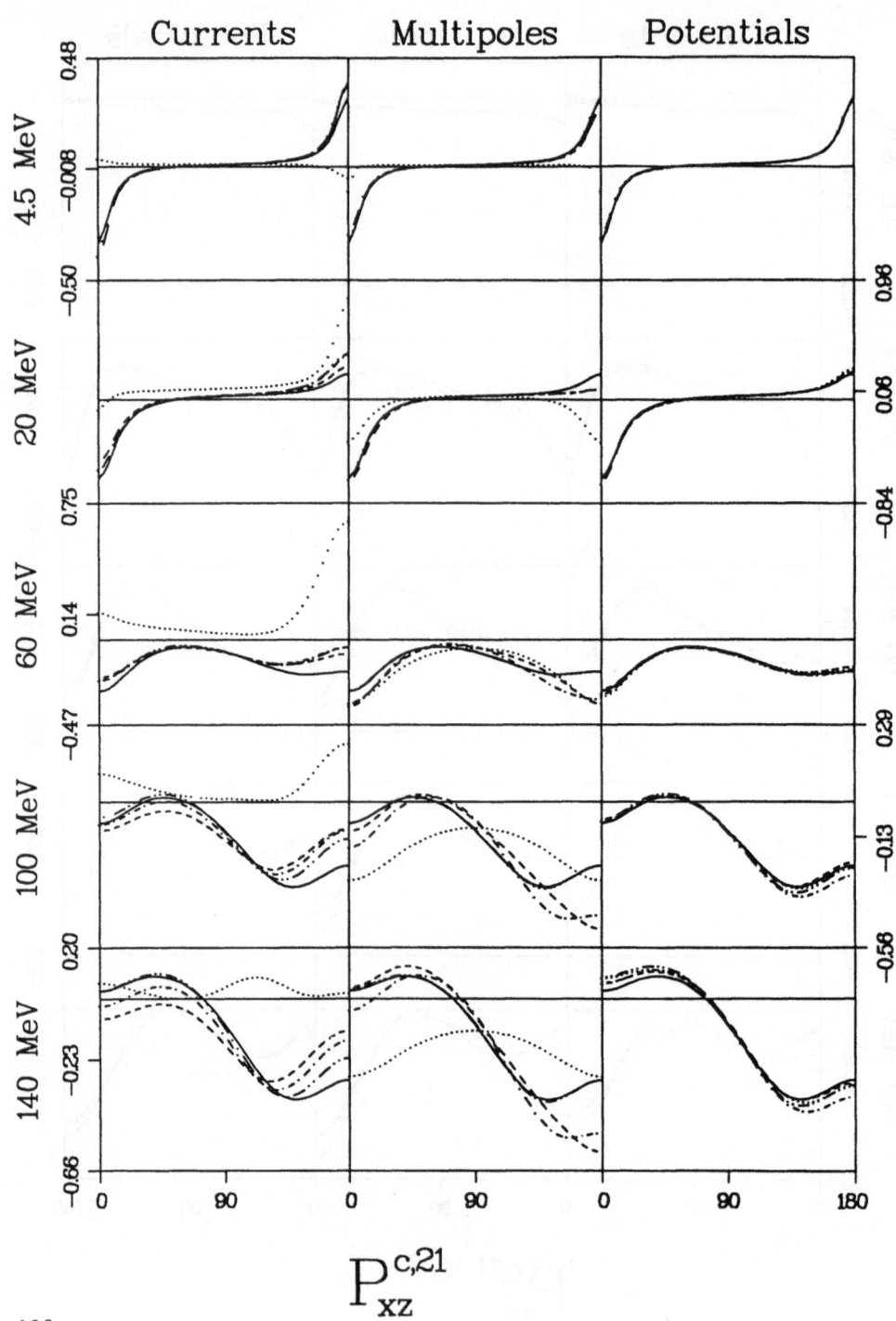

$$P^{c,21}_{xz}$$

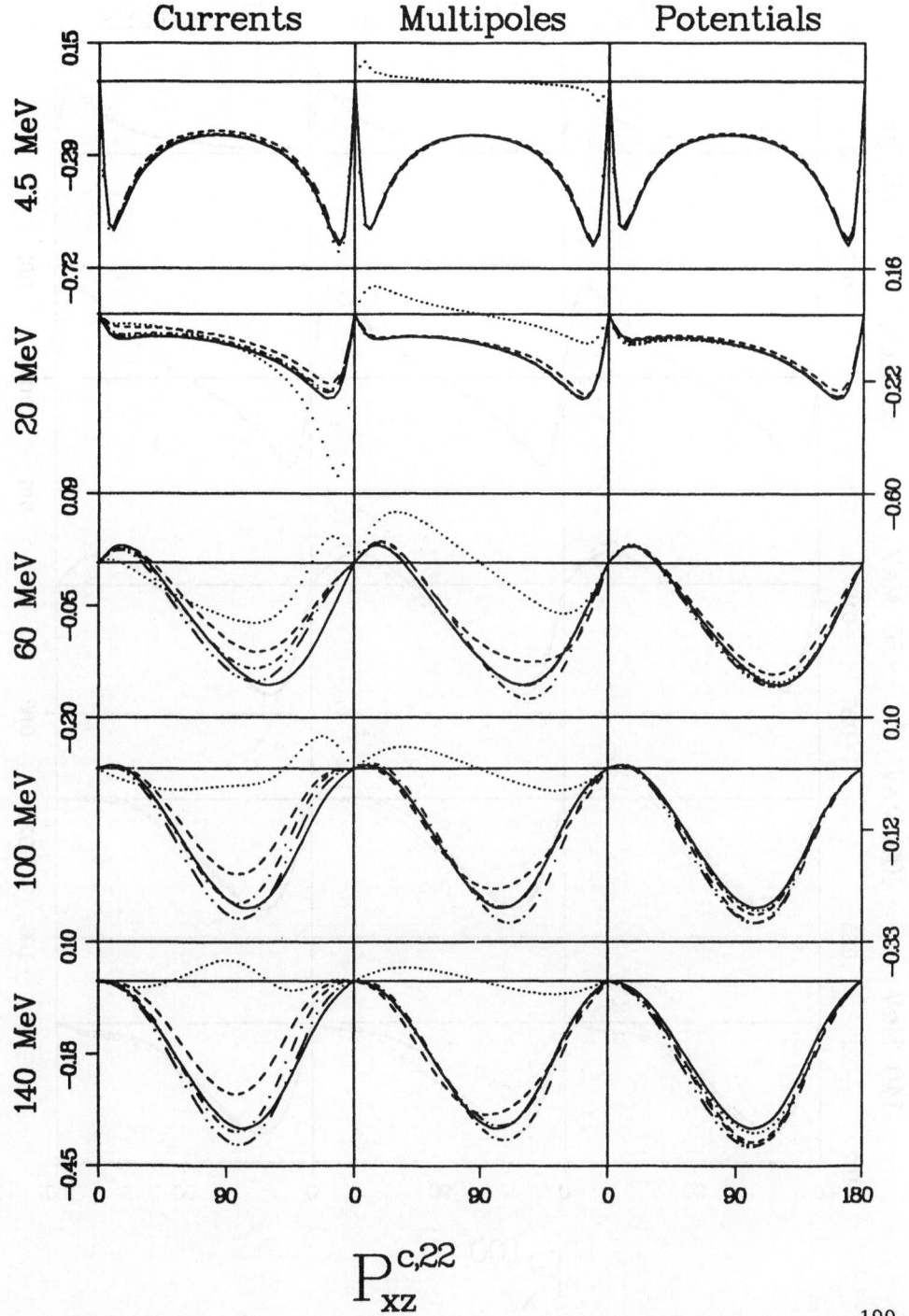

$$P_{xz}^{c,22}$$

5.34 Proton-Neutron Spin Correlation $P_{zz}^{l,IM}$

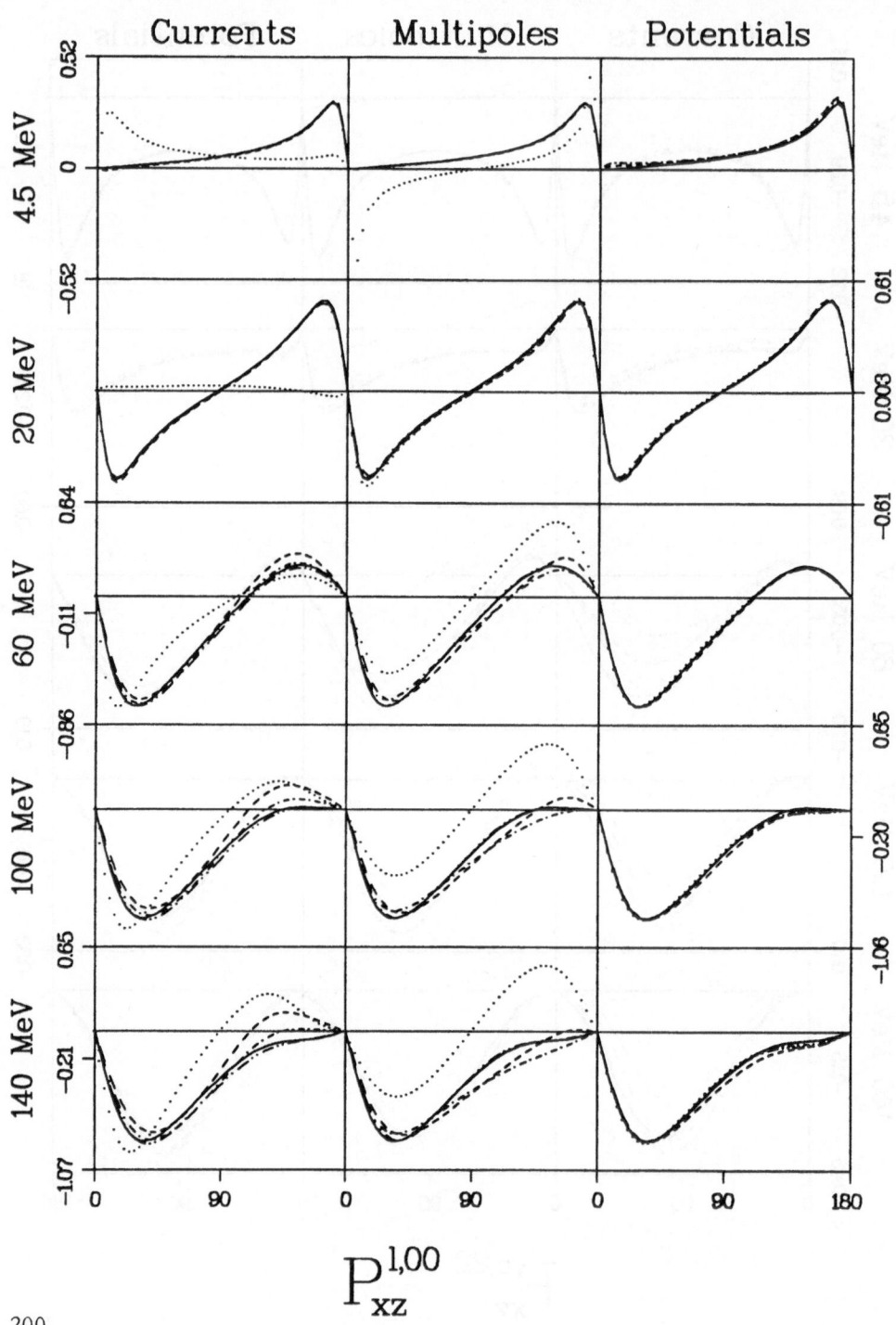

$$P_{xz}^{1,00}$$

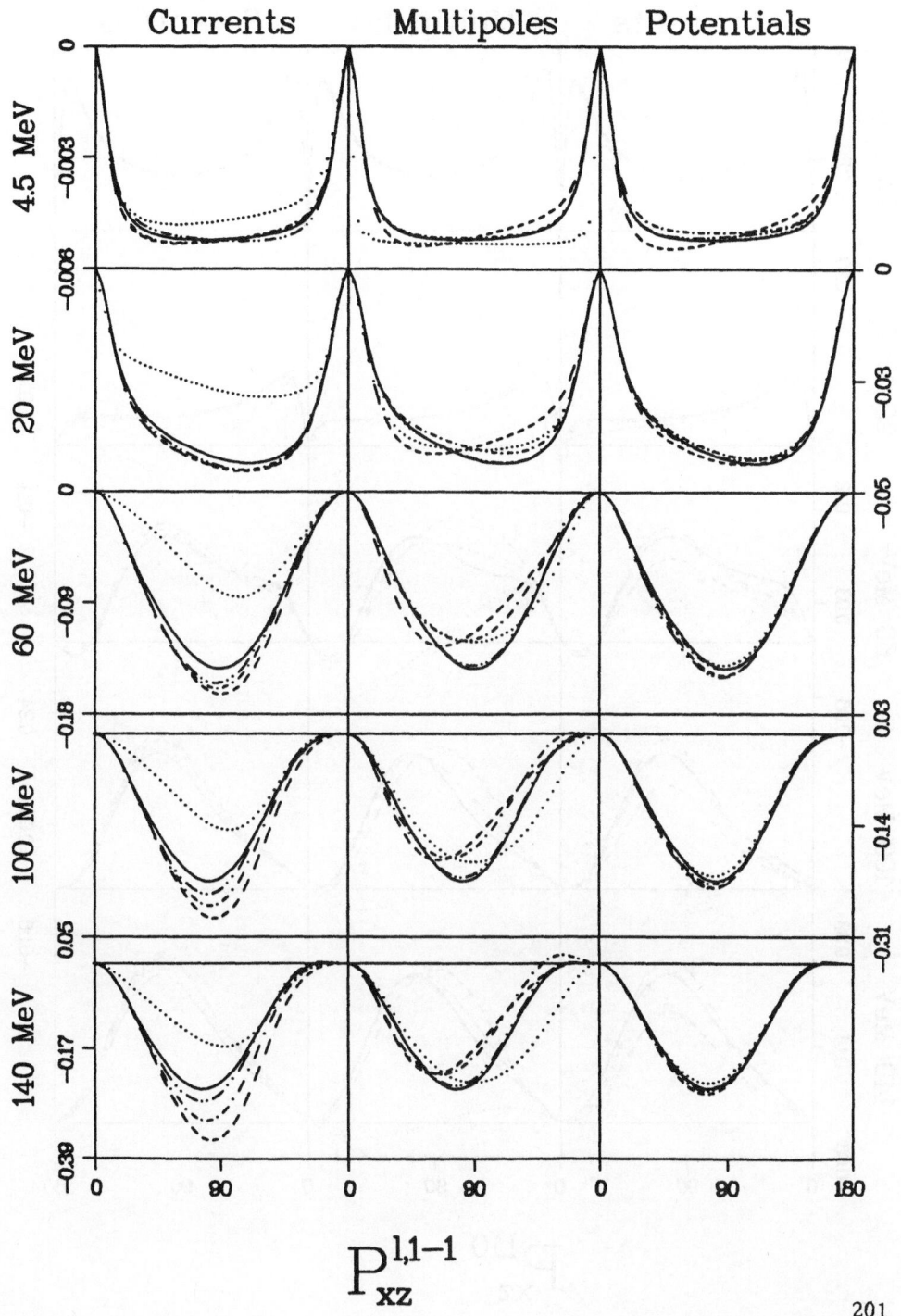

Currents   Multipoles   Potentials

4.5 MeV

20 MeV

60 MeV

100 MeV

140 MeV

$$P_{xz}^{1,1-1}$$

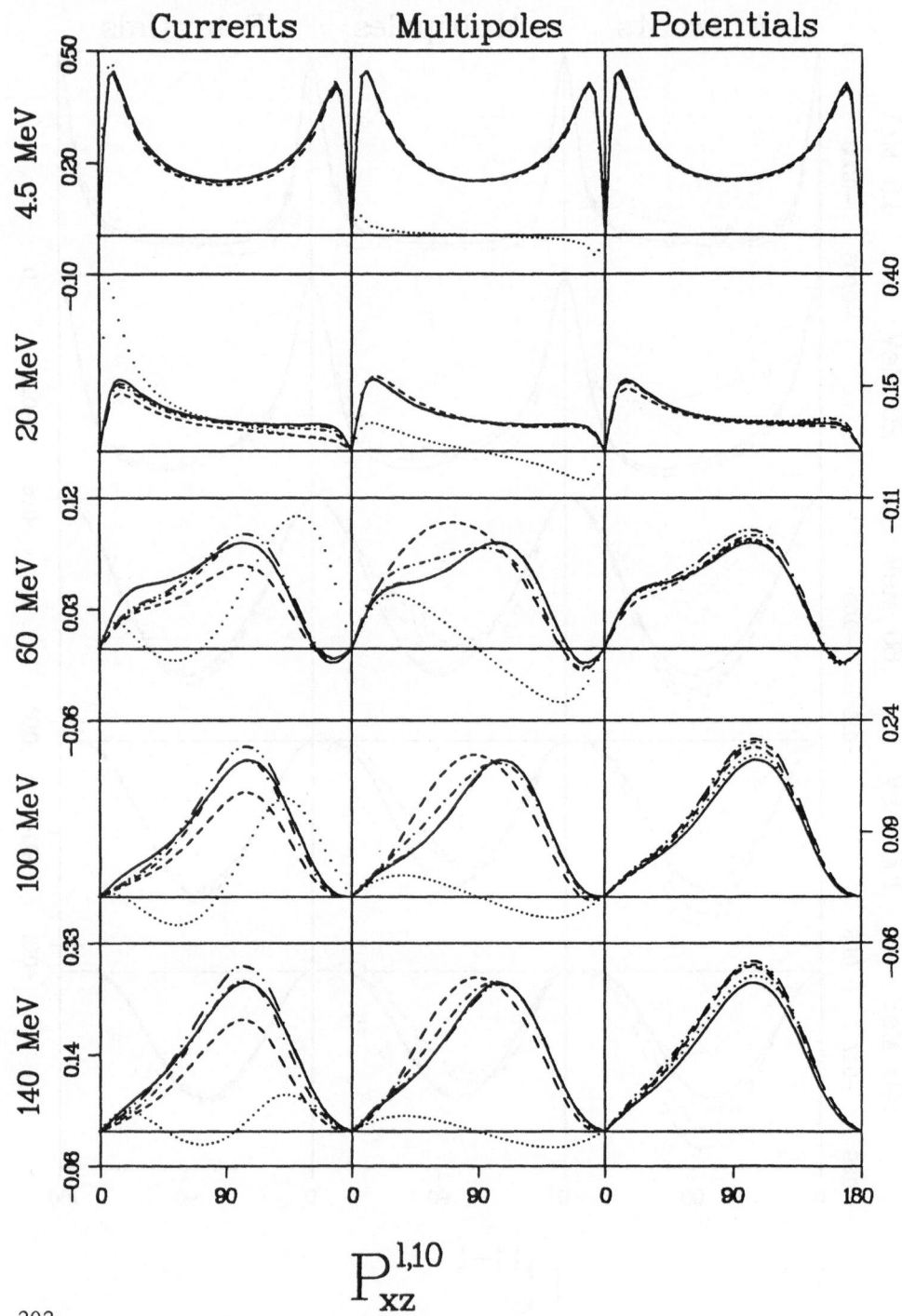

Currents   Multipoles   Potentials

4.5 MeV

20 MeV

60 MeV

100 MeV

140 MeV

$$P_{xz}^{1,10}$$

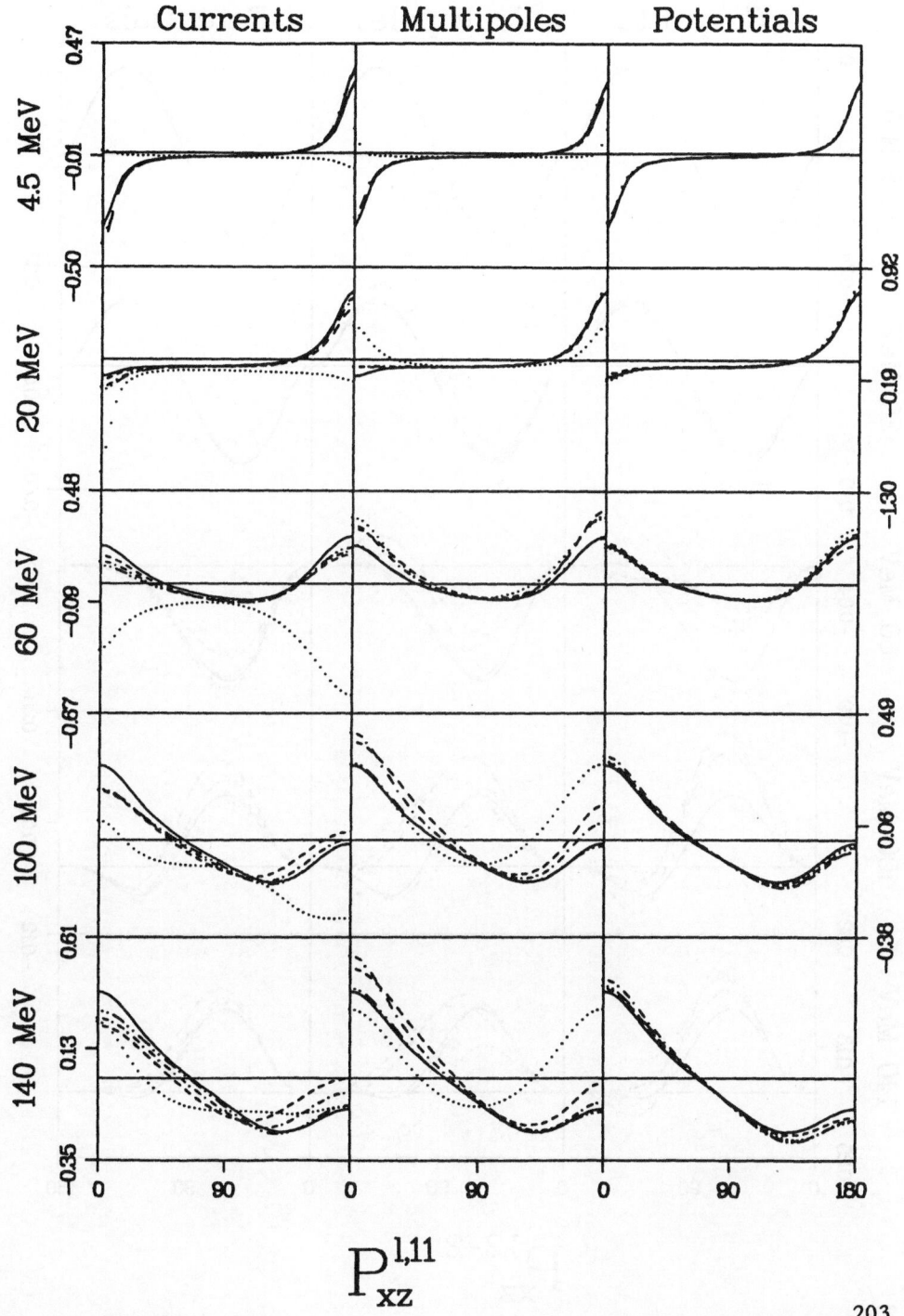

Currents   Multipoles   Potentials

$$P_{xz}^{l,11}$$

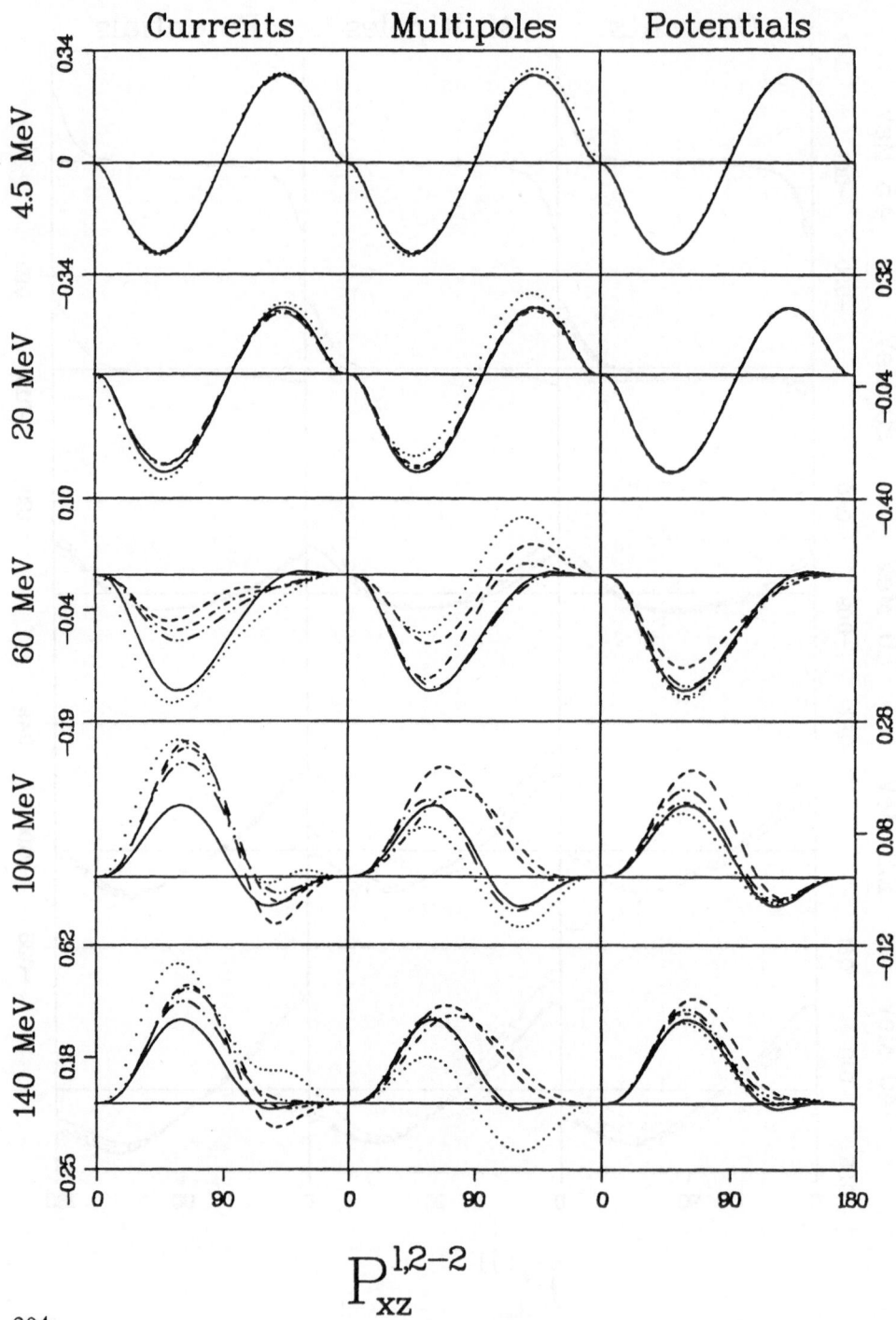

$$P_{xz}^{1,2-2}$$

Proton-Neutron Spin Correlation $P_{xz}^{l,IM}$

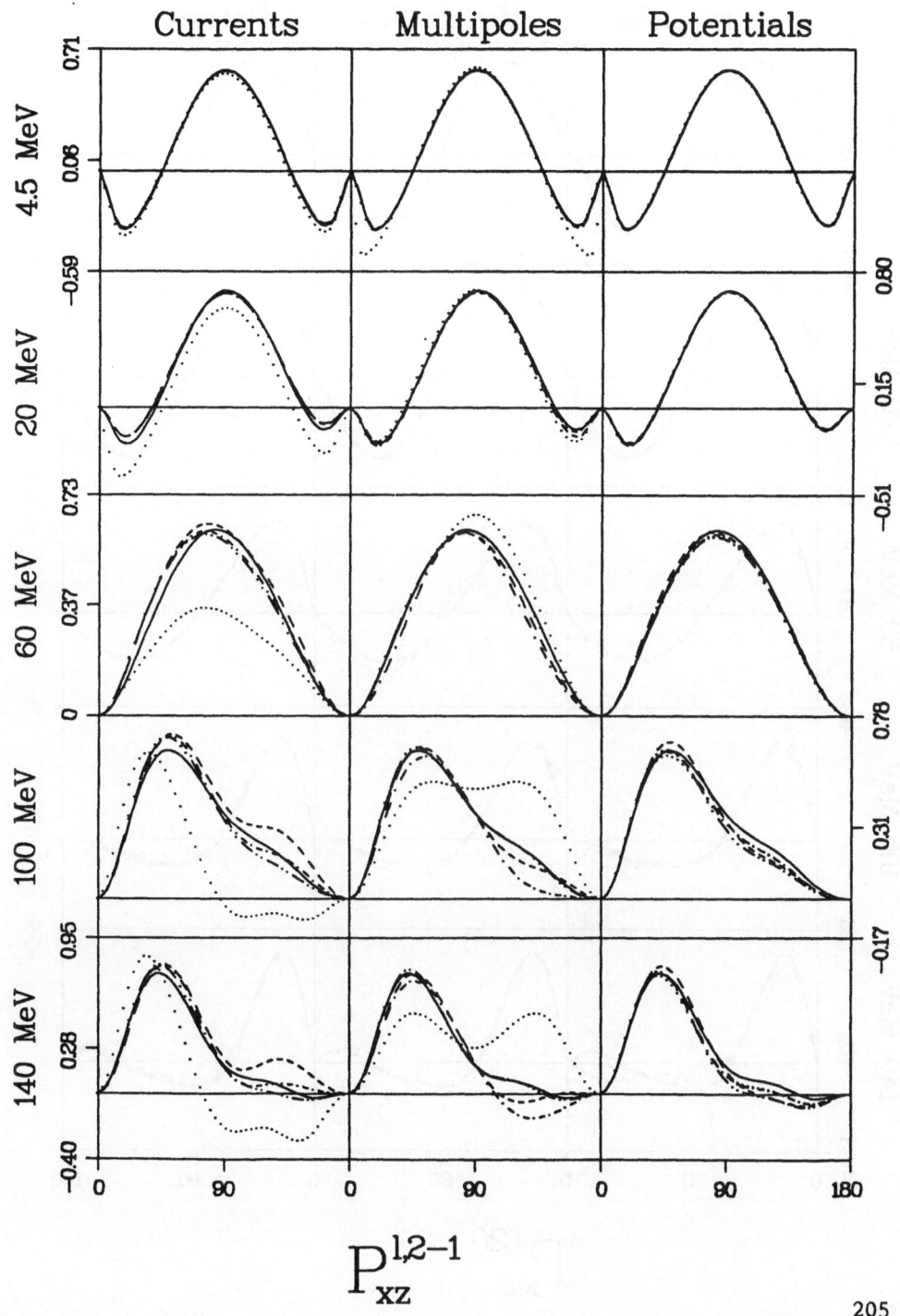

$$P_{xz}^{1,2-1}$$

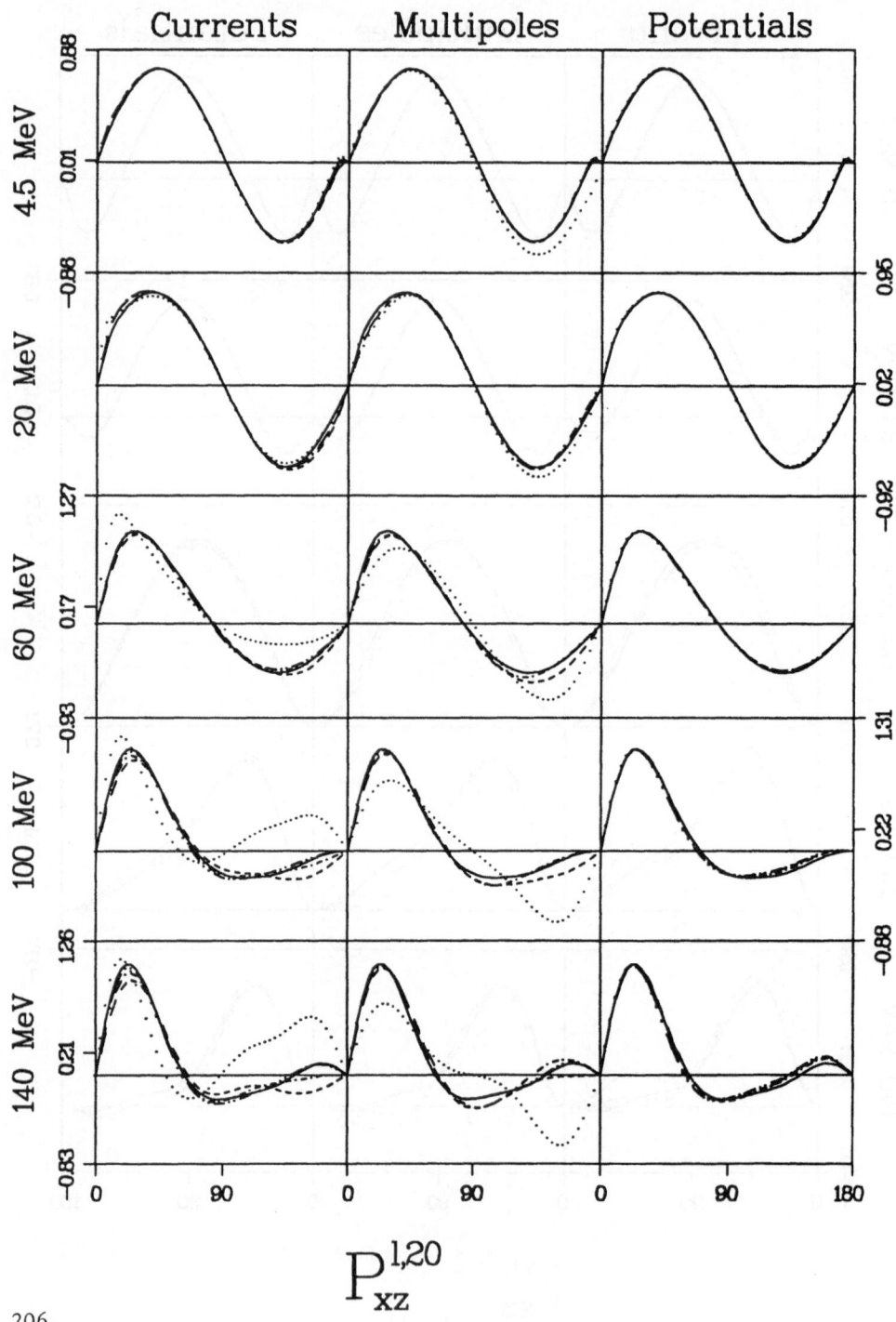

$$P_{xz}^{1,20}$$

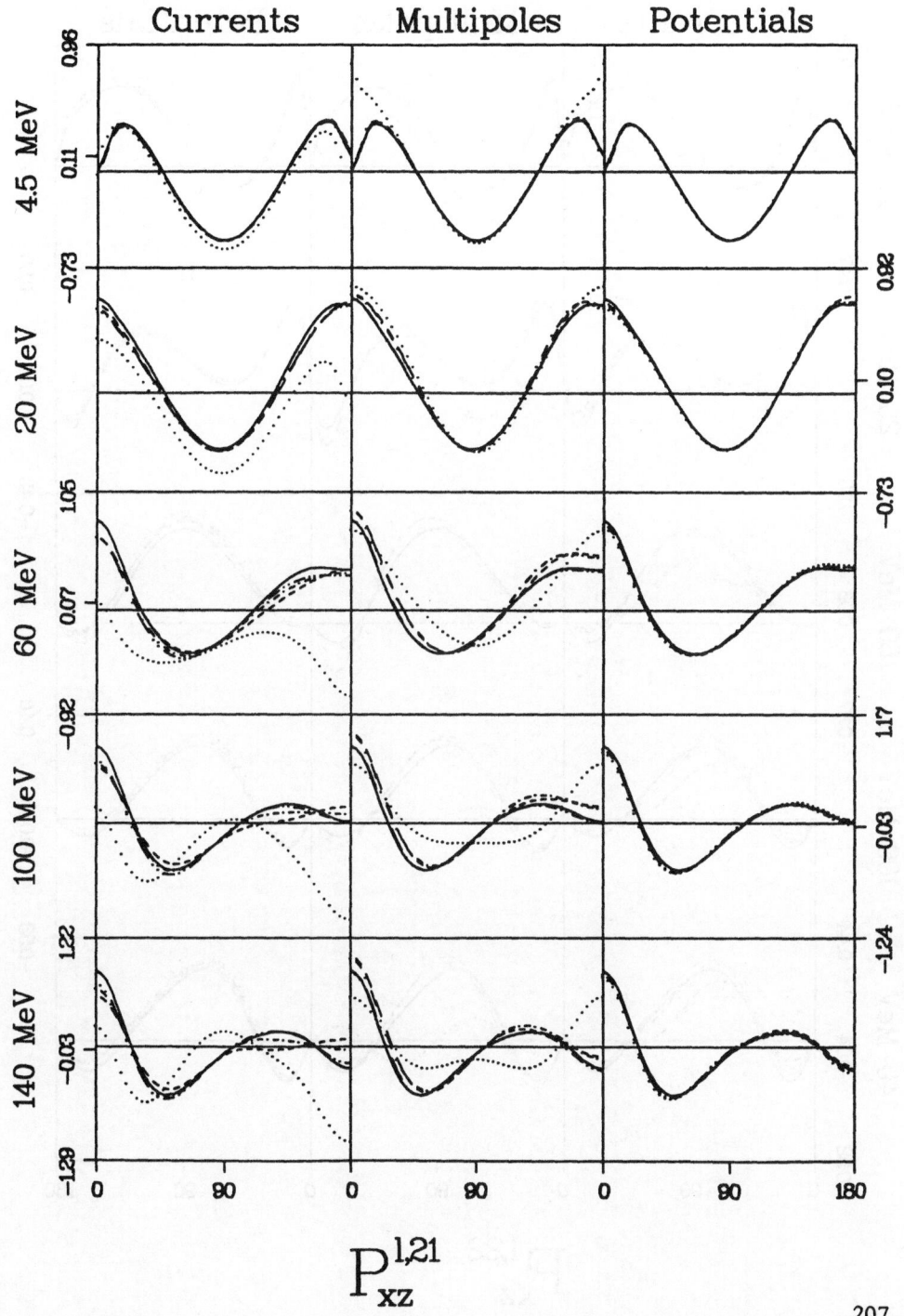

$$P_{xz}^{1,21}$$

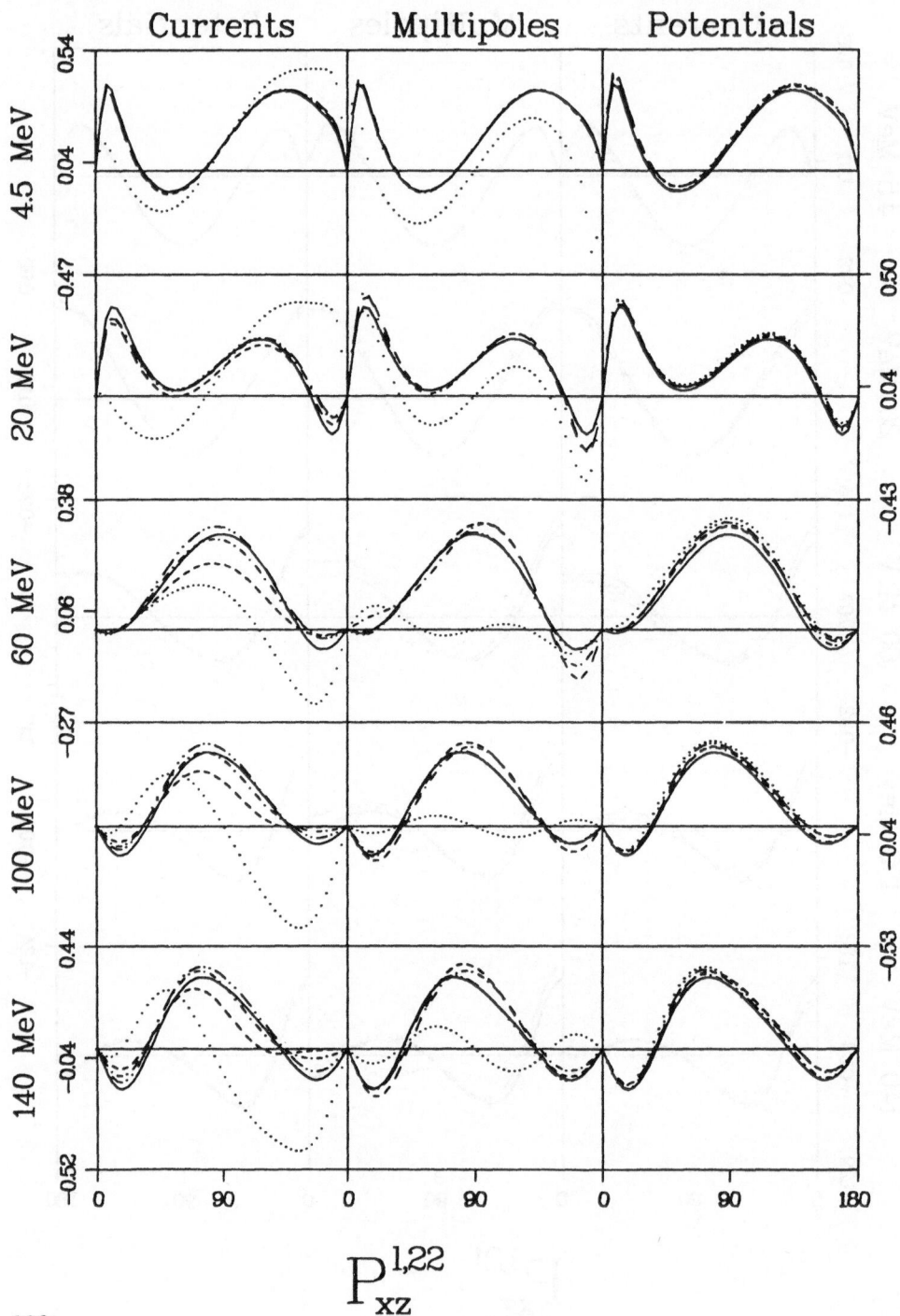

Currents Multipoles Potentials

$$P_{xz}^{1,22}$$

5.35 Proton-Neutron Spin Correlation $P_{zz}^{0,IM}$

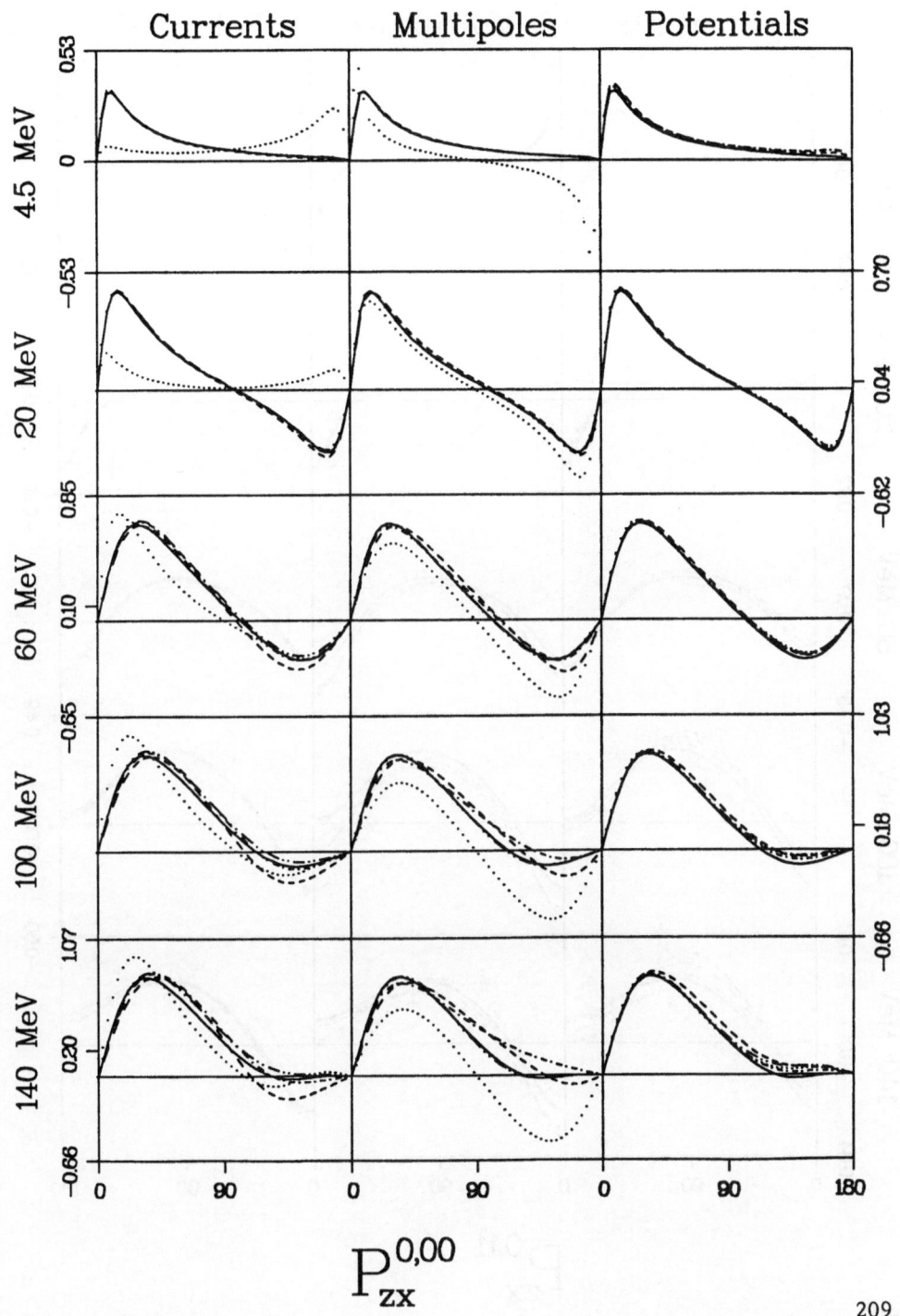

$$P_{zx}^{0,00}$$

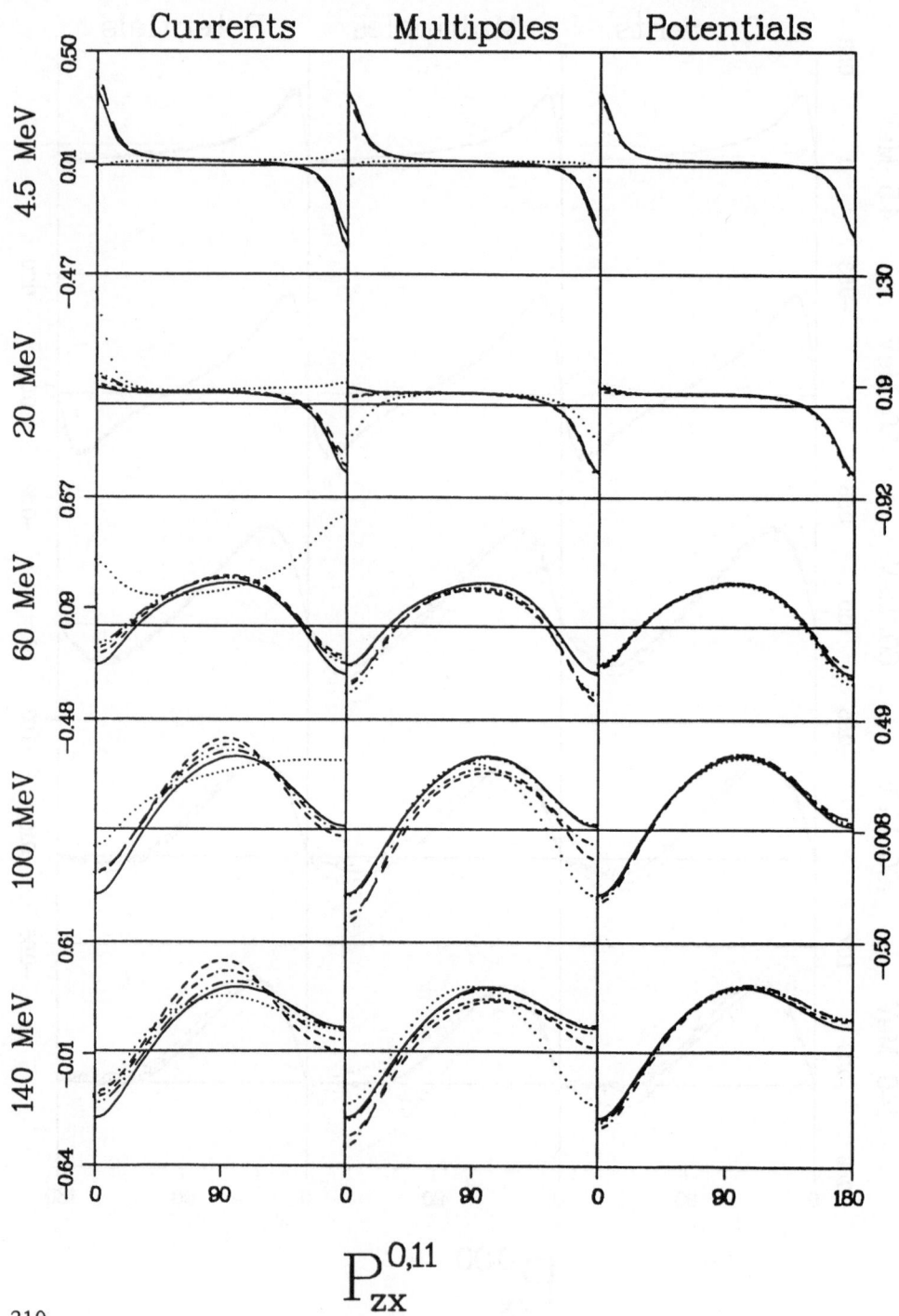

Currents   Multipoles   Potentials

4.5 MeV
20 MeV
60 MeV
100 MeV
140 MeV

$$P_{zx}^{0,11}$$

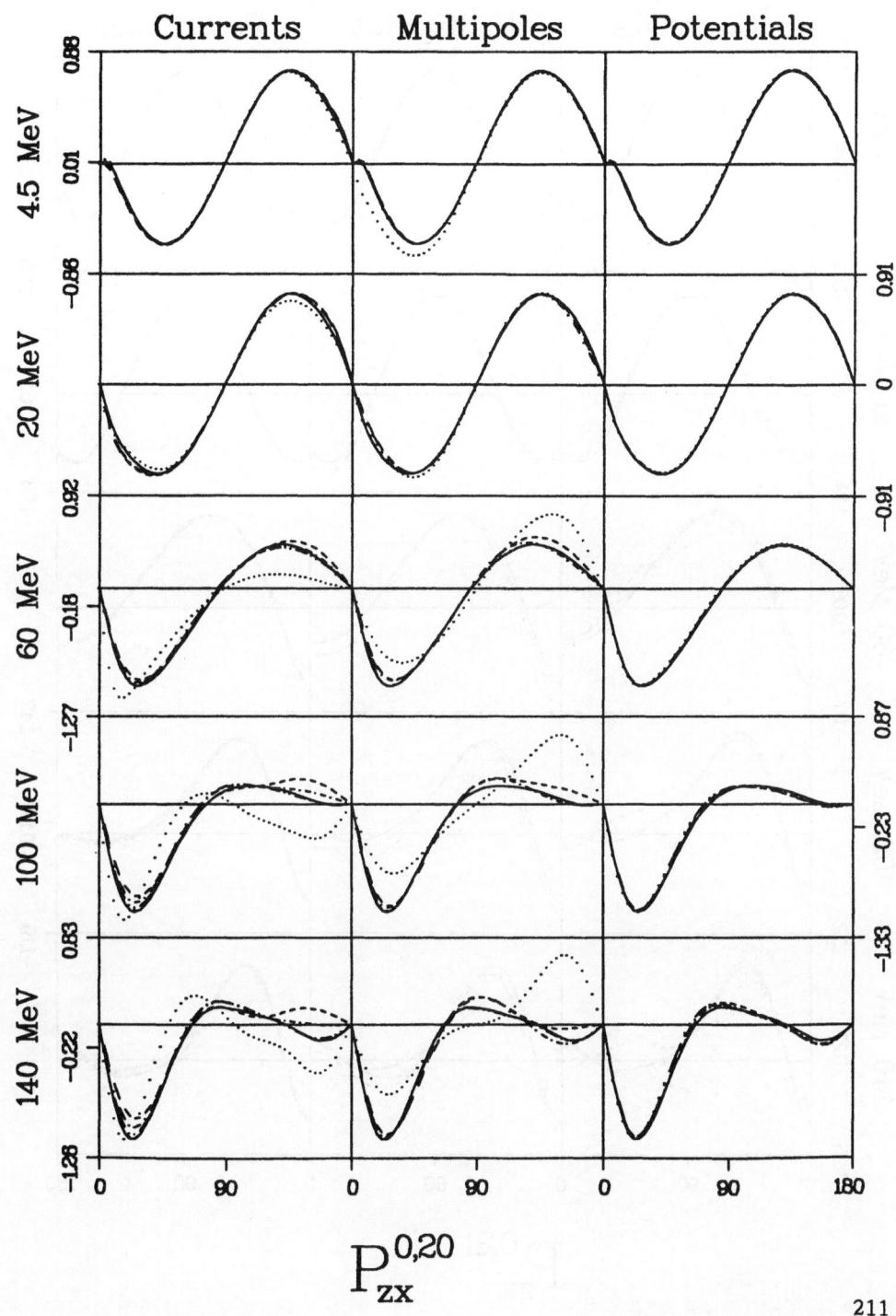

$$P_{zx}^{0,20}$$

Proton-Neutron Spin Correlation $P_{zx}^{0,IM}$

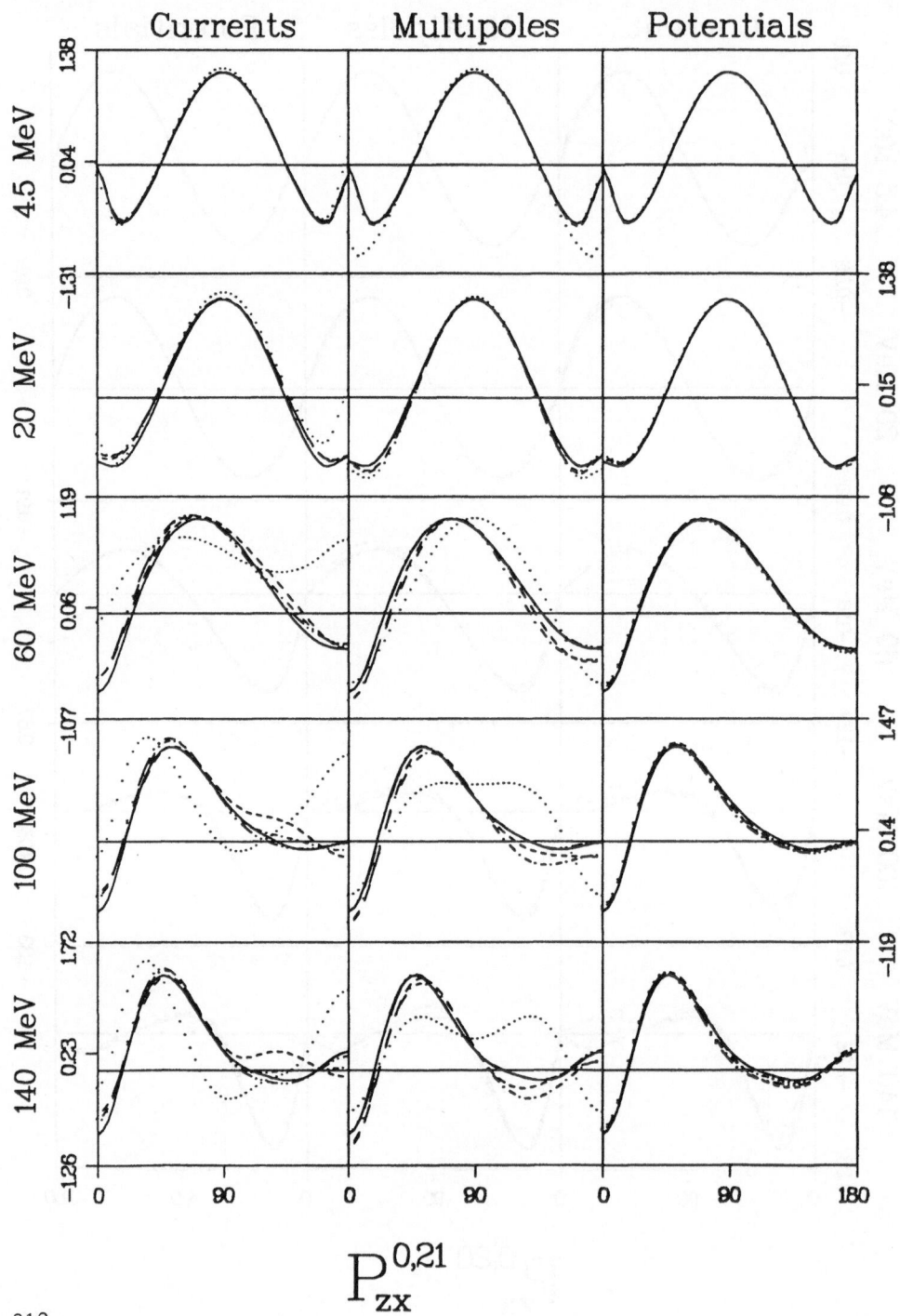

$$P_{zx}^{0,21}$$

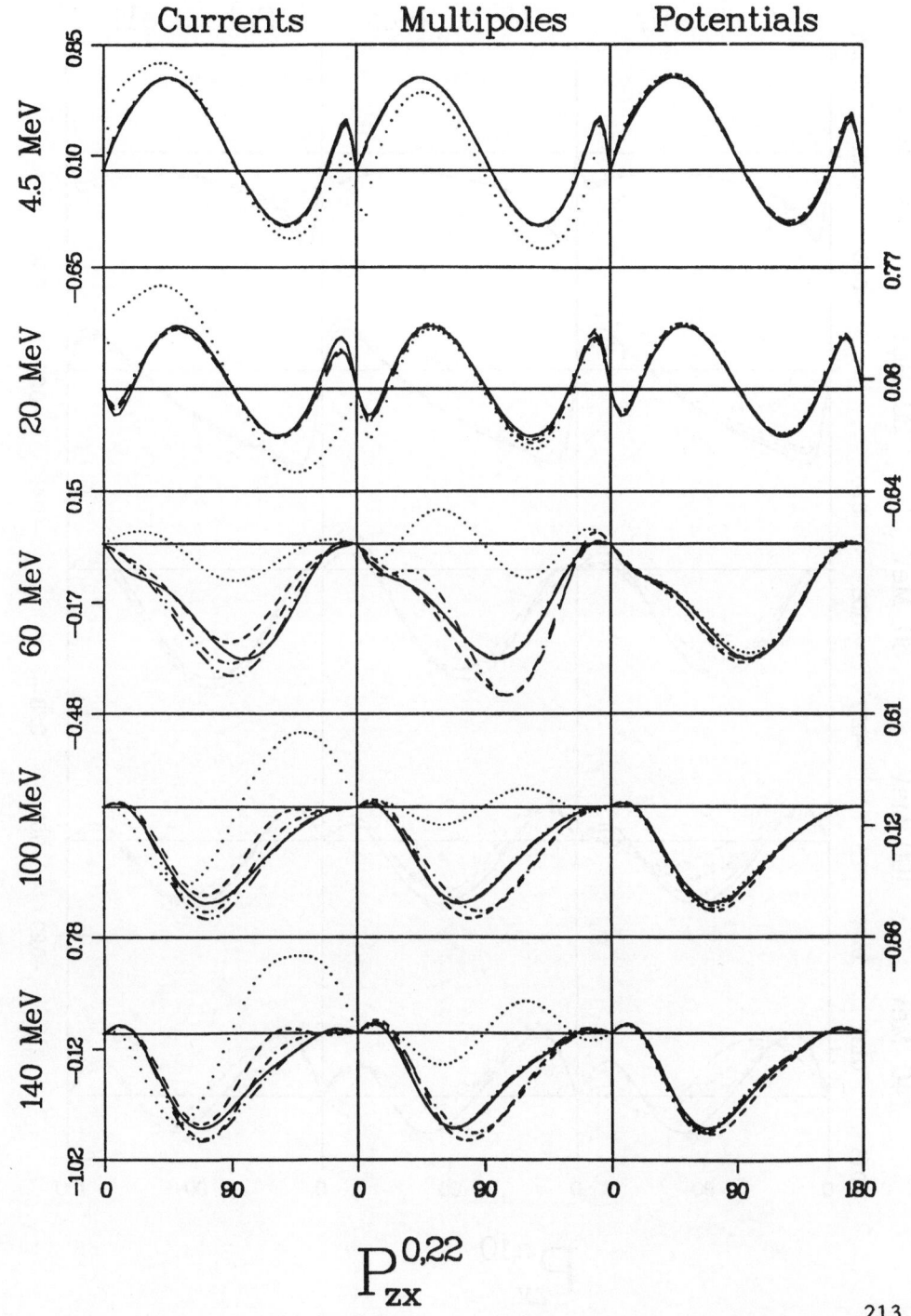

$$P_{zx}^{0,22}$$

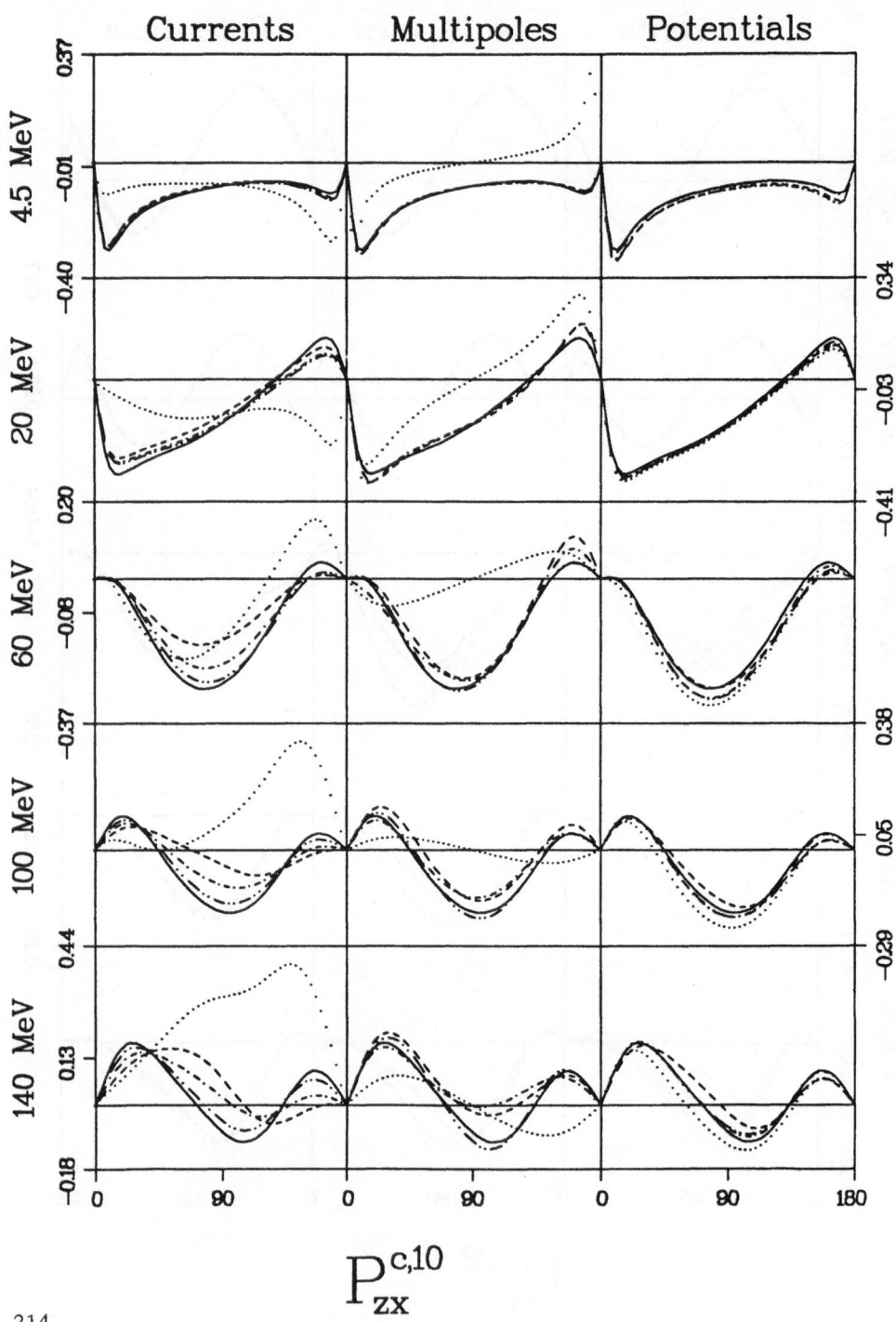

Currents   Multipoles   Potentials

4.5 MeV

20 MeV

60 MeV

100 MeV

140 MeV

$$P^{c,10}_{zx}$$

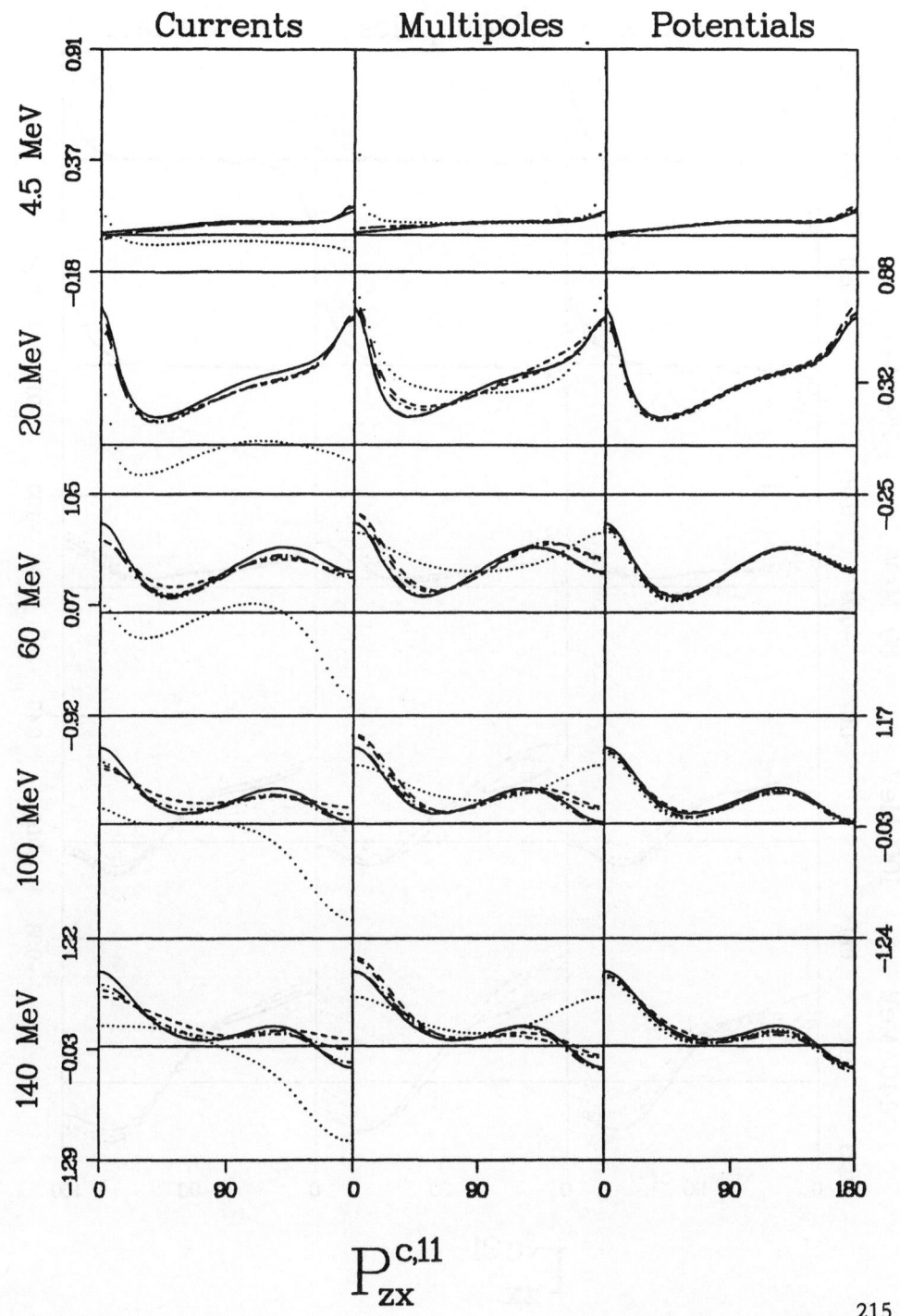

$$P_{zx}^{c,11}$$

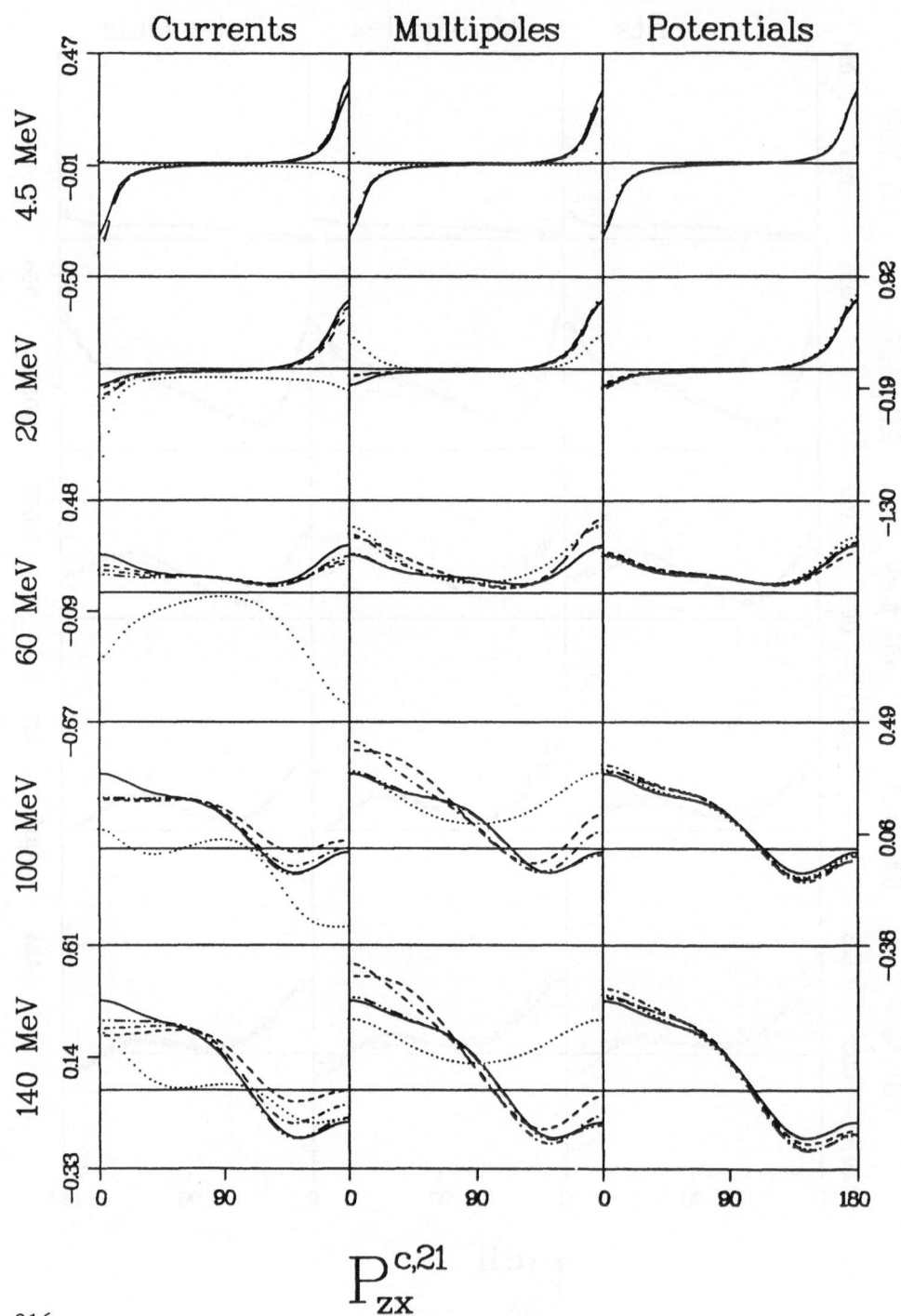

Currents   Multipoles   Potentials

$$P_{zx}^{c,21}$$

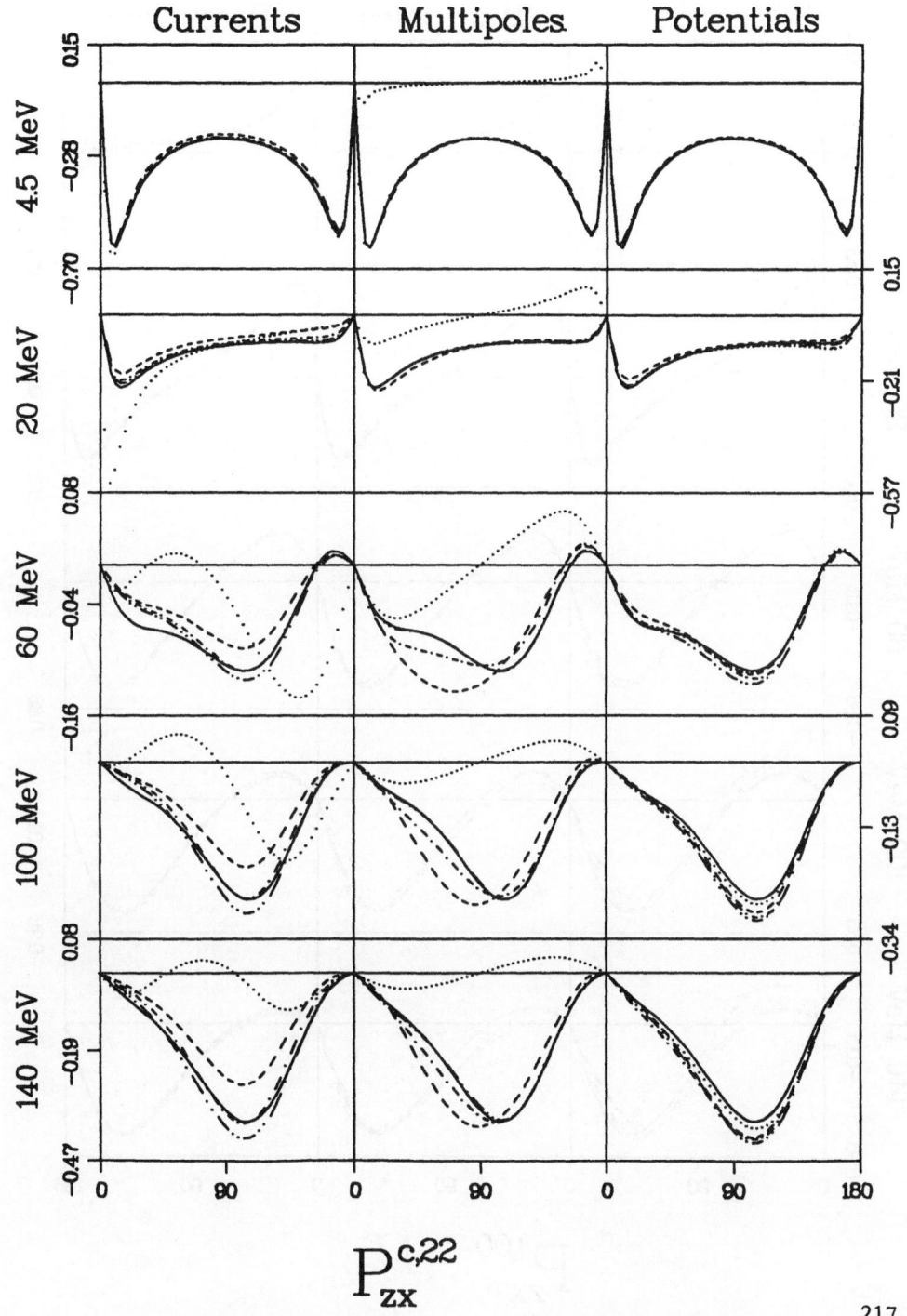

Currents — Multipoles — Potentials

4.5 MeV — 20 MeV — 60 MeV — 100 MeV — 140 MeV

$$P_{zx}^{c,22}$$

5.37 Proton-Neutron Spin Correlation $P_{zx}^{l,IM}$

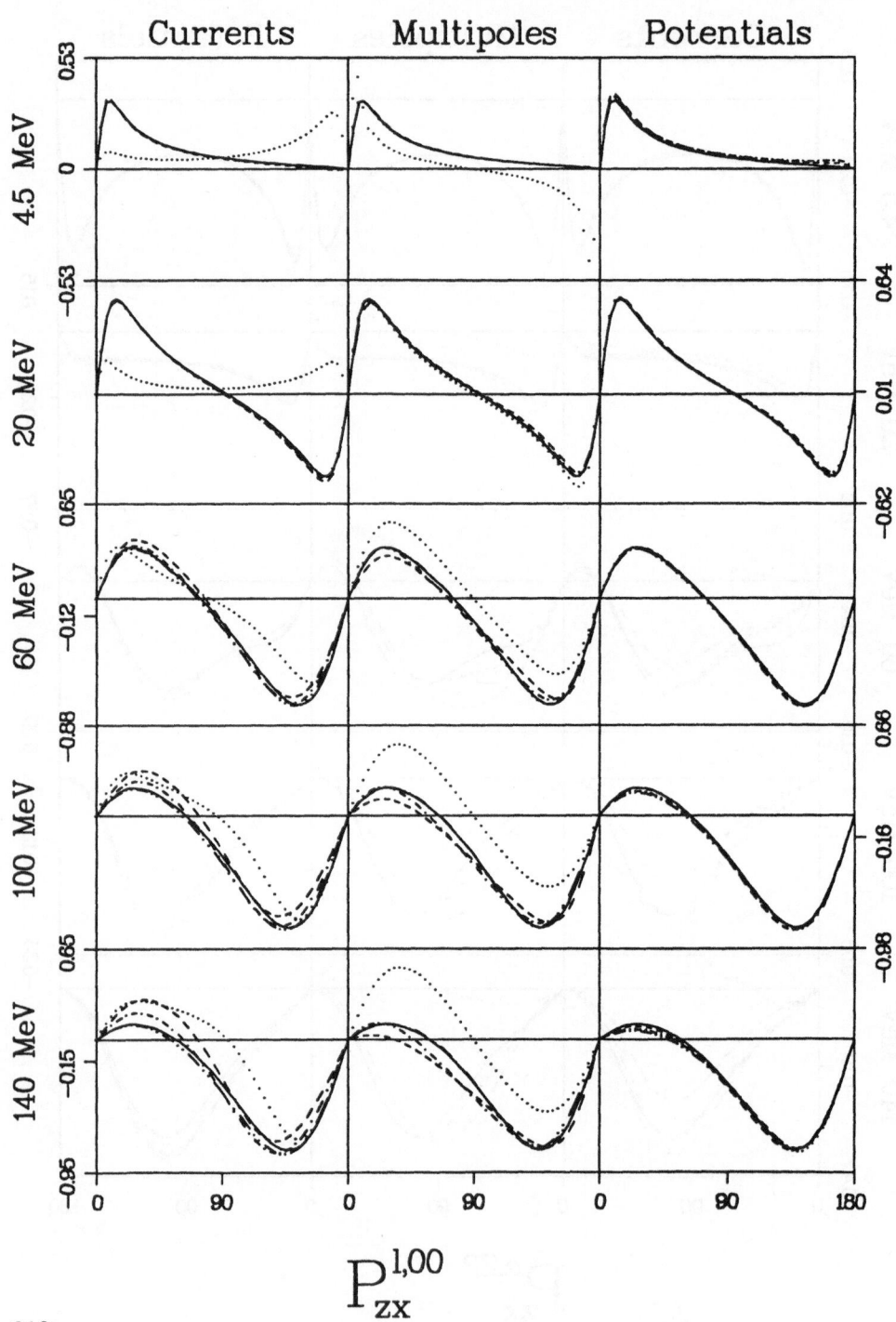

$$P_{zx}^{1,00}$$

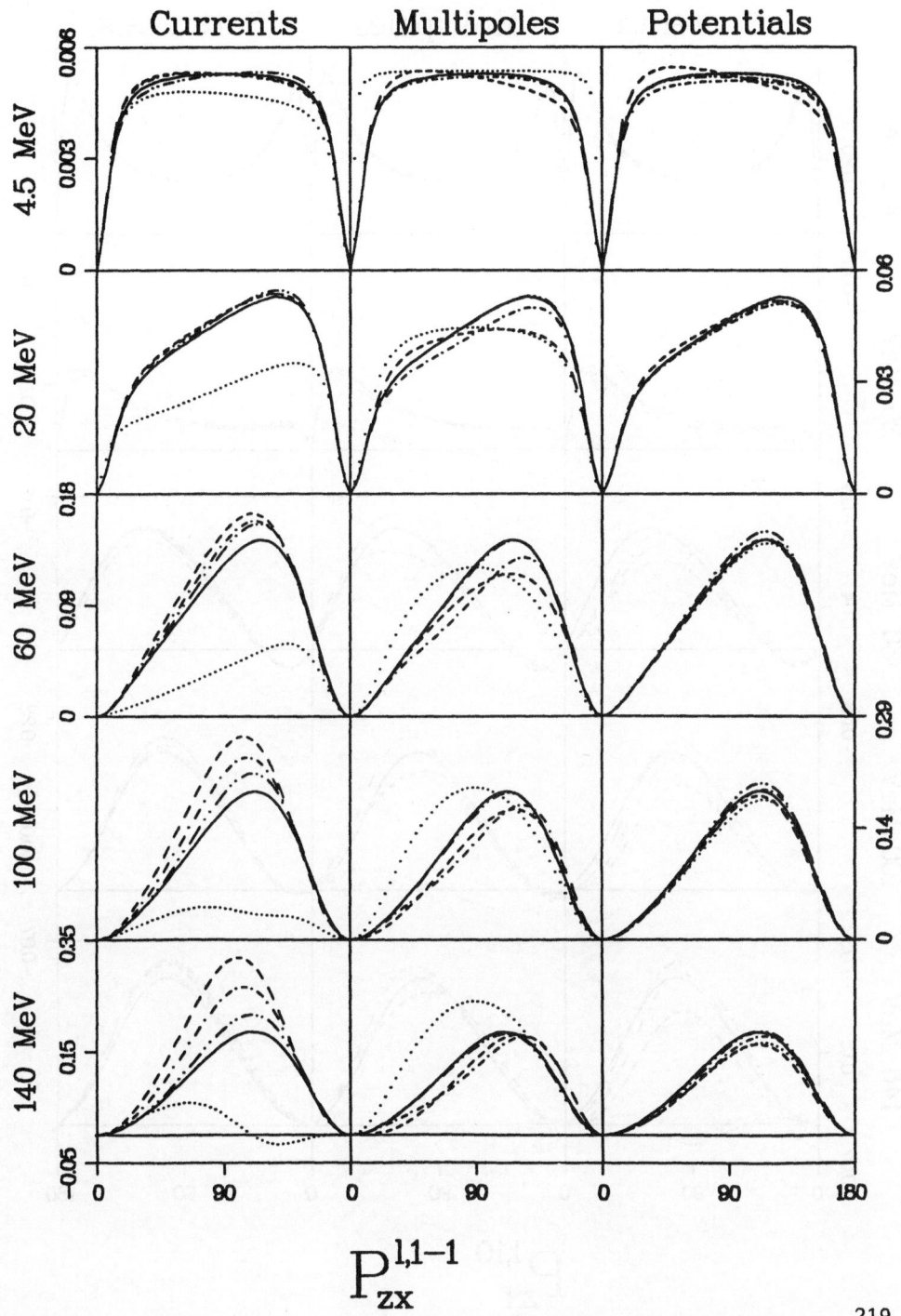

$$P_{zx}^{l,1-1}$$

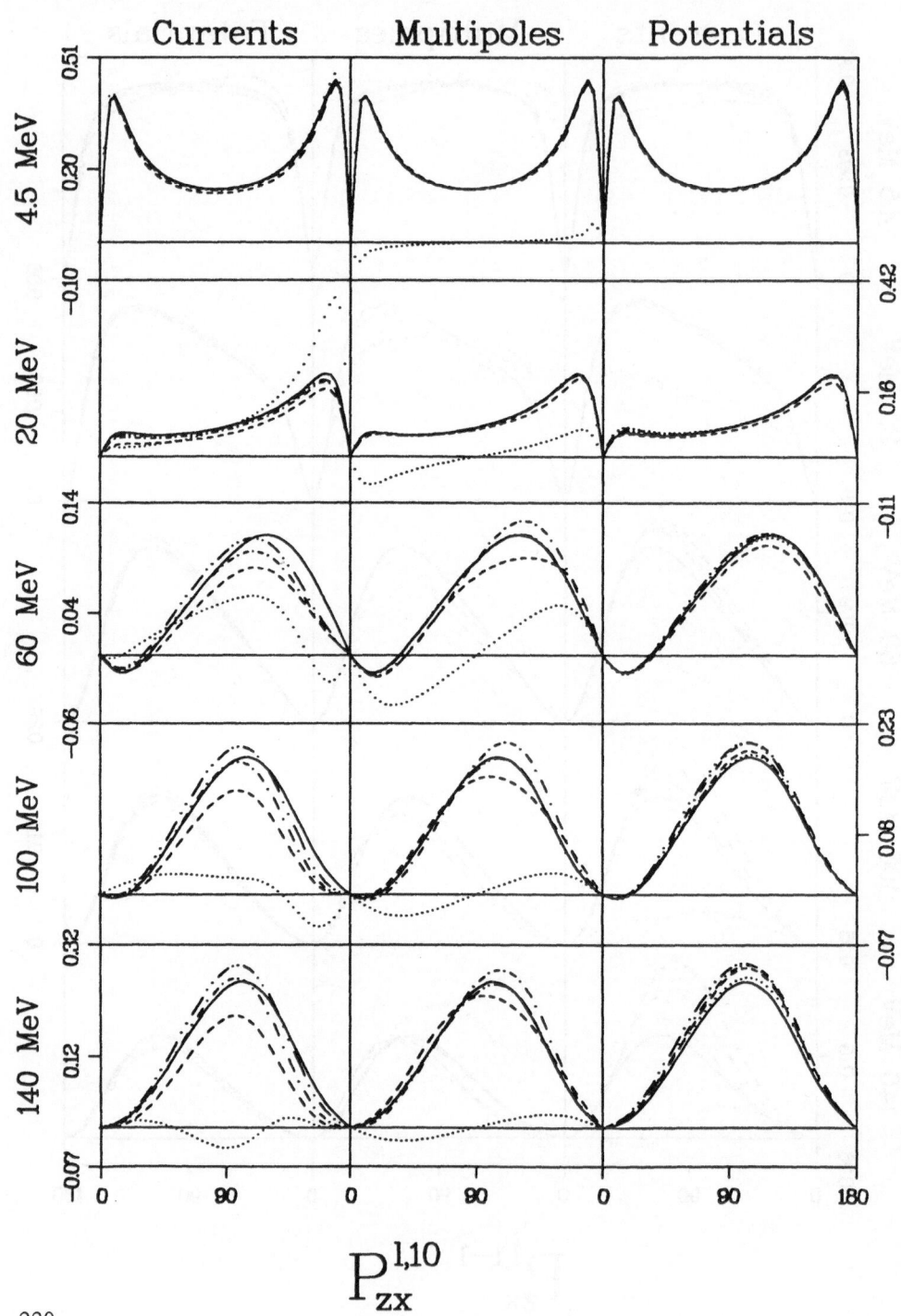

Currents   Multipoles   Potentials

4.5 MeV   20 MeV   60 MeV   100 MeV   140 MeV

$$P_{zx}^{1,10}$$

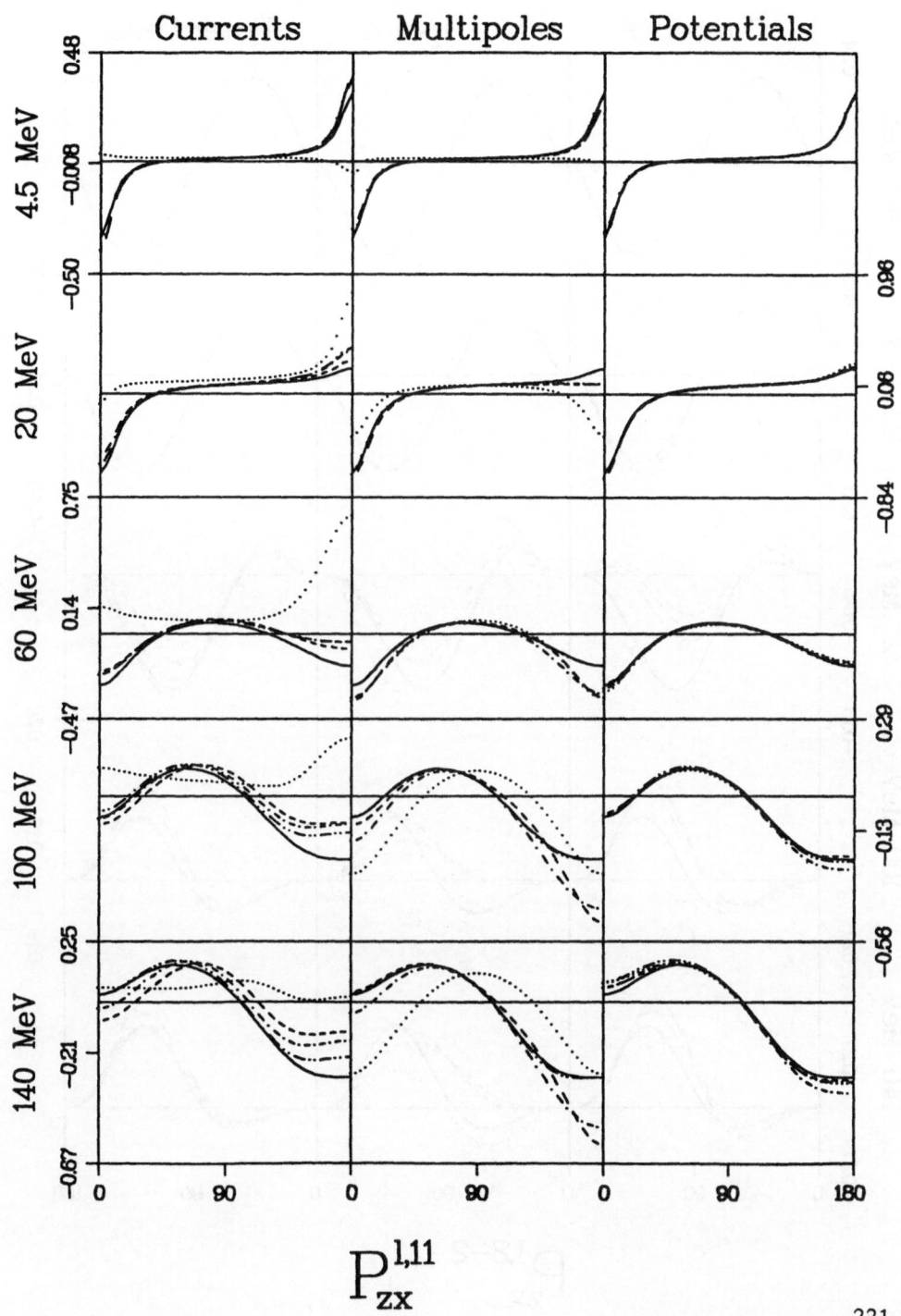

$$P_{zx}^{l,11}$$

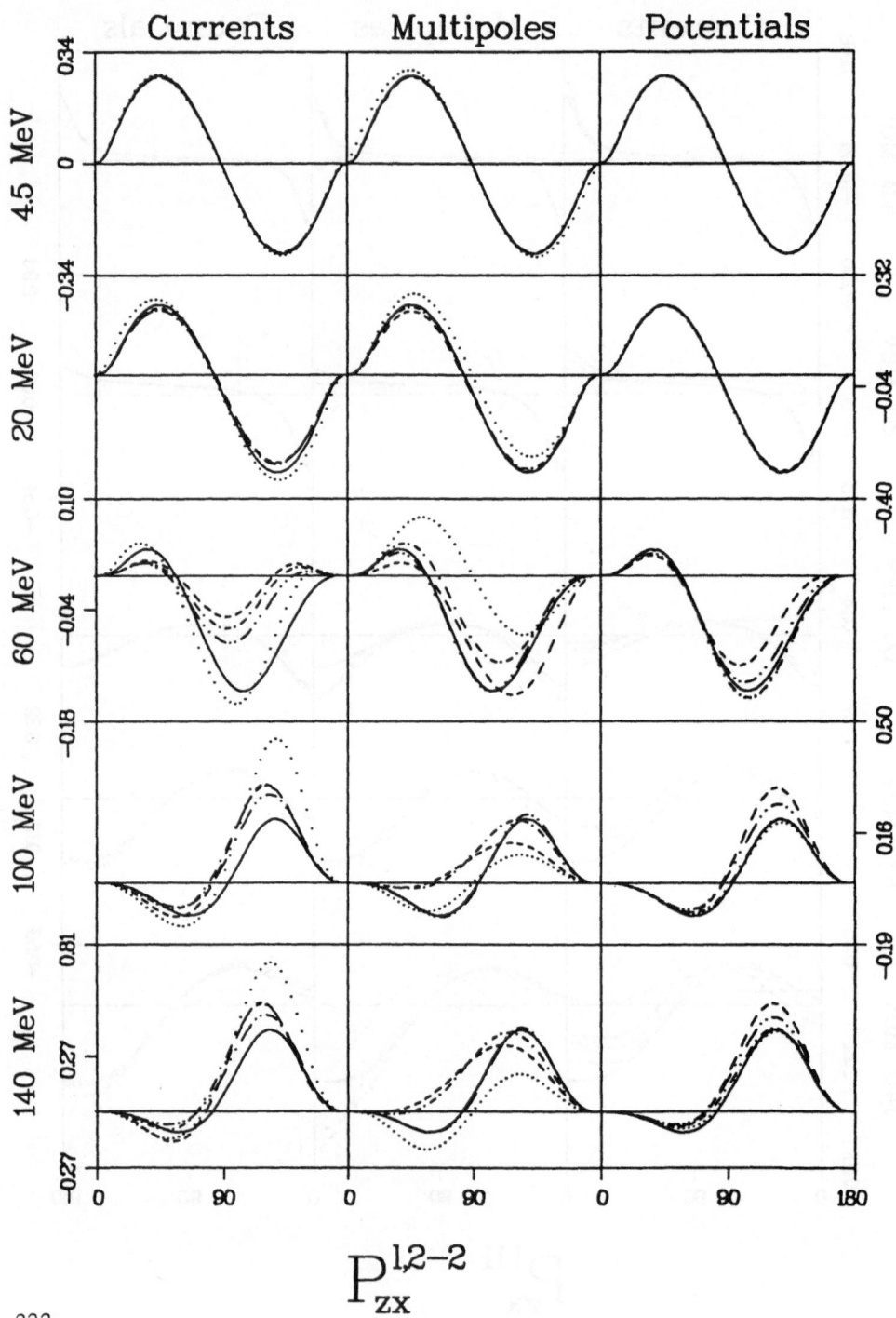

$$P_{zx}^{1,2-2}$$

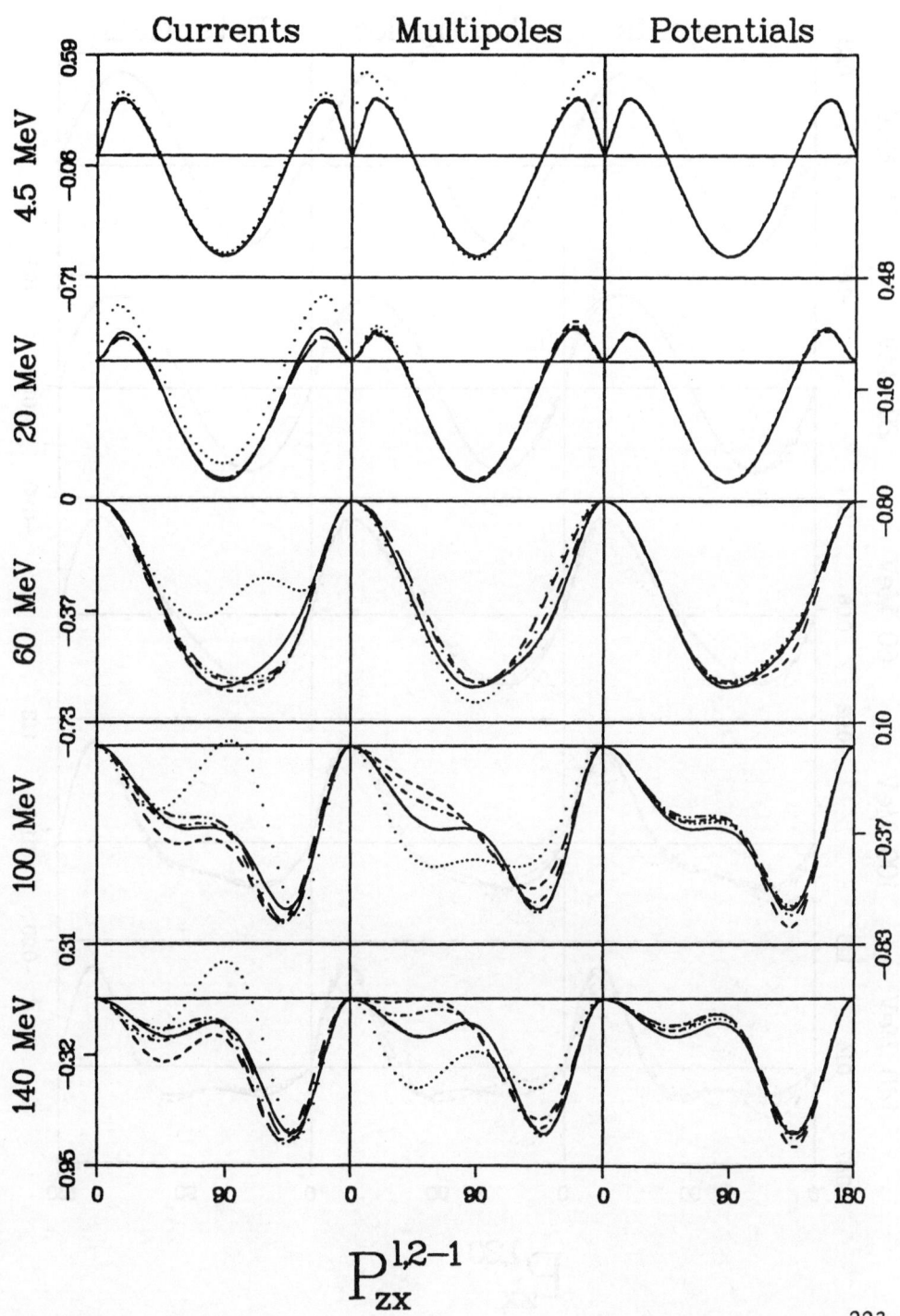

$$P_{zx}^{1,2-1}$$

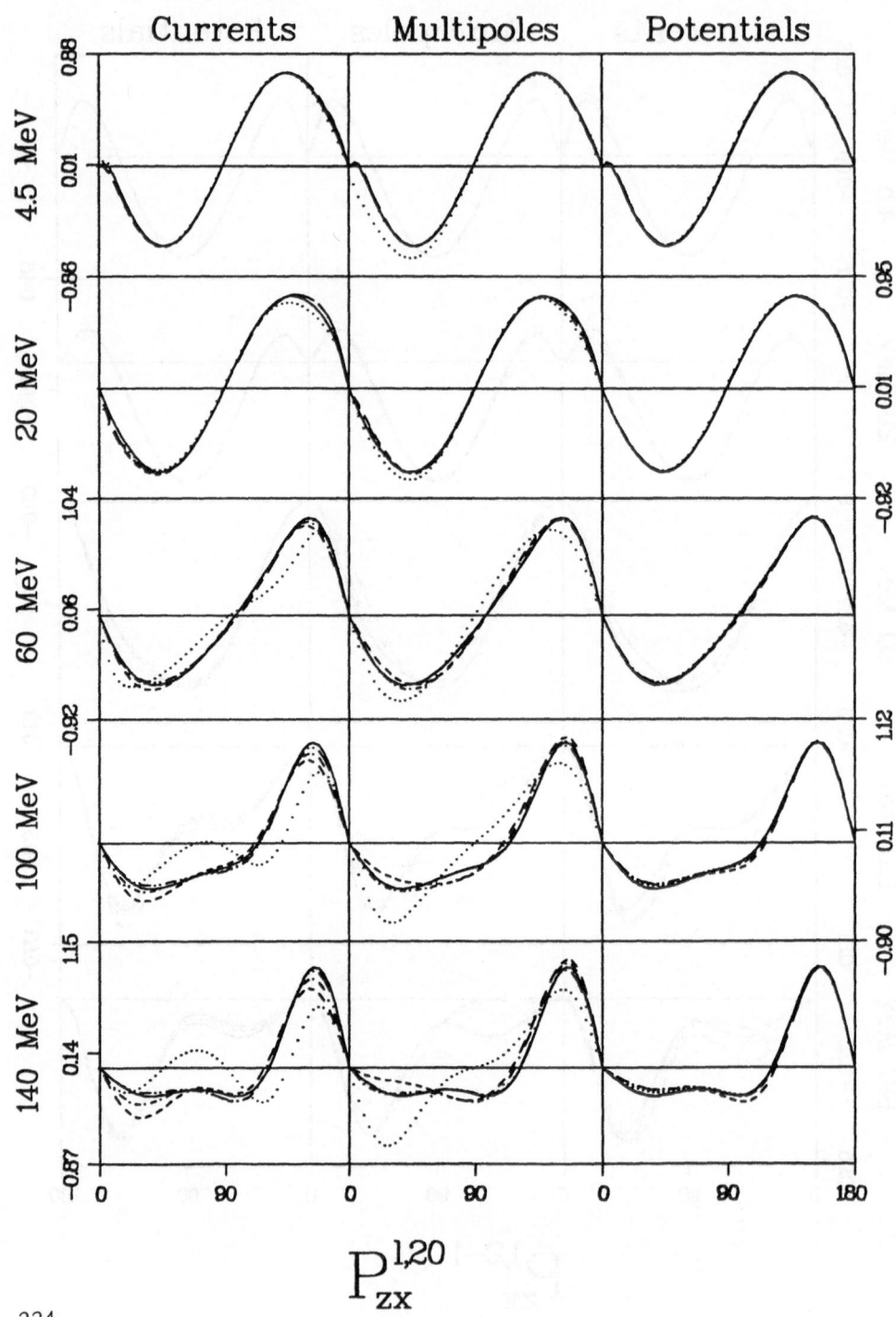

$$P_{zx}^{1,20}$$

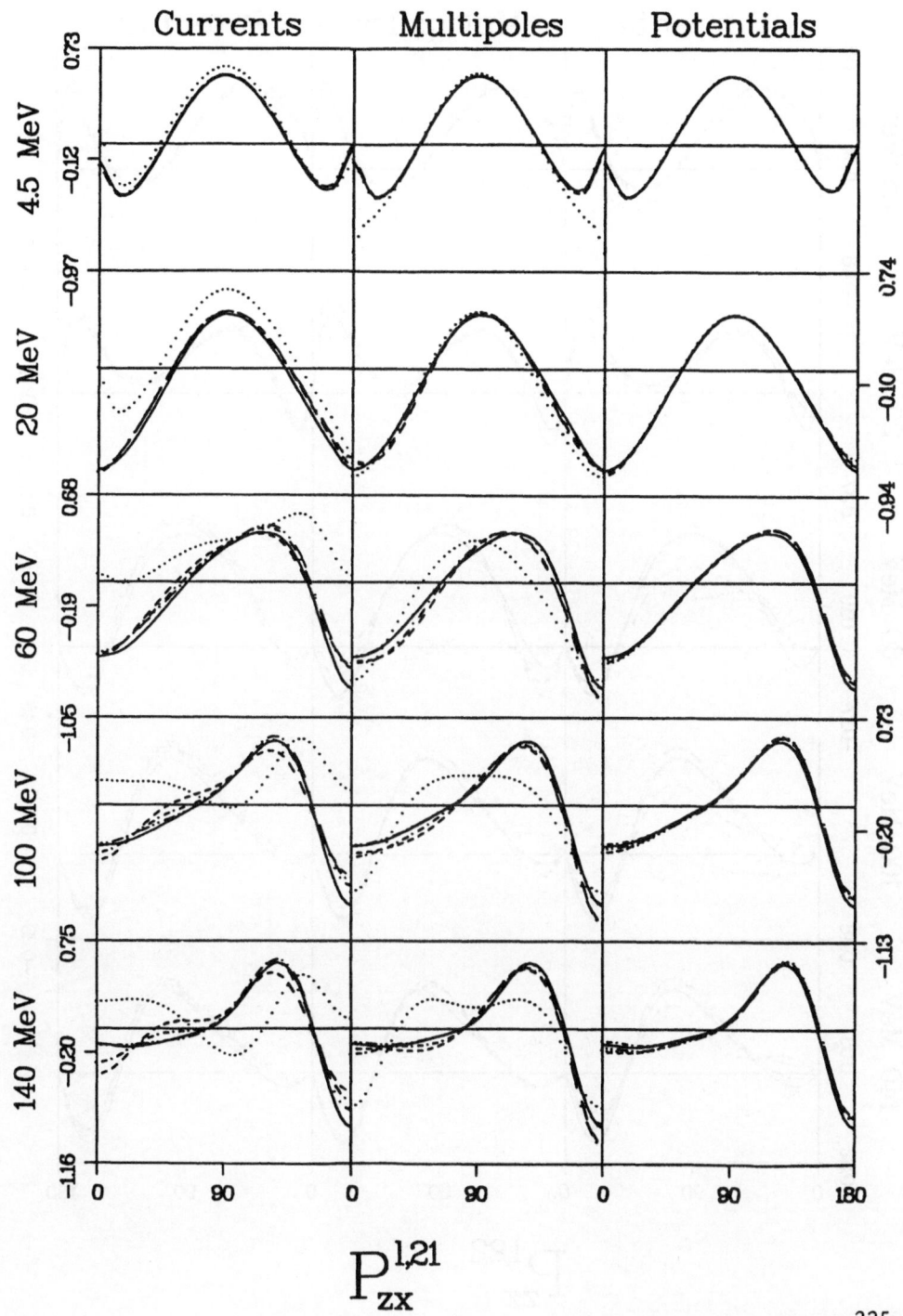

Currents   Multipoles   Potentials

$$P_{zx}^{1,21}$$

225

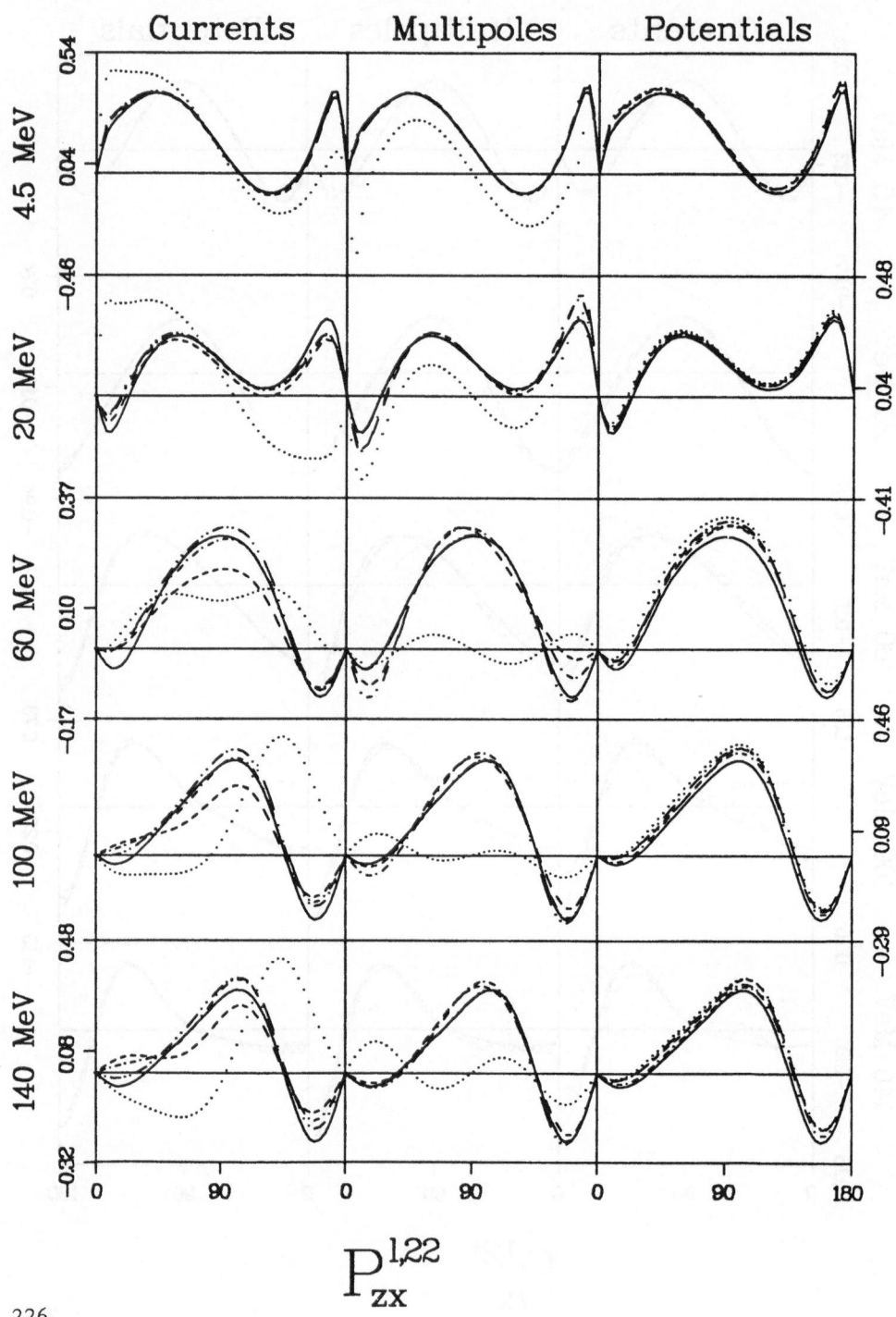

Currents   Multipoles   Potentials

4.5 MeV

20 MeV

60 MeV

100 MeV

140 MeV

$$P_{zx}^{1,22}$$

5.38 Proton-Neutron Spin Correlation $P_{zy}^{0,IM}$

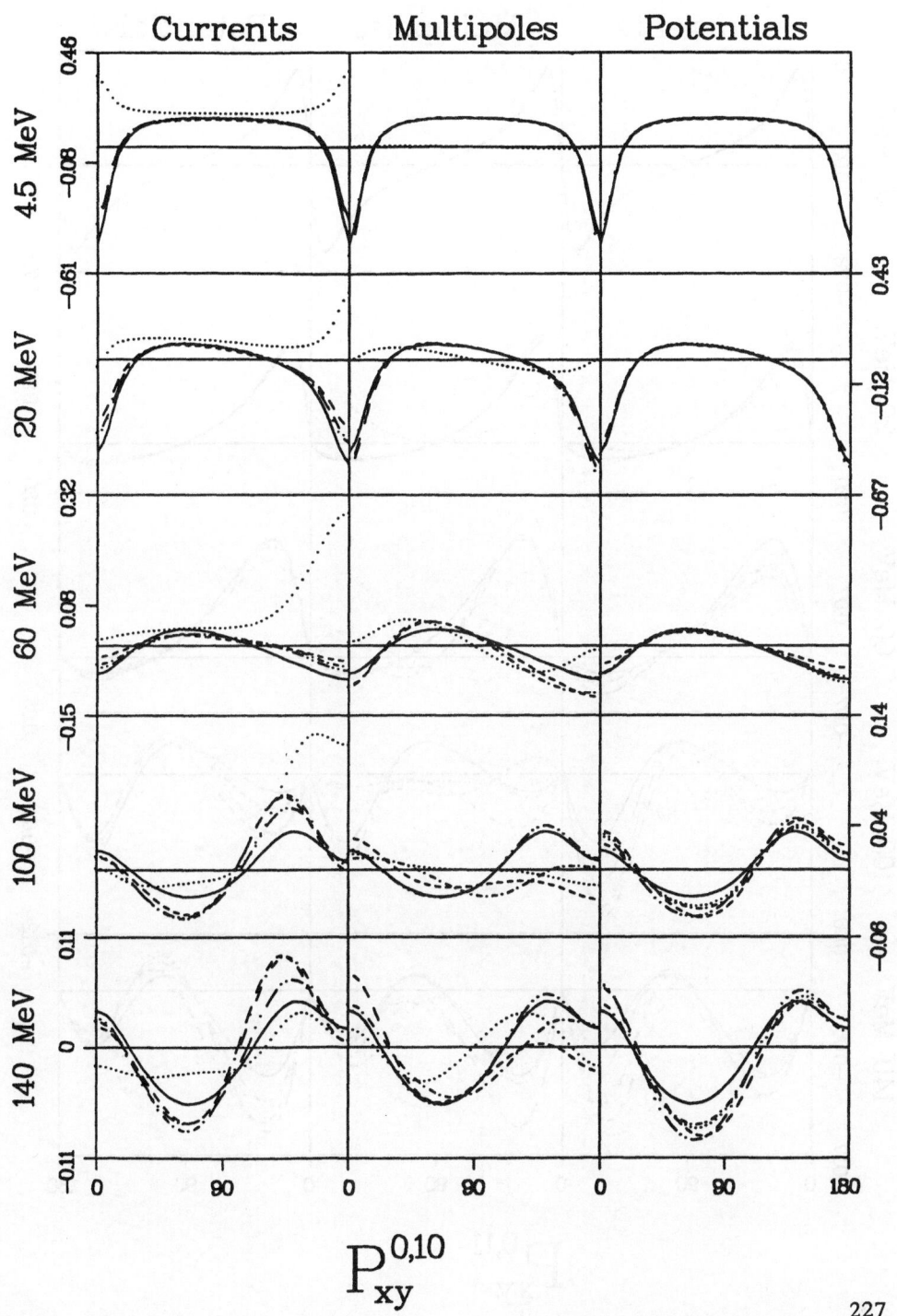

$$P_{xy}^{0,10}$$

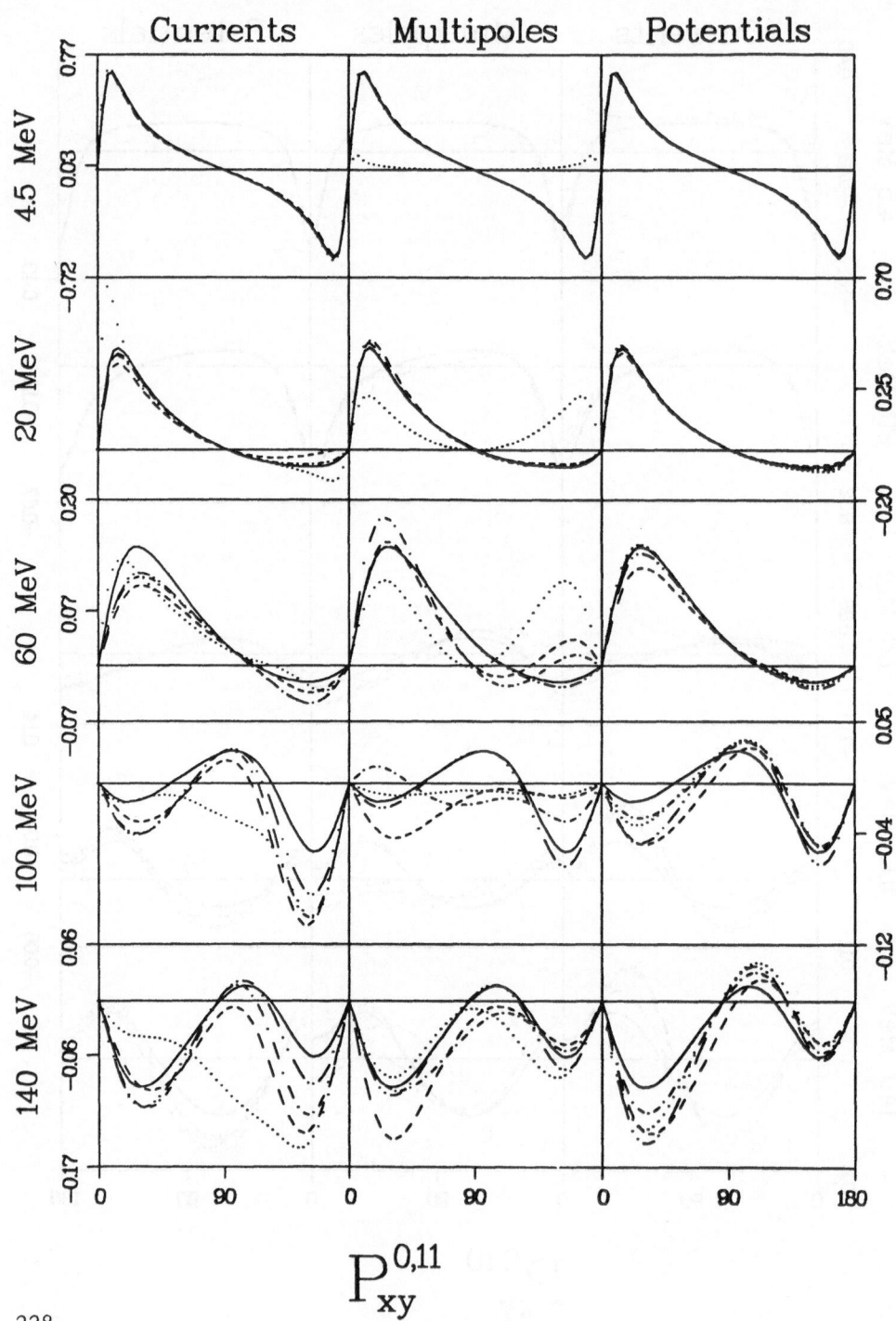

Currents   Multipoles   Potentials

4.5 MeV
20 MeV
60 MeV
100 MeV
140 MeV

$$P_{xy}^{0,11}$$

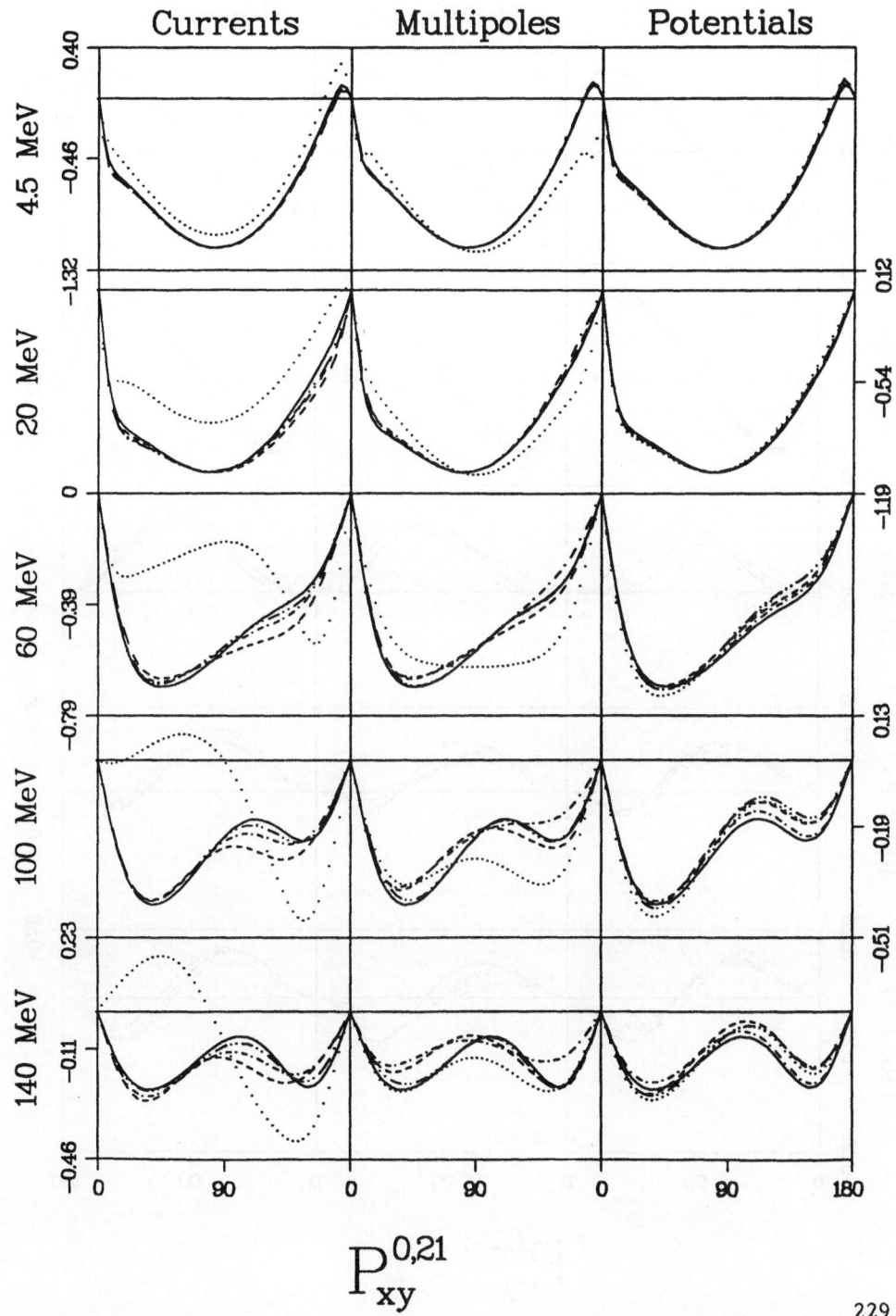

$$P_{xy}^{0,21}$$

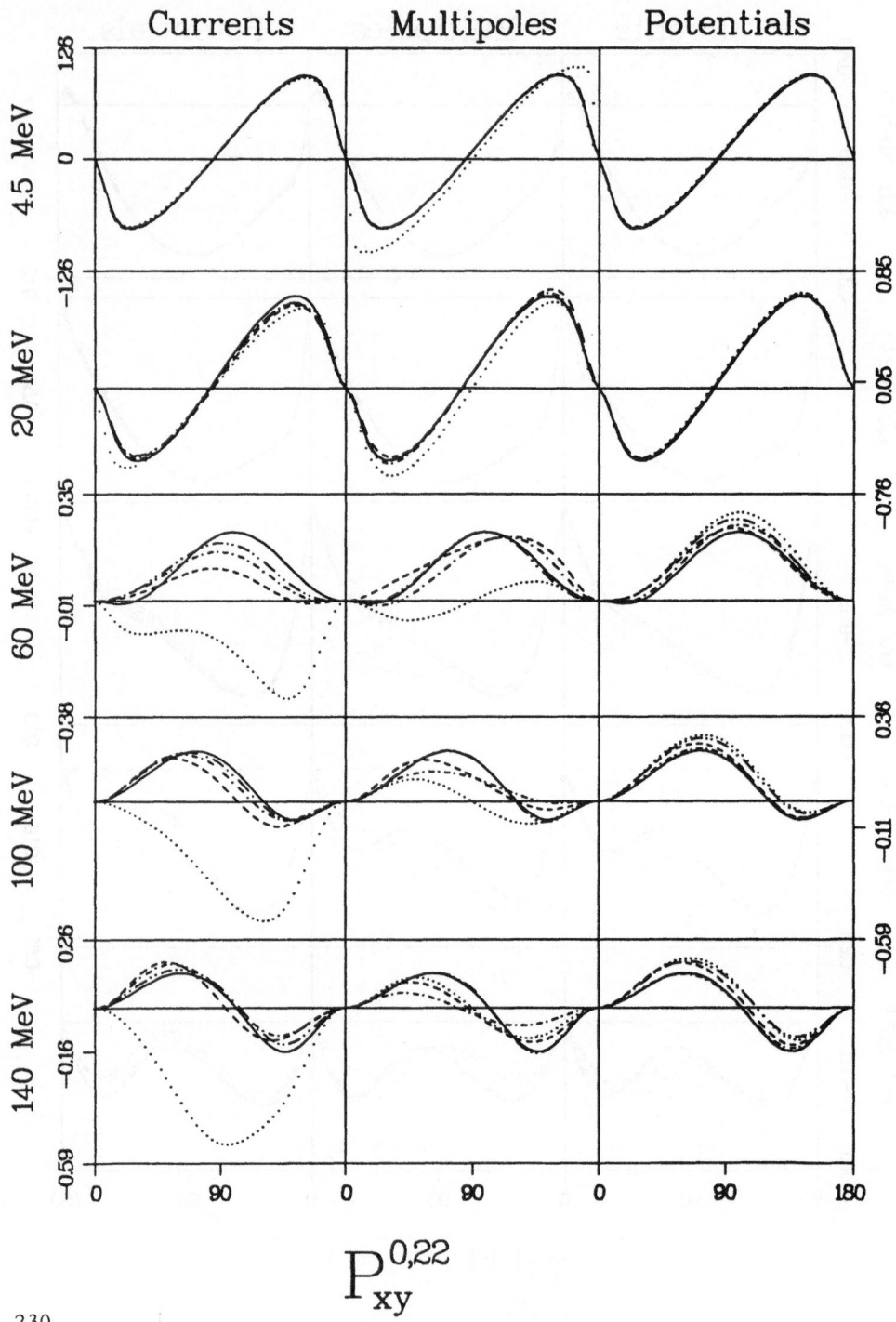

$$P_{xy}^{0,22}$$

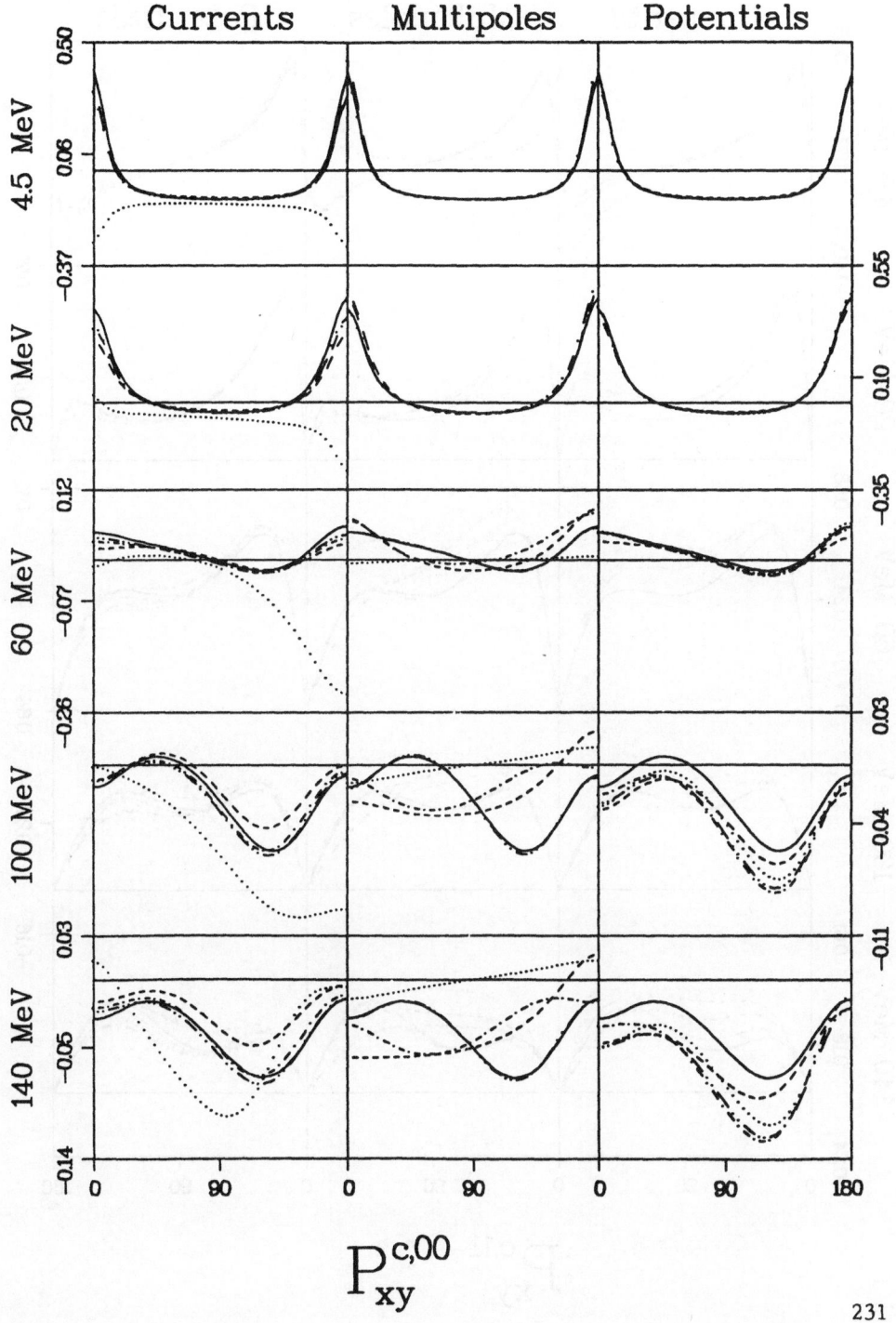

$$P_{xy}^{c,00}$$

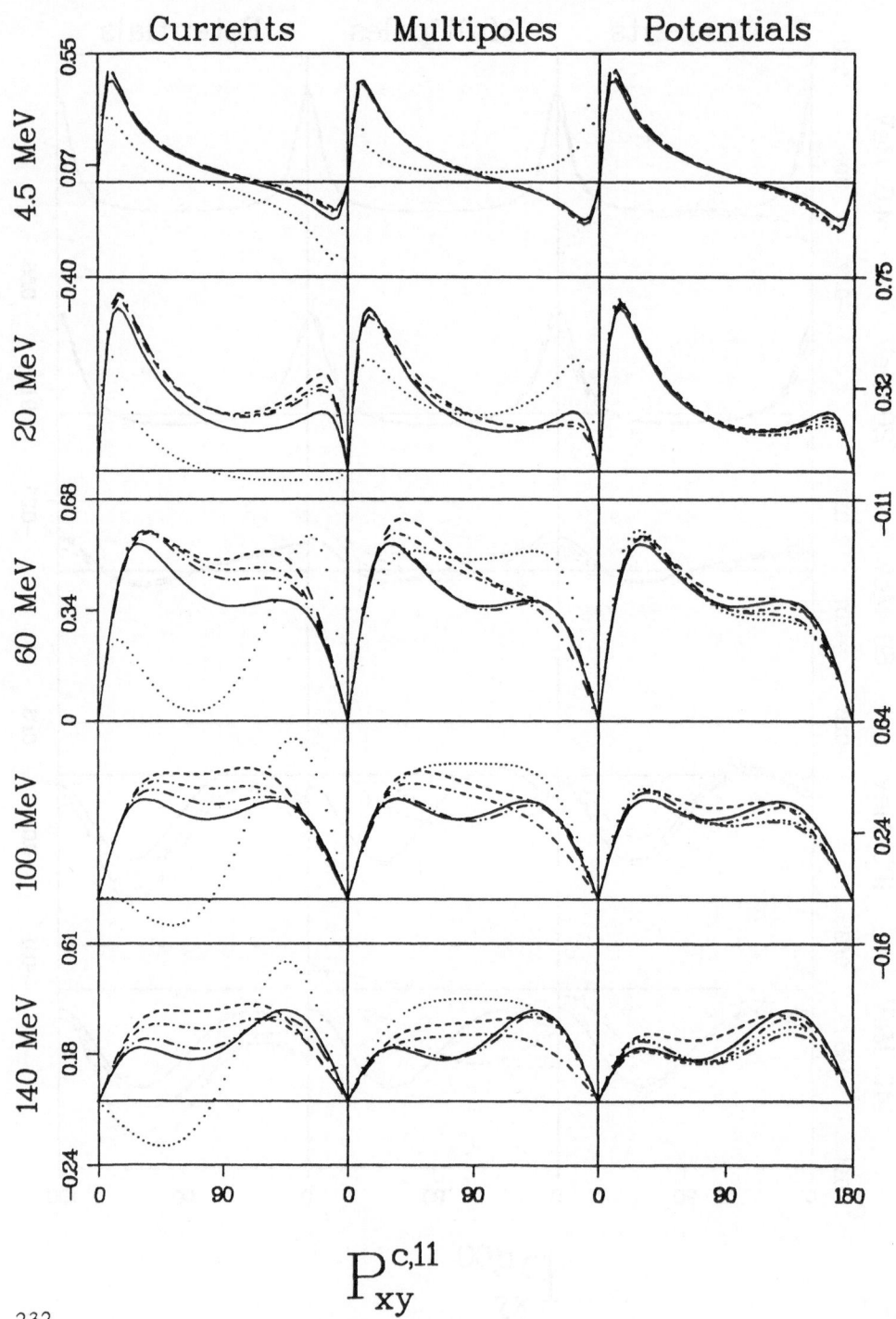

$$P_{xy}^{c,11}$$

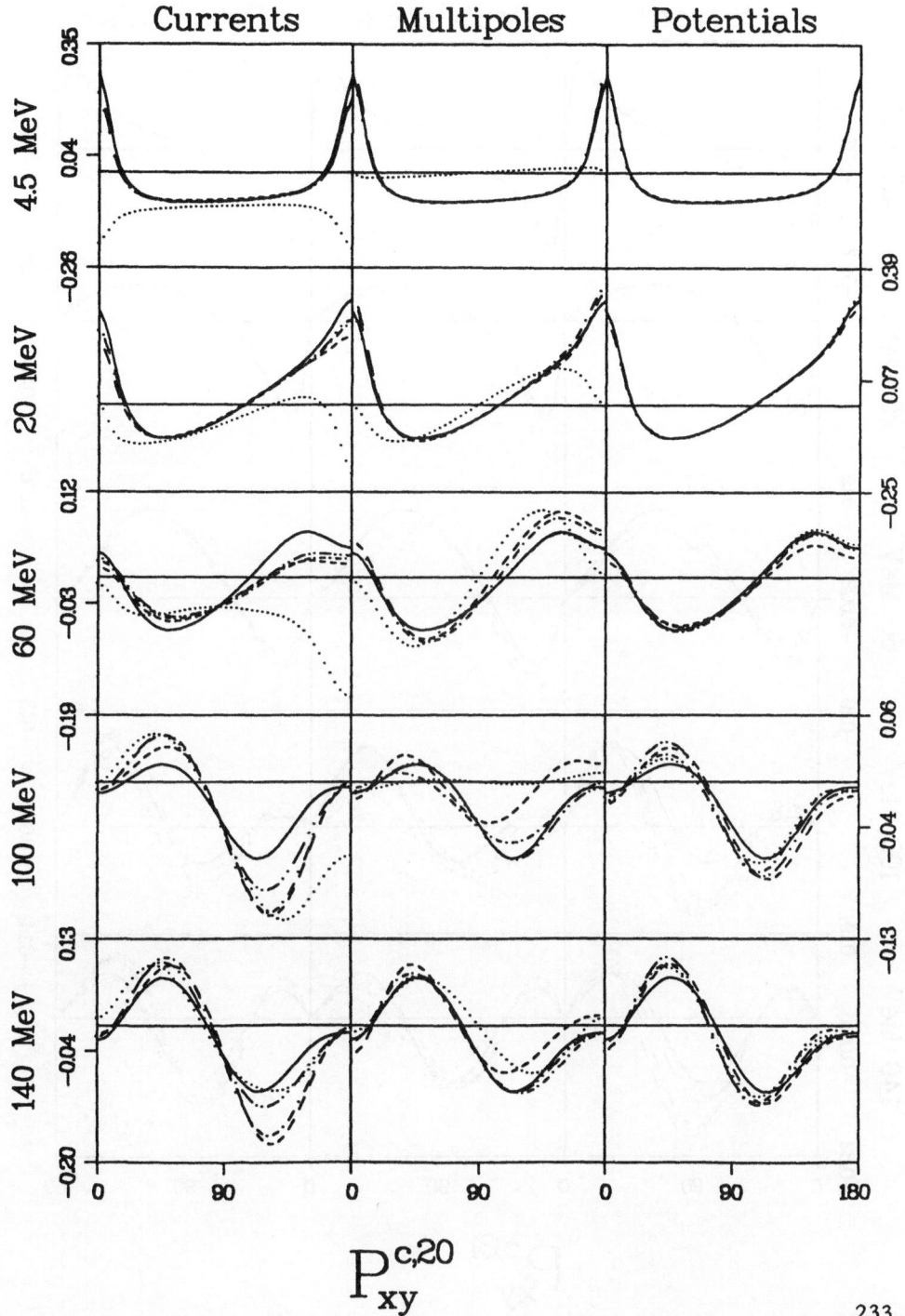

$$P_{xy}^{c,20}$$

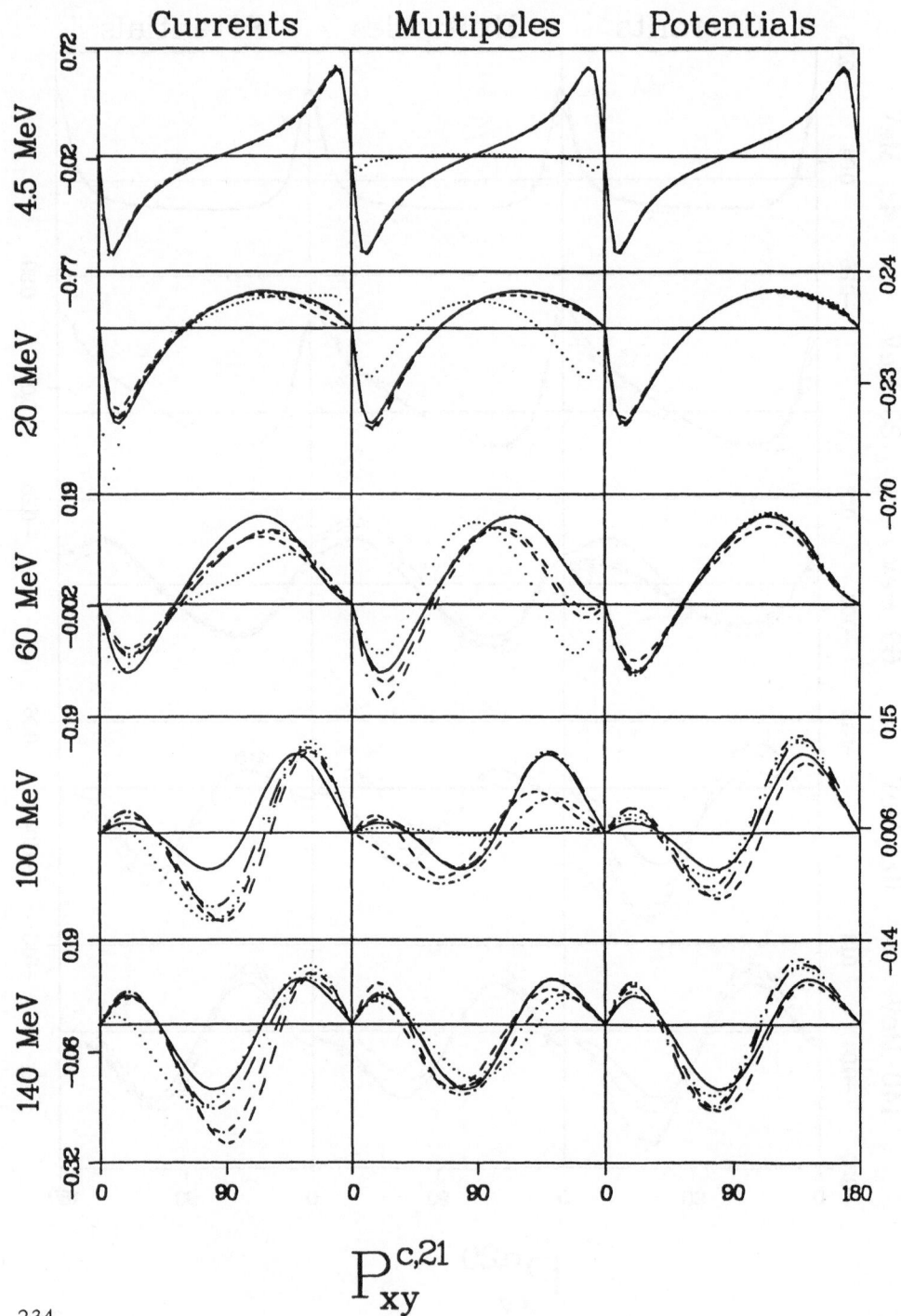

$$P_{xy}^{c,21}$$

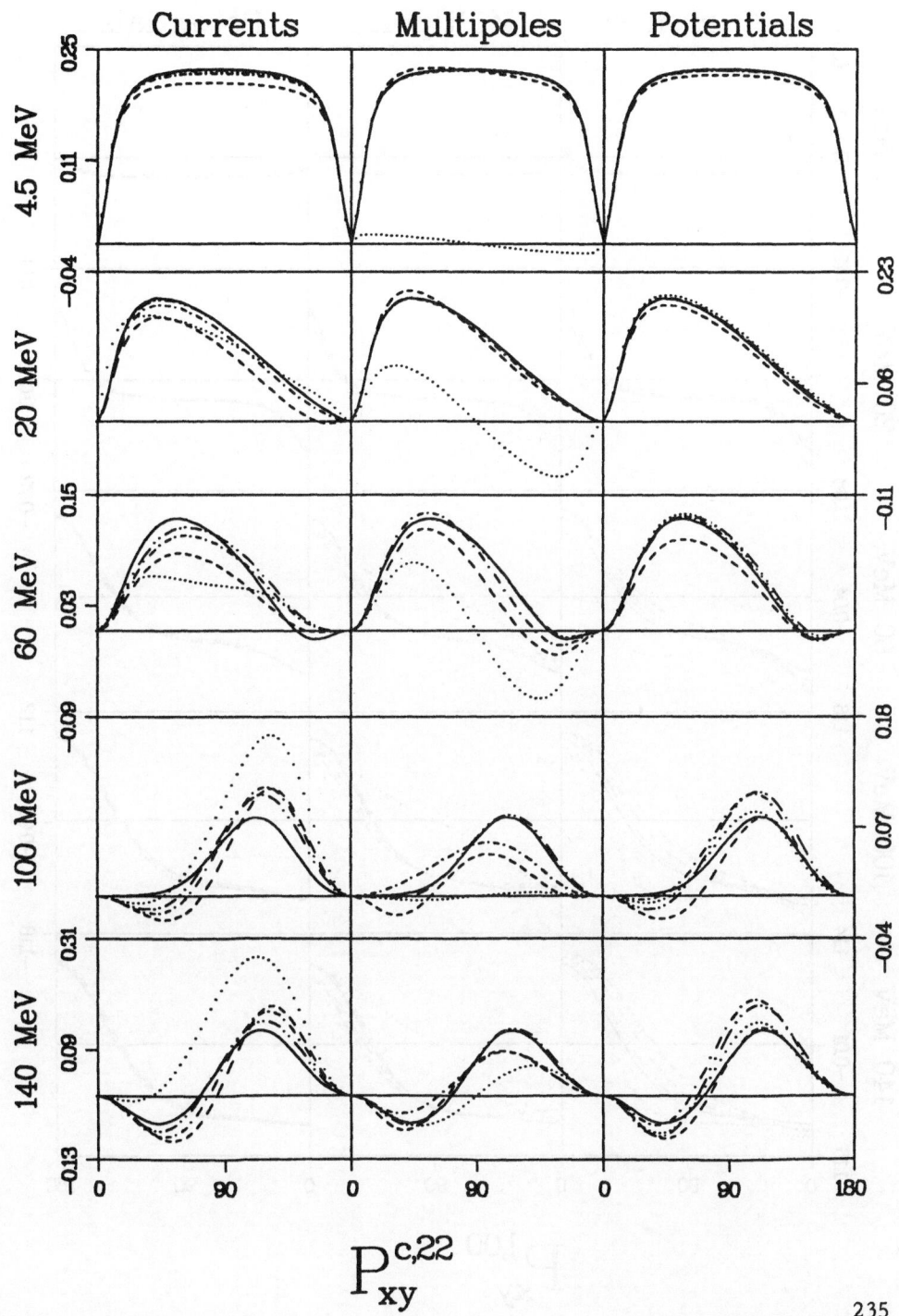

$$P^{c,22}_{xy}$$

235

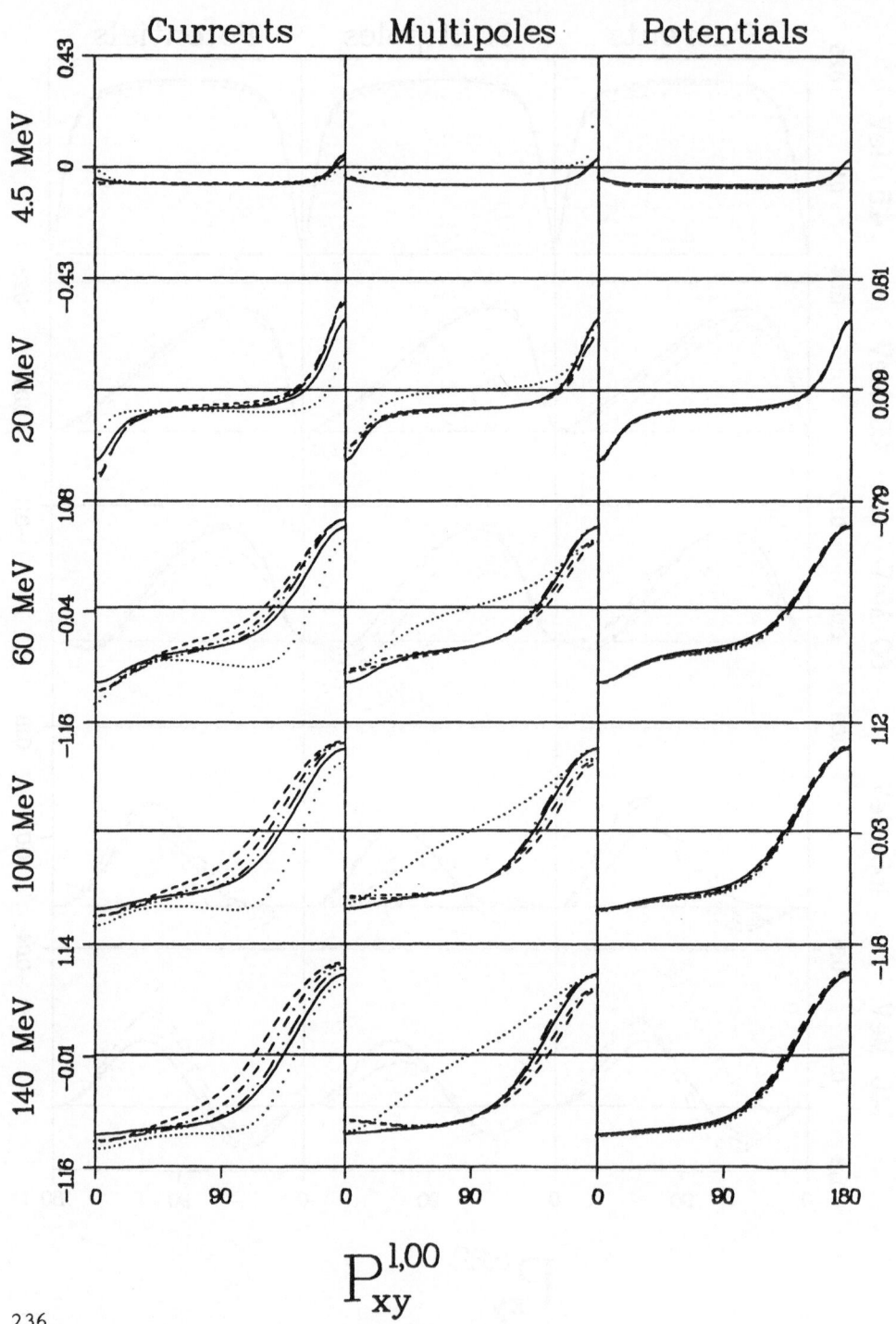

$$P_{xy}^{1,00}$$

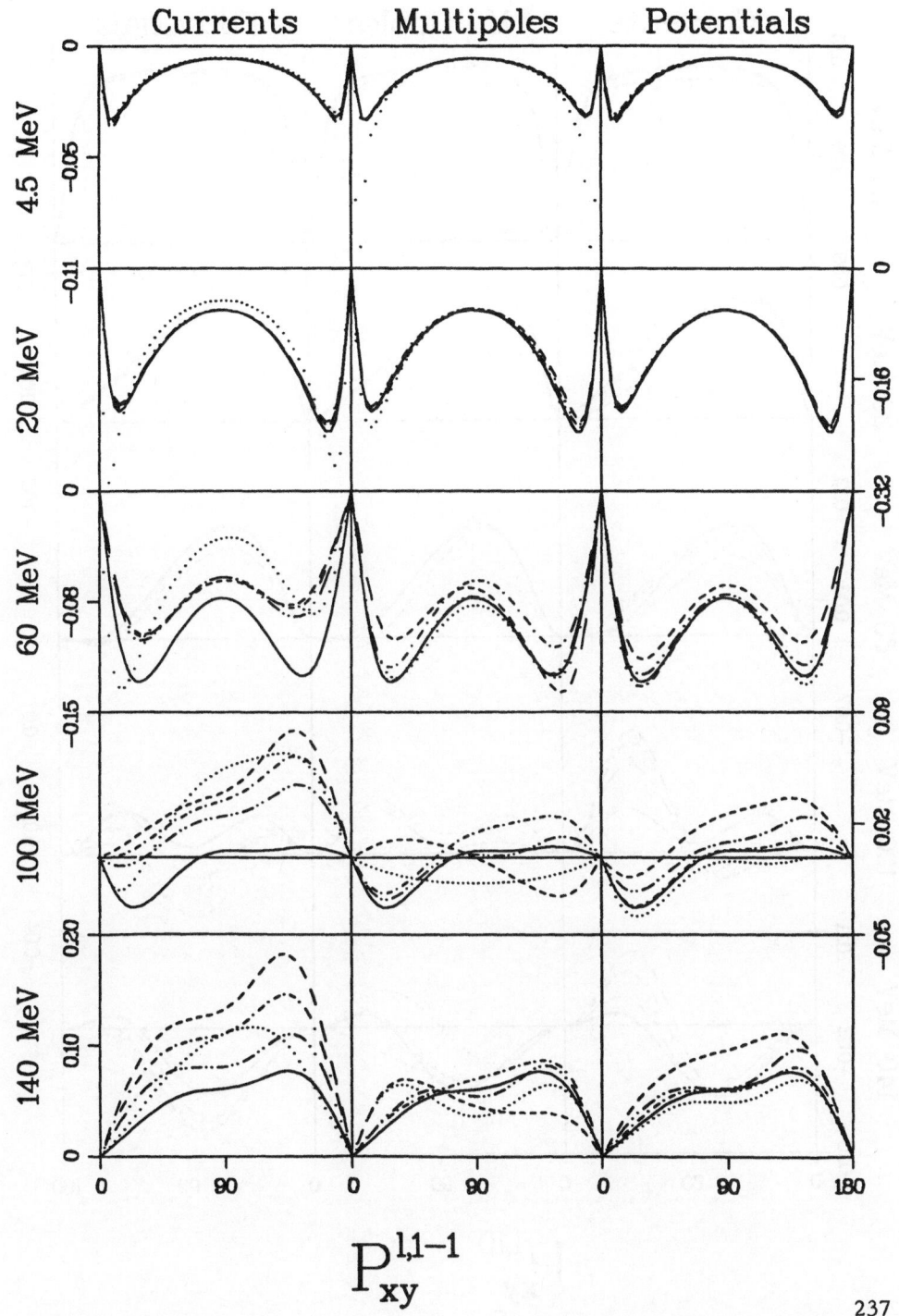

Currents   Multipoles   Potentials

4.5 MeV   20 MeV   60 MeV   100 MeV   140 MeV

$$P_{xy}^{1,1-1}$$

237

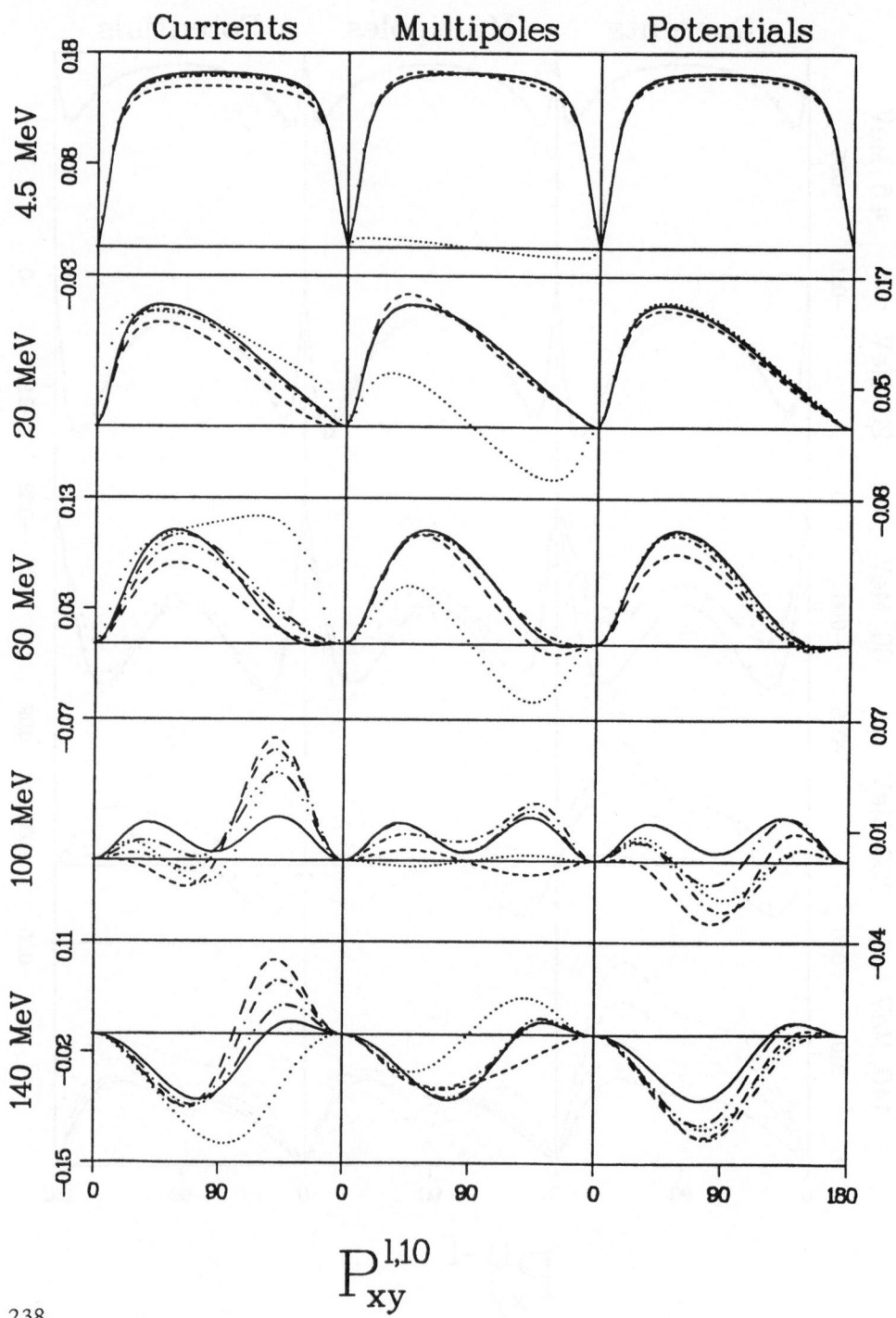

Currents   Multipoles   Potentials

4.5 MeV
20 MeV
60 MeV
100 MeV
140 MeV

$$P_{xy}^{1,10}$$

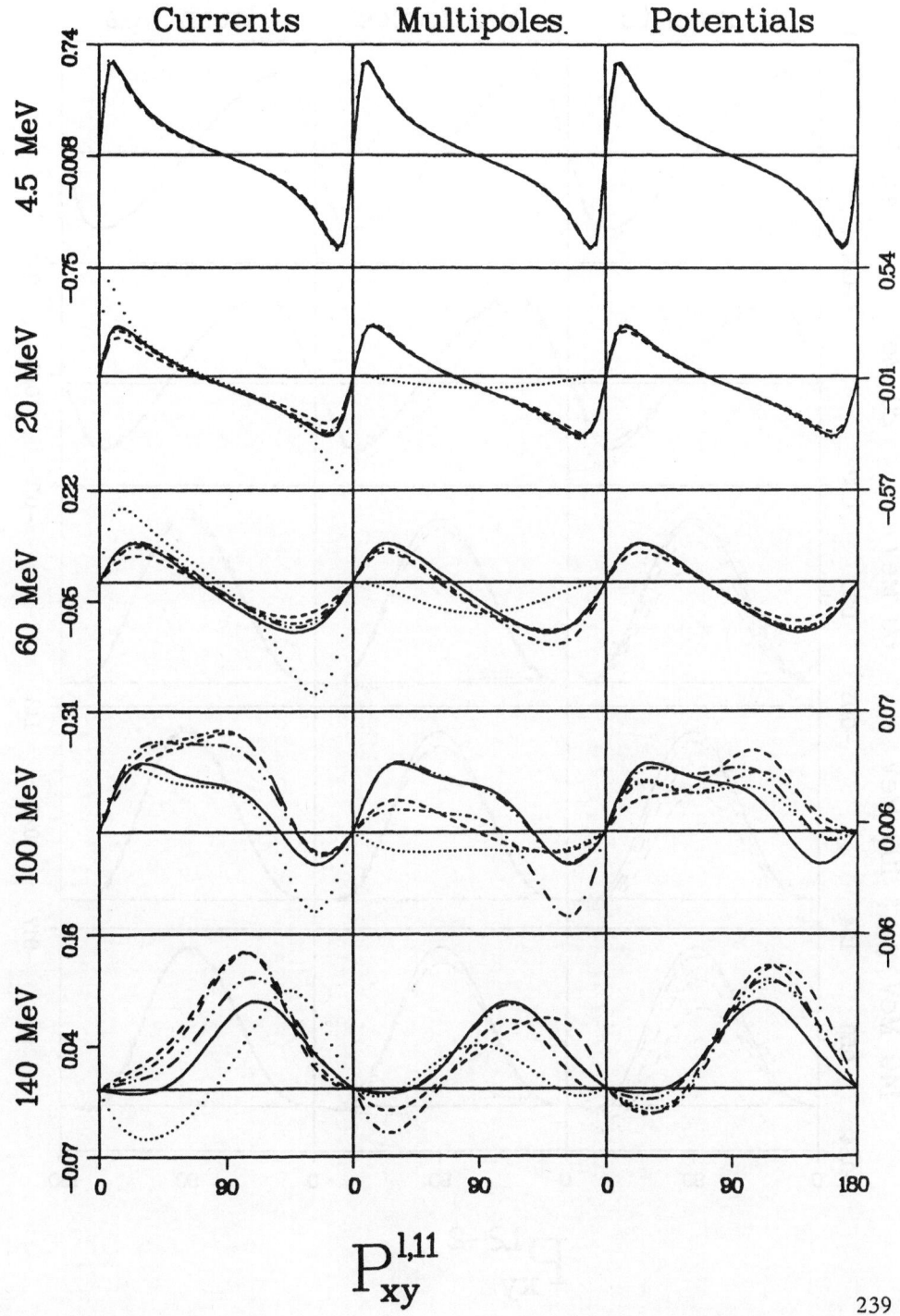

Currents   Multipoles   Potentials

$$P_{xy}^{l,11}$$

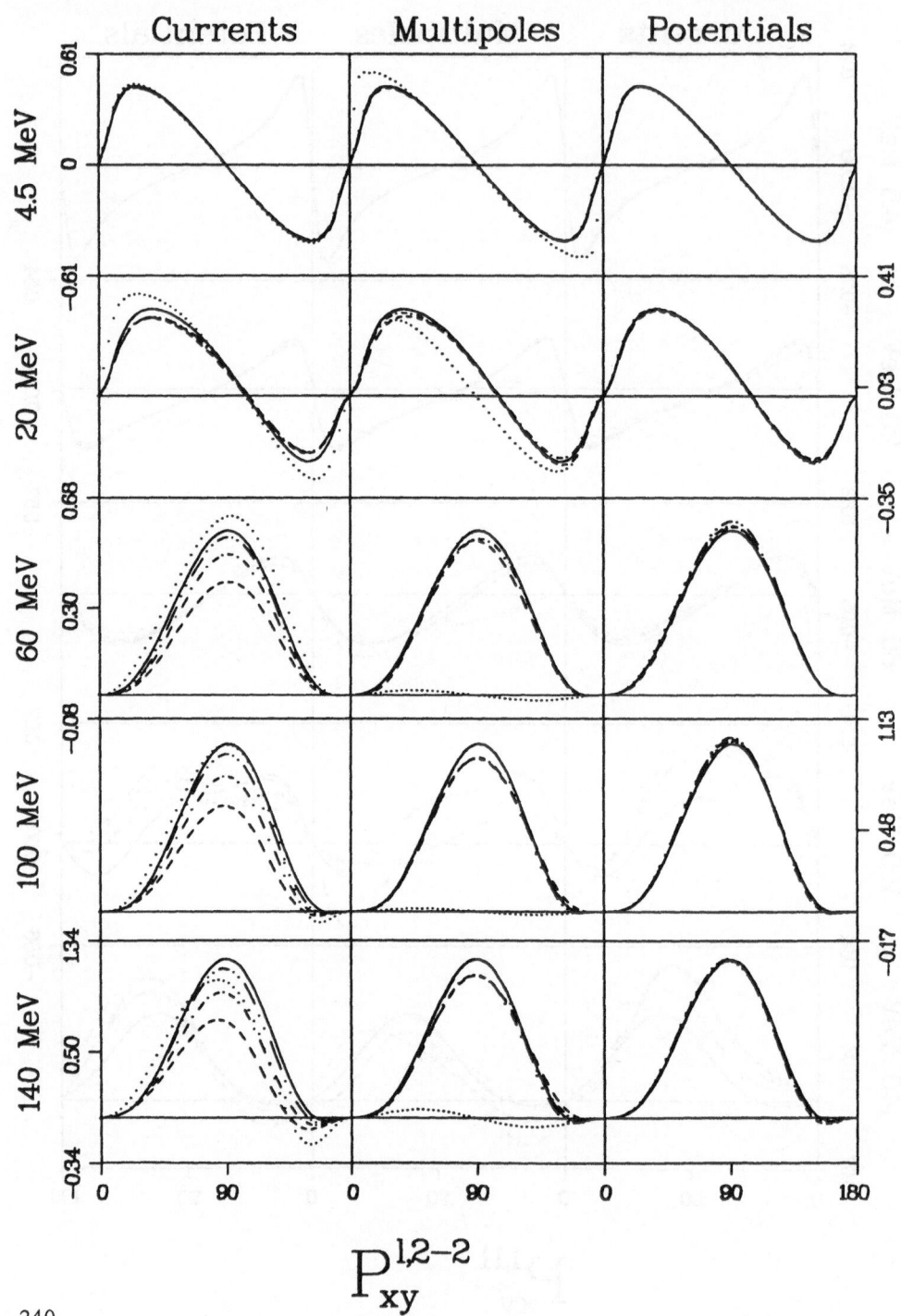

$$P_{xy}^{1,2-2}$$

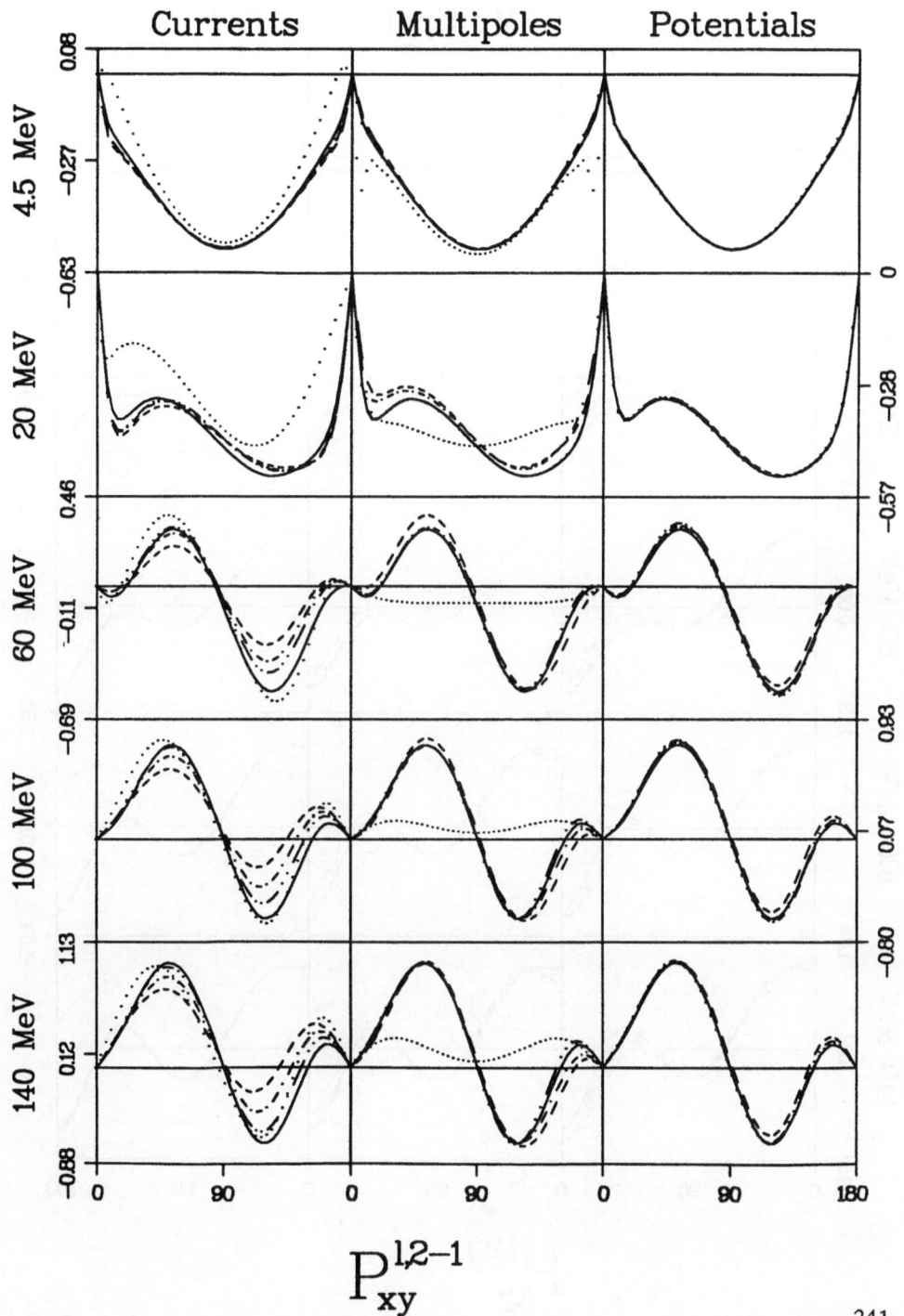

Currents Multipoles Potentials

4.5 MeV

20 MeV

60 MeV

100 MeV

140 MeV

$$P_{xy}^{1,2-1}$$

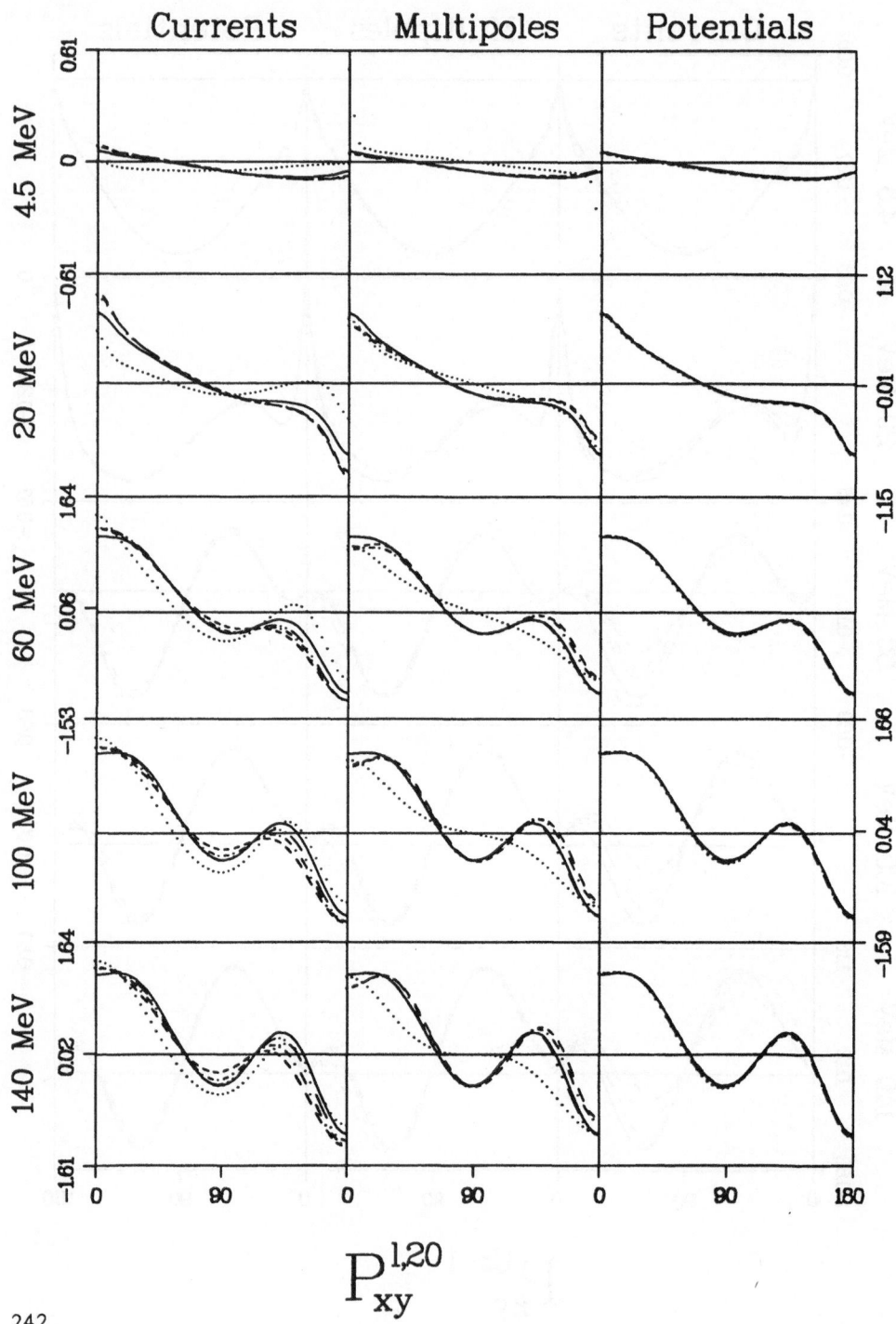

$$P_{xy}^{1,20}$$

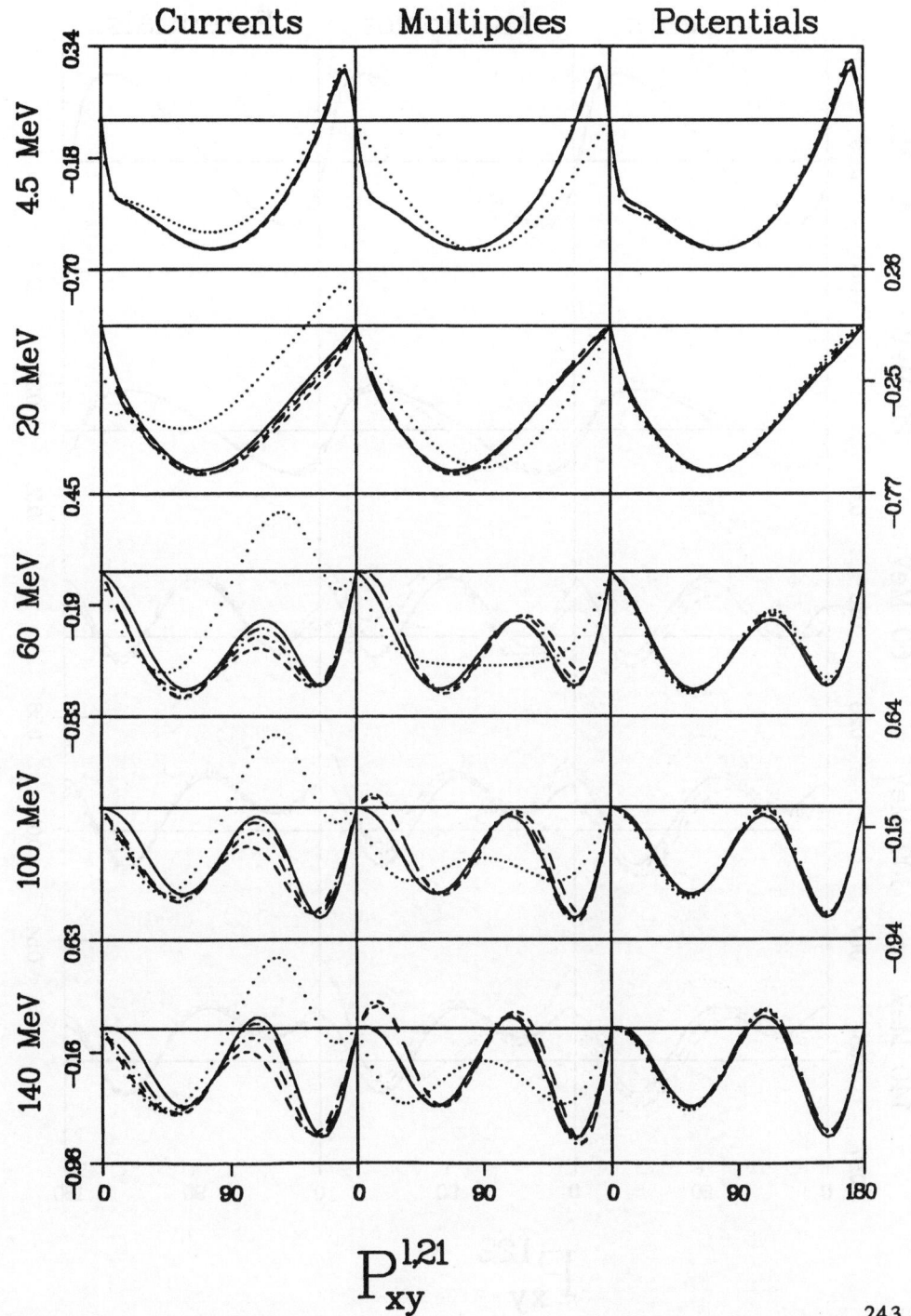

$$P_{xy}^{1,21}$$

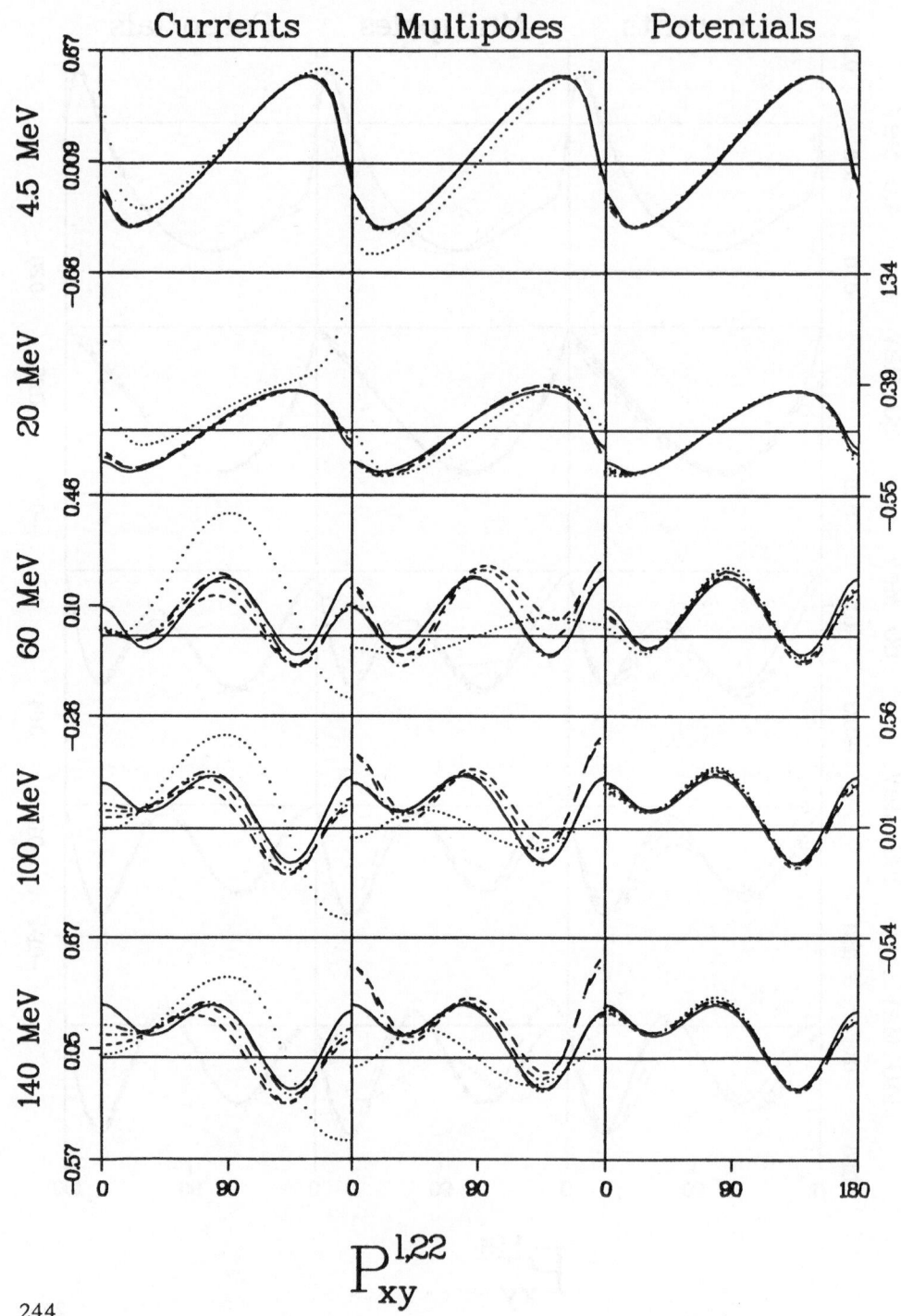

$$P_{xy}^{1,22}$$

5.41 Proton-Neutron Spin Correlation $P_{yx}^{0,IM}$

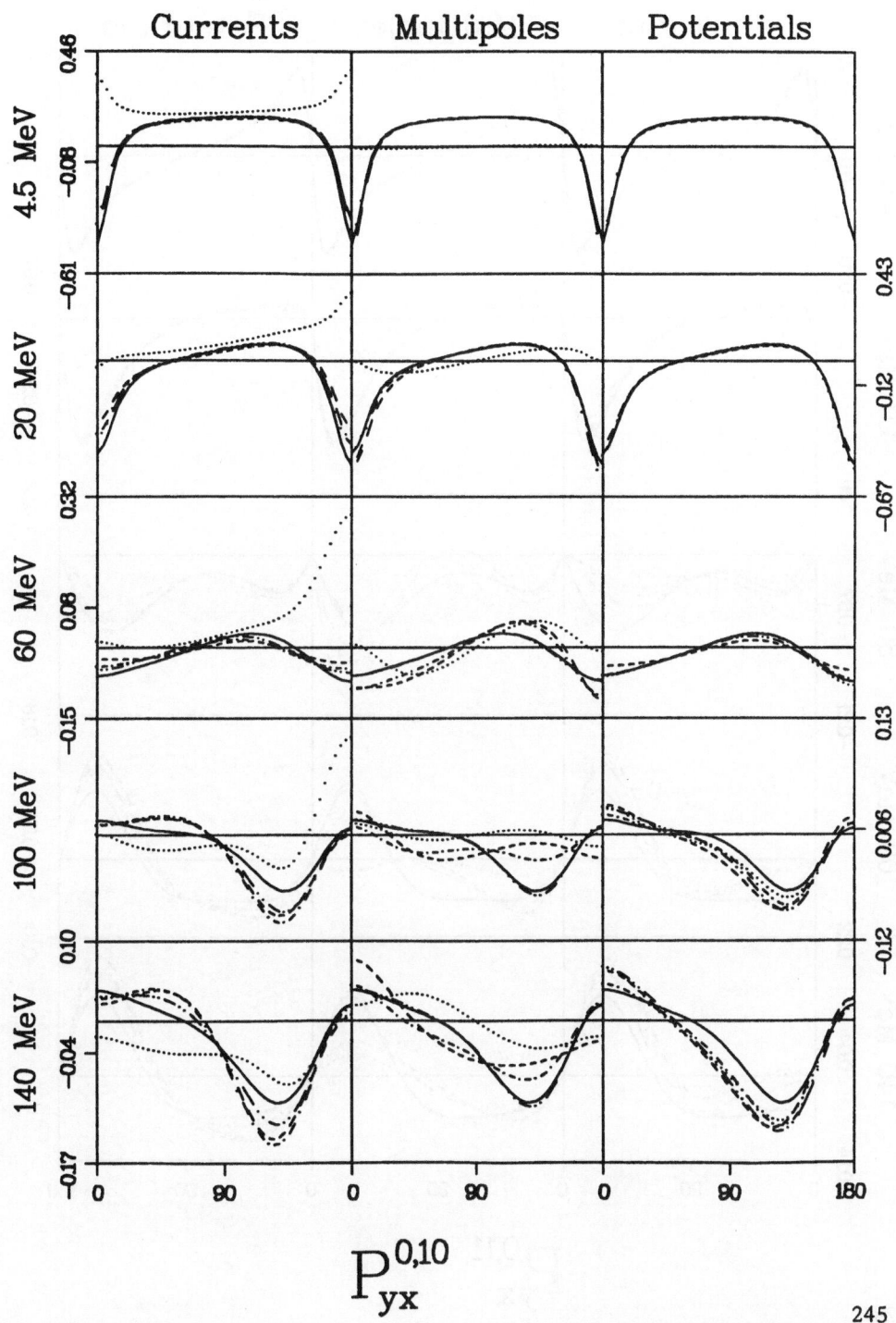

$$P_{yx}^{0,10}$$

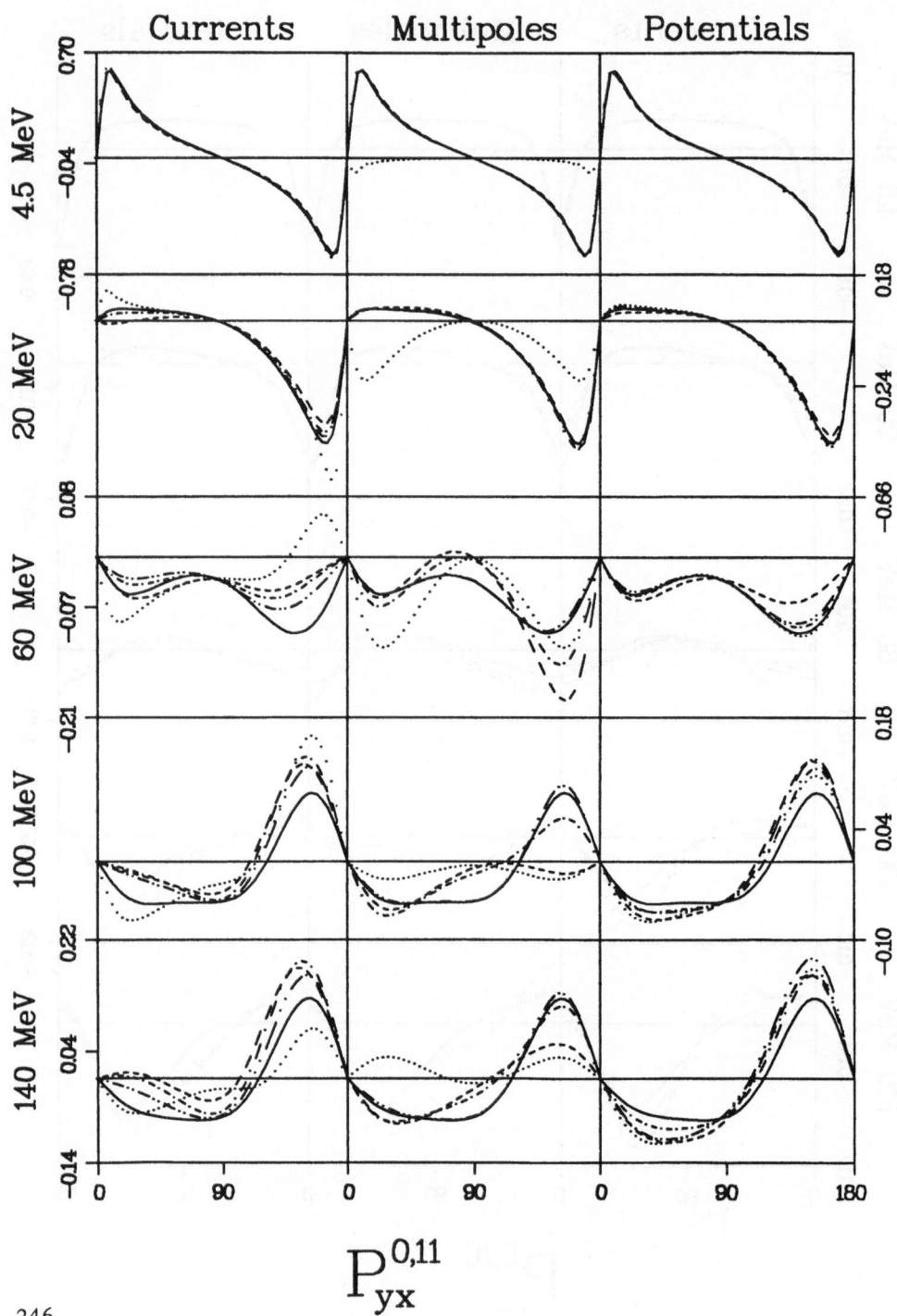

$$P_{yx}^{0,11}$$

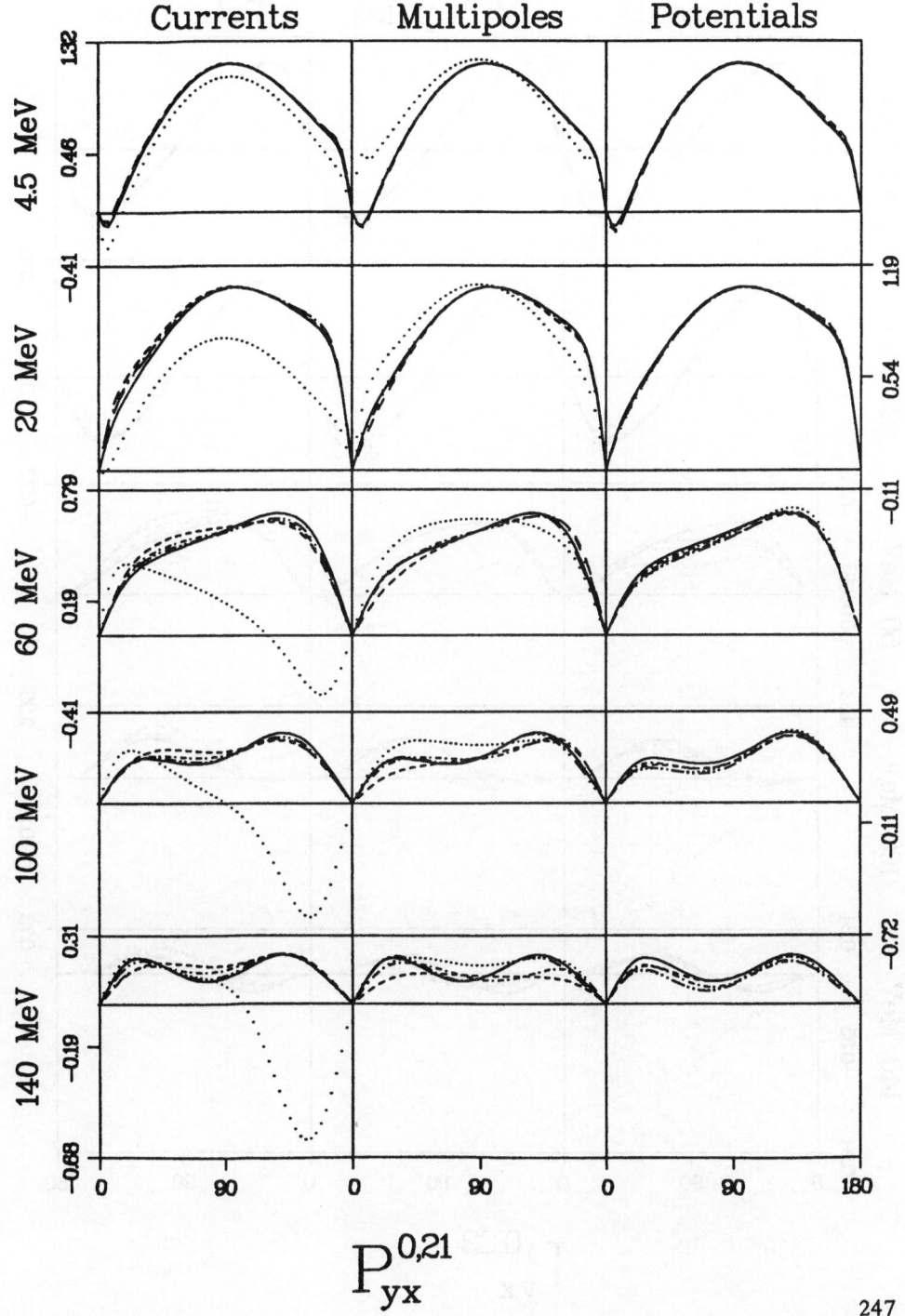

$$P_{yx}^{0,21}$$

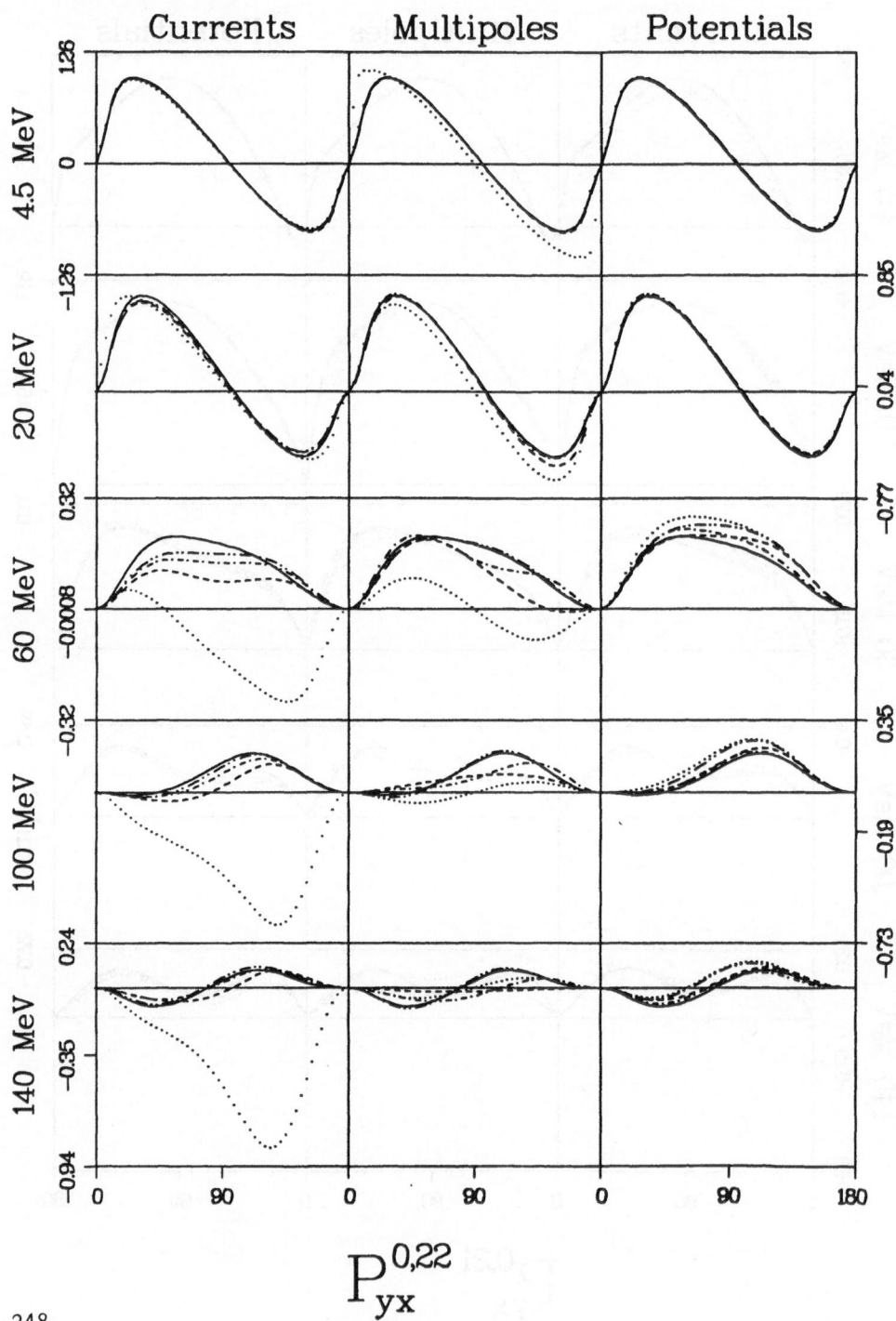

$$P_{yx}^{0,22}$$

5.42 Proton-Neutron Spin Correlation $P_{yx}^{c,IM}$

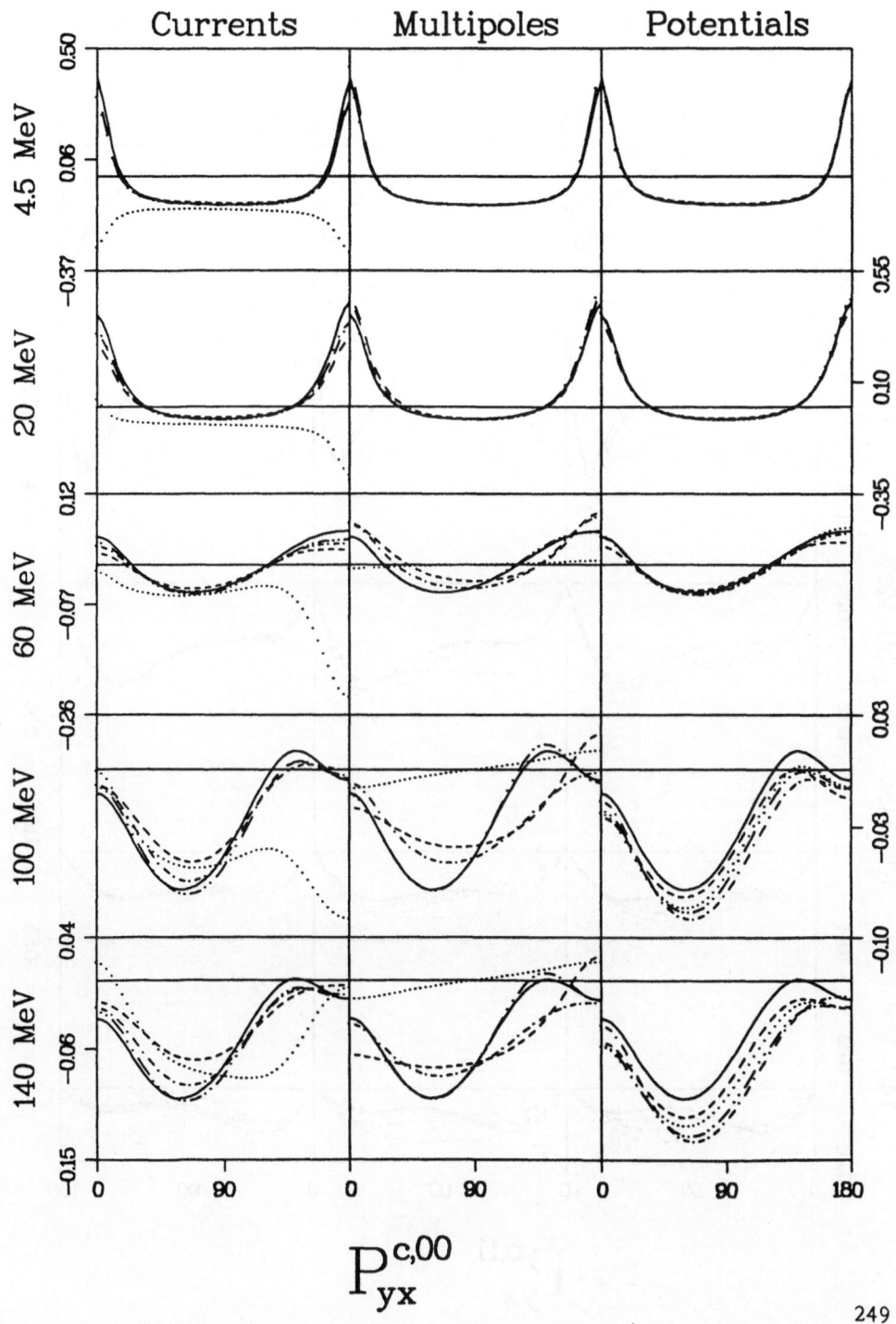

$$P_{yx}^{c,00}$$

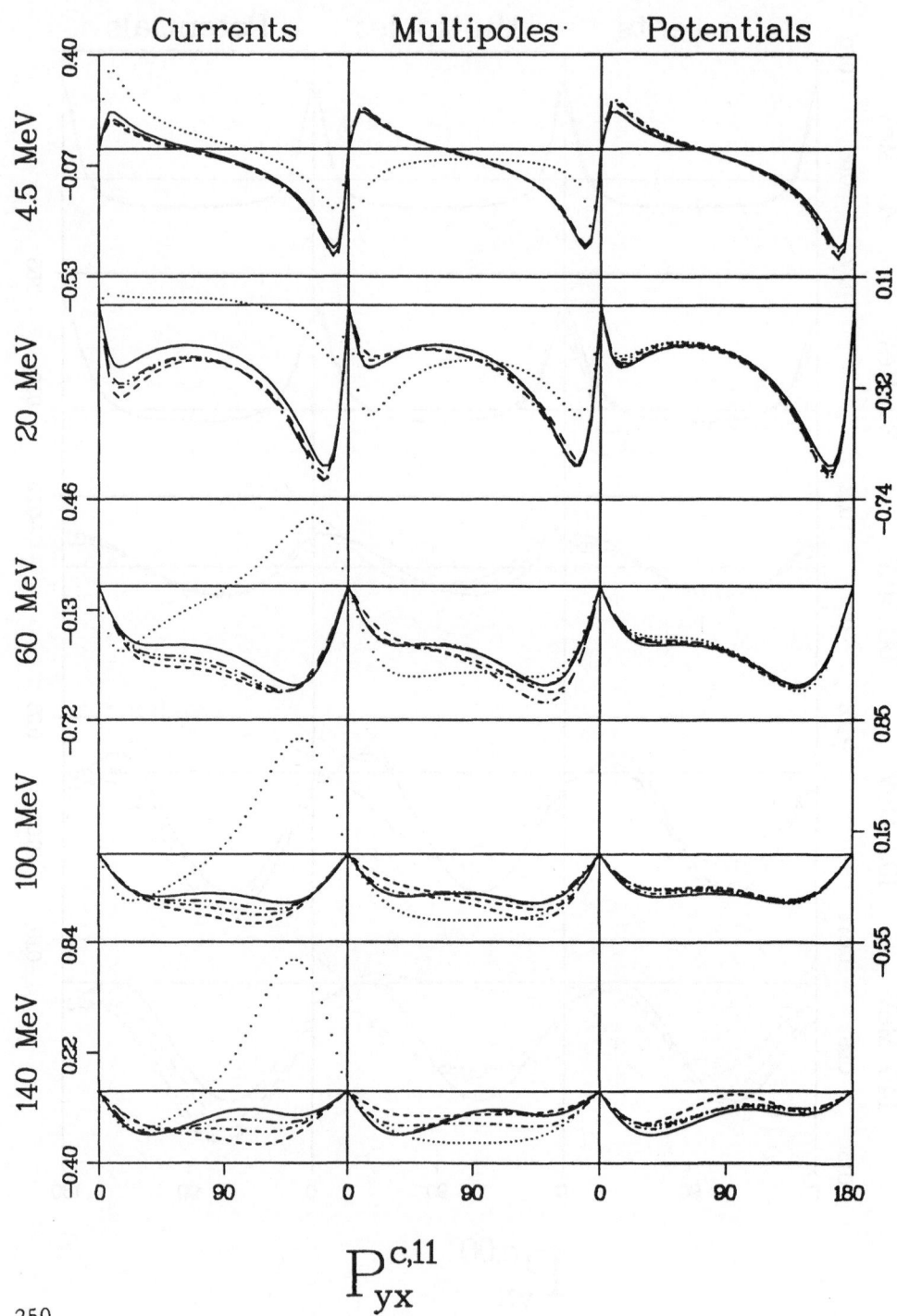

$$P_{yx}^{c,11}$$

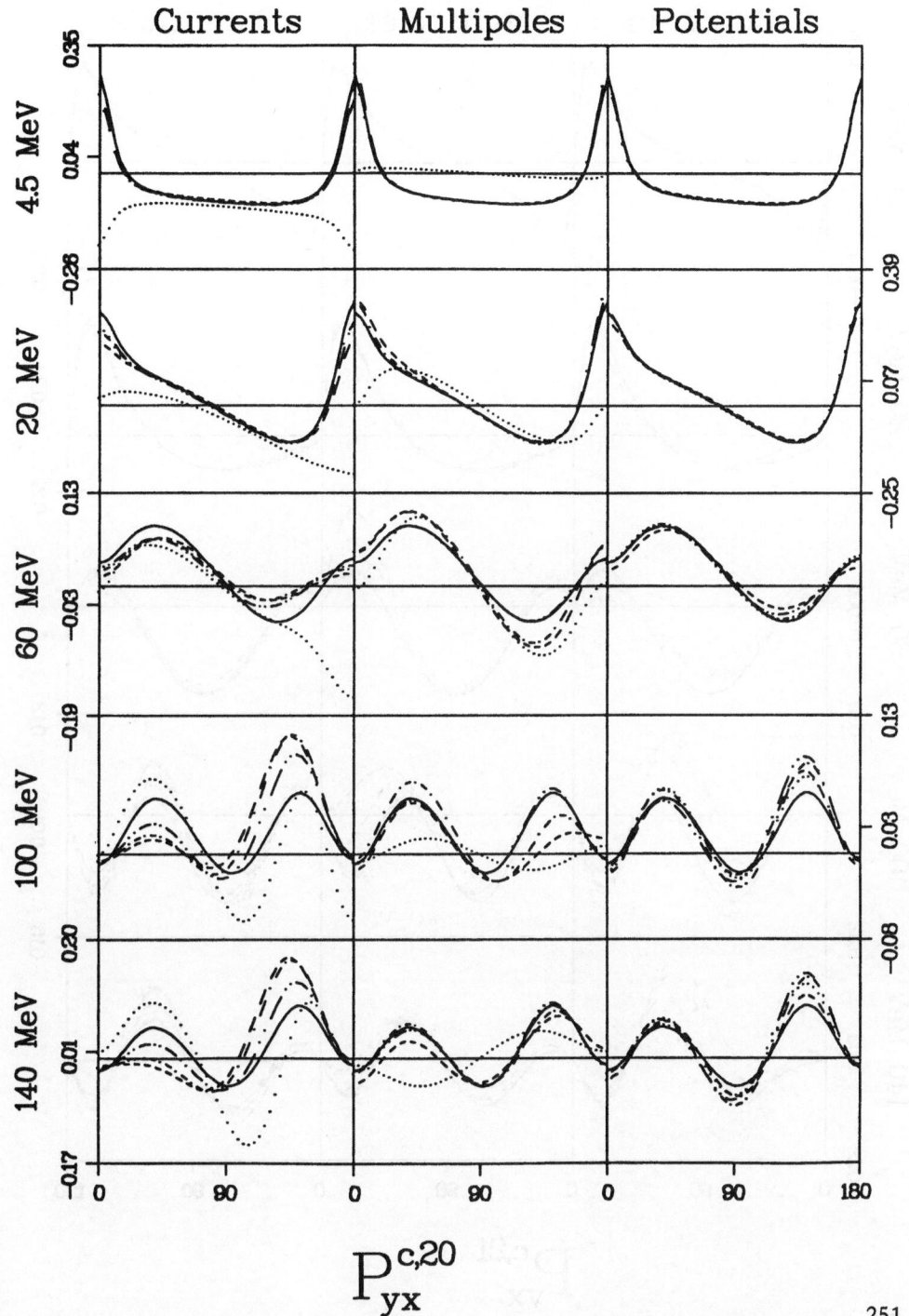

Currents   Multipoles   Potentials

$$P_{yx}^{c,20}$$

251

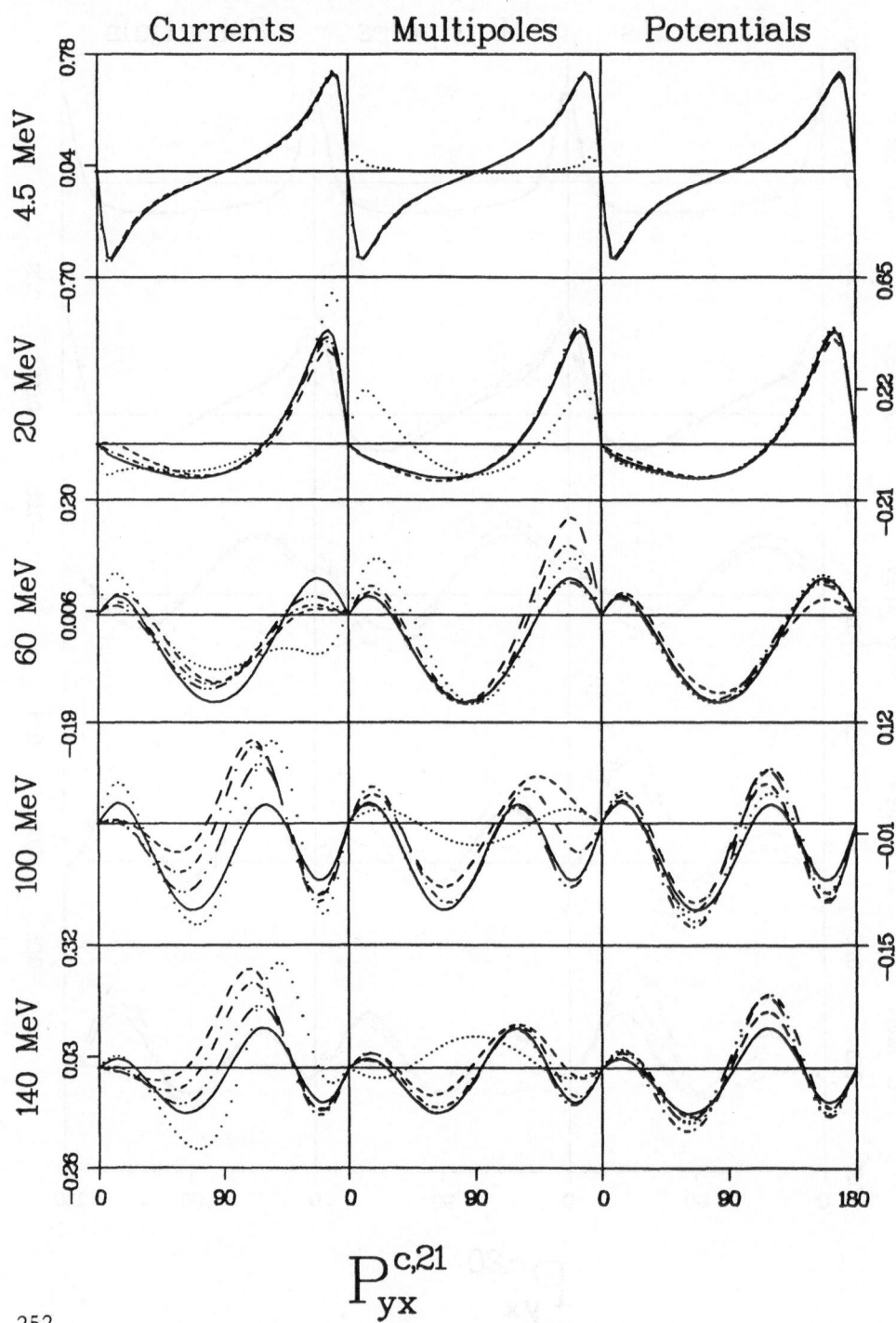

$$P_{yx}^{c,21}$$

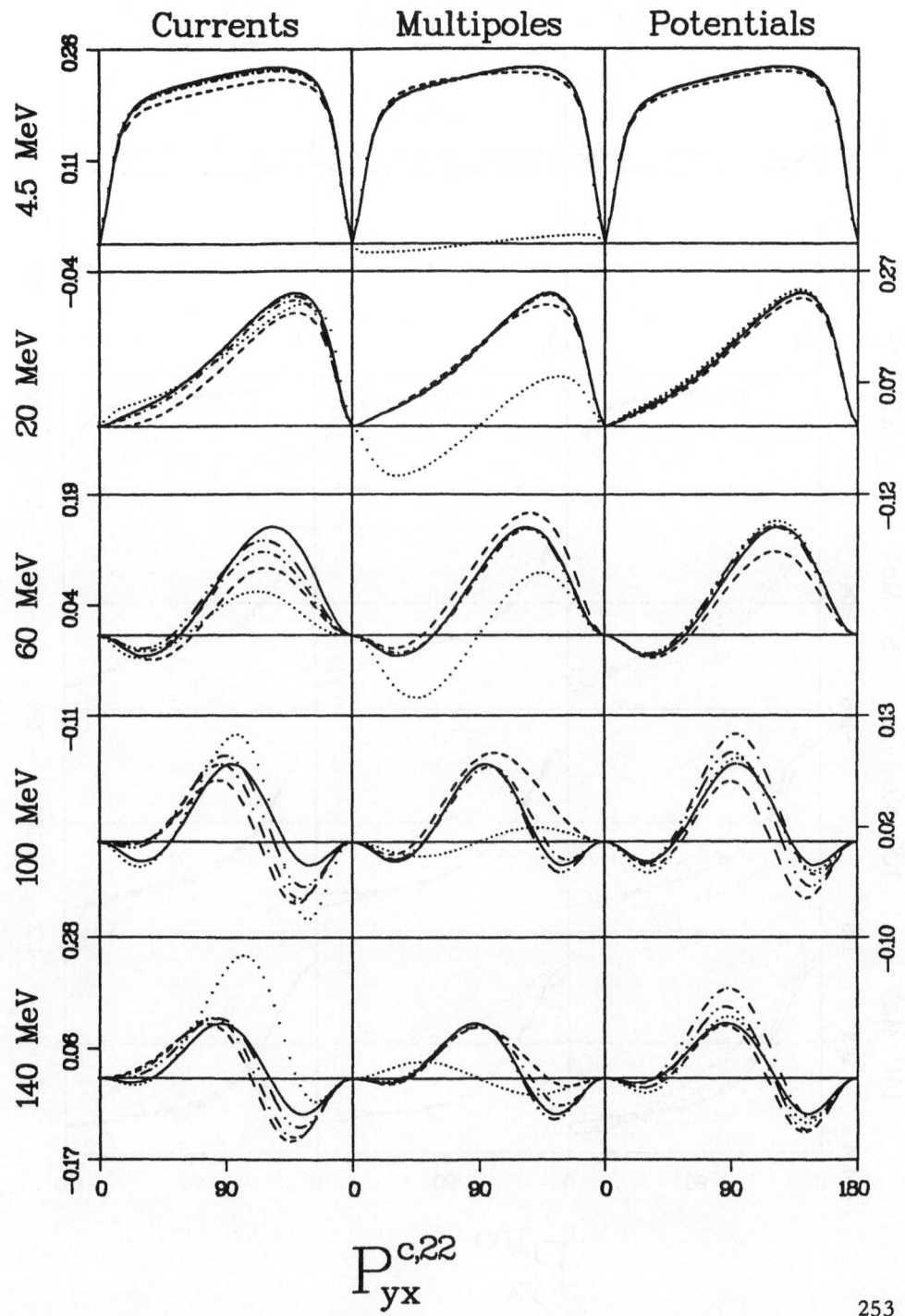

Currents   Multipoles   Potentials

4.5 MeV

20 MeV

60 MeV

100 MeV

140 MeV

$$P_{yx}^{c,22}$$

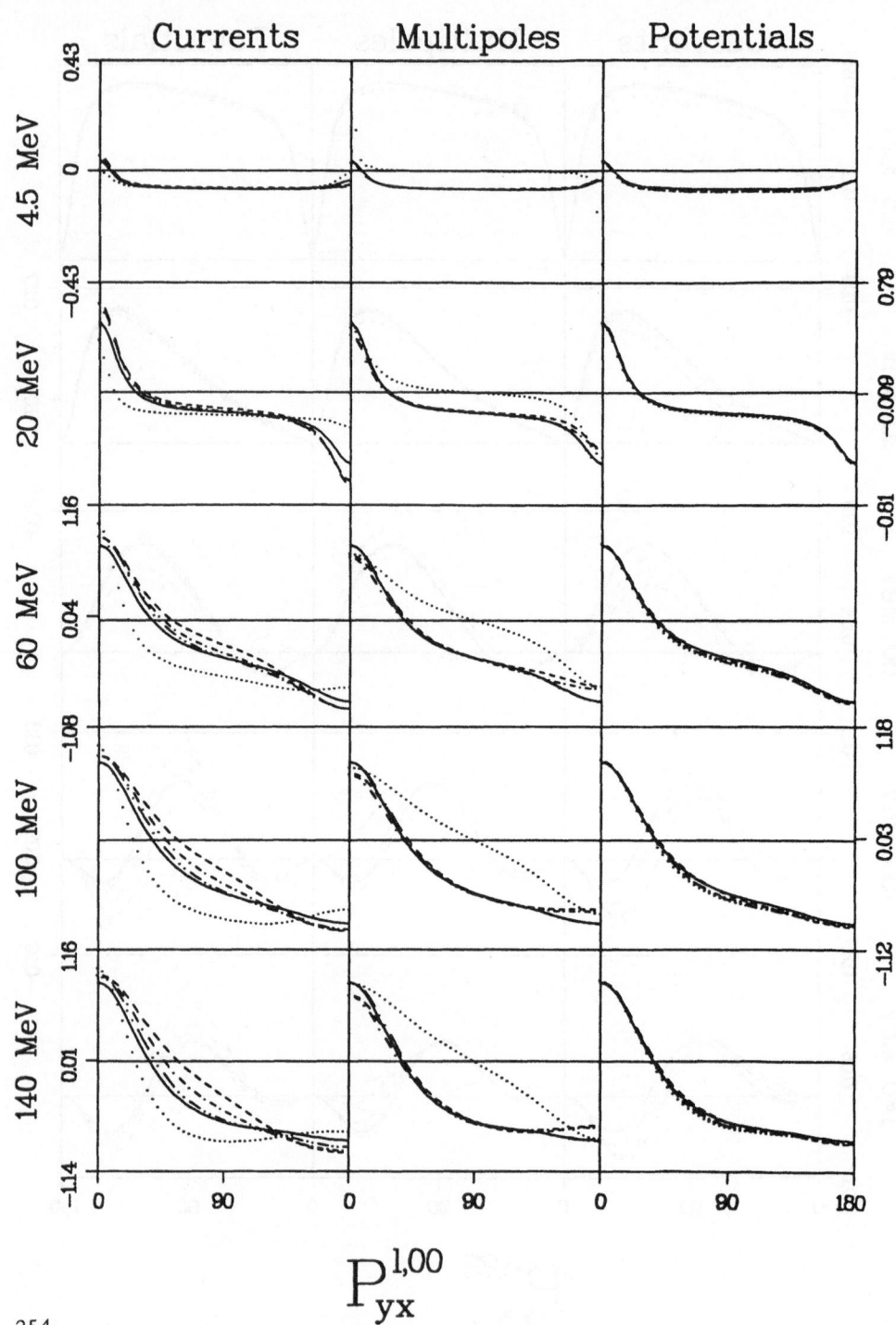

$$P_{yx}^{1,00}$$

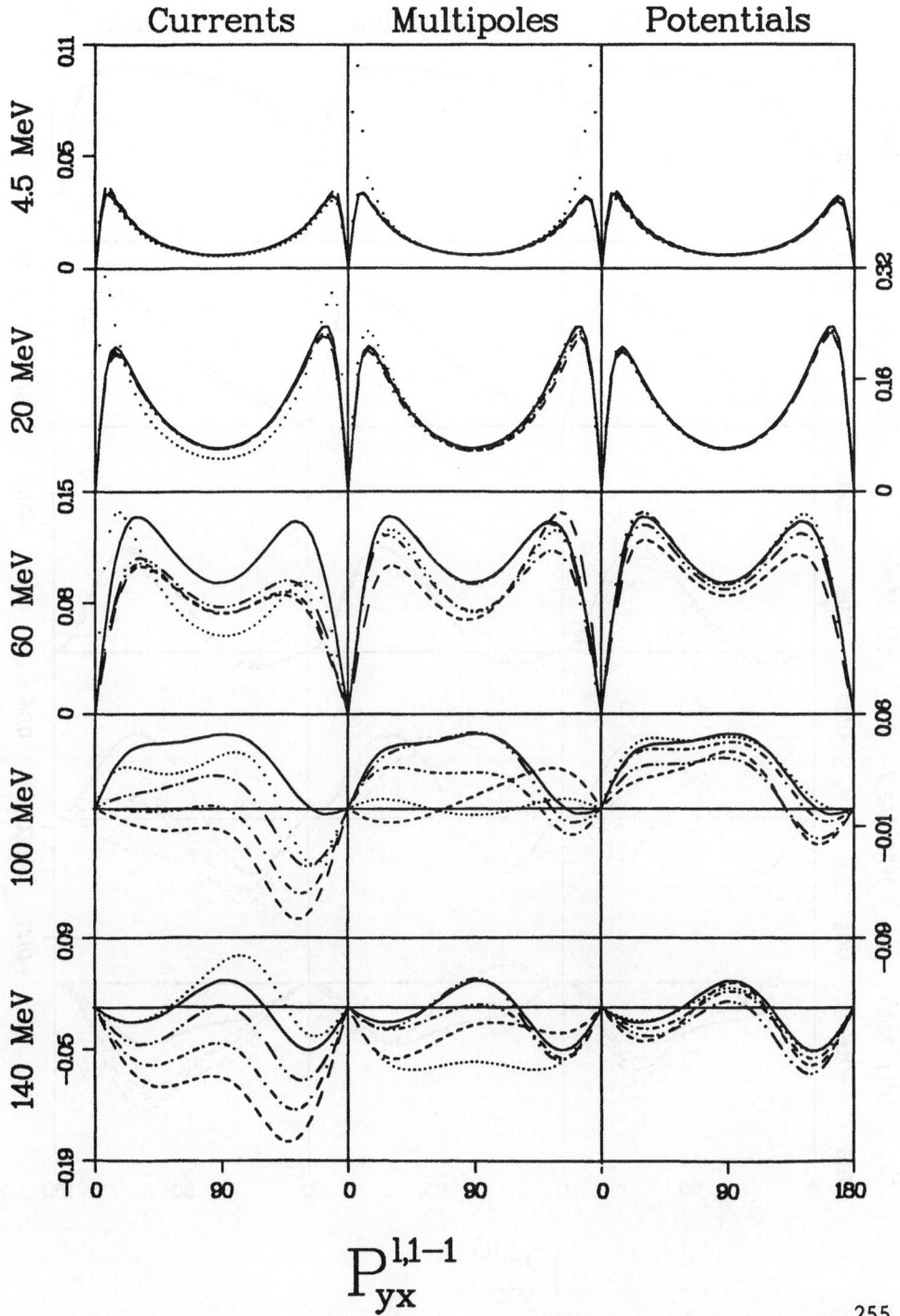

$$P_{yx}^{1,1-1}$$

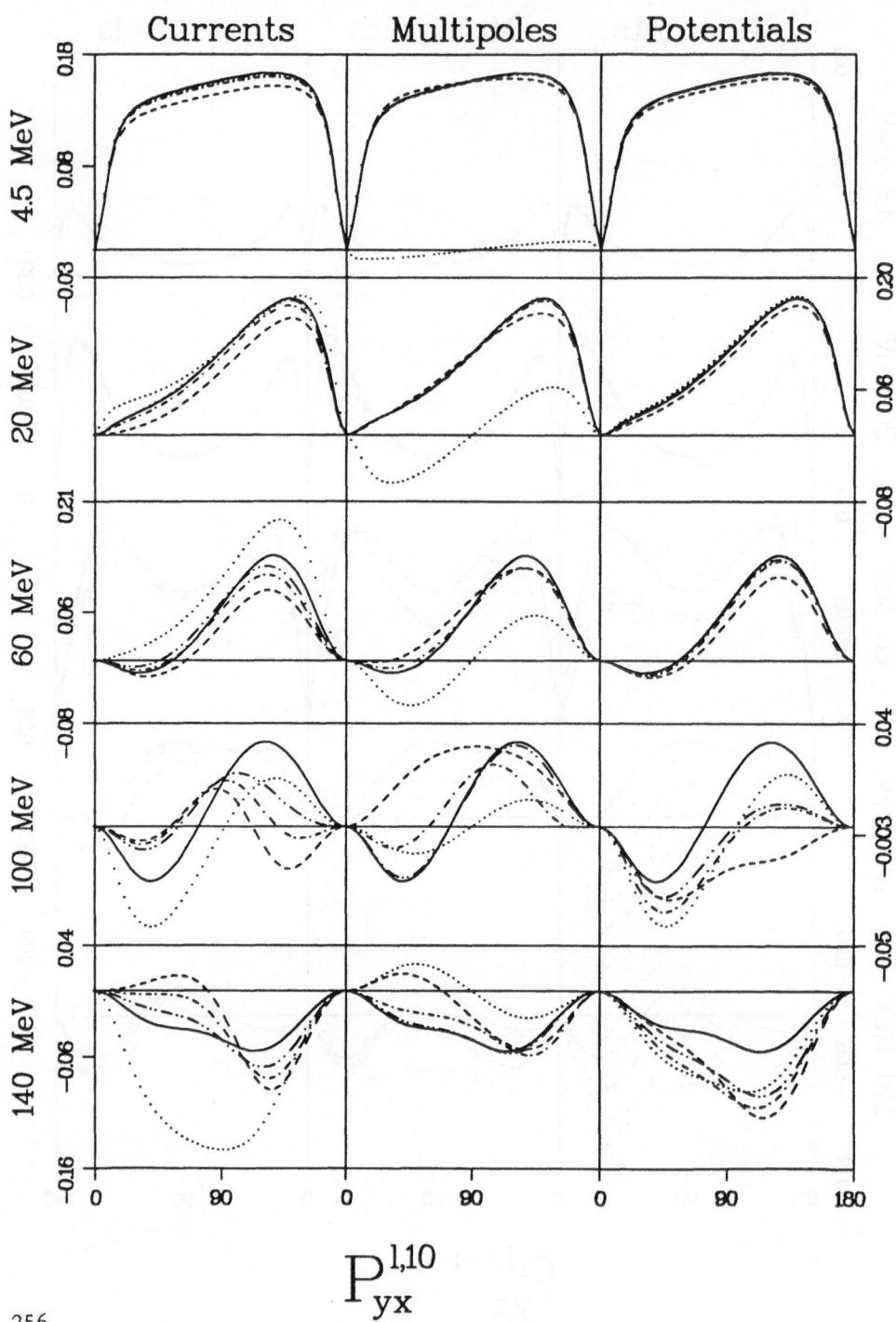

$$\mathrm{P}_{yx}^{1,10}$$

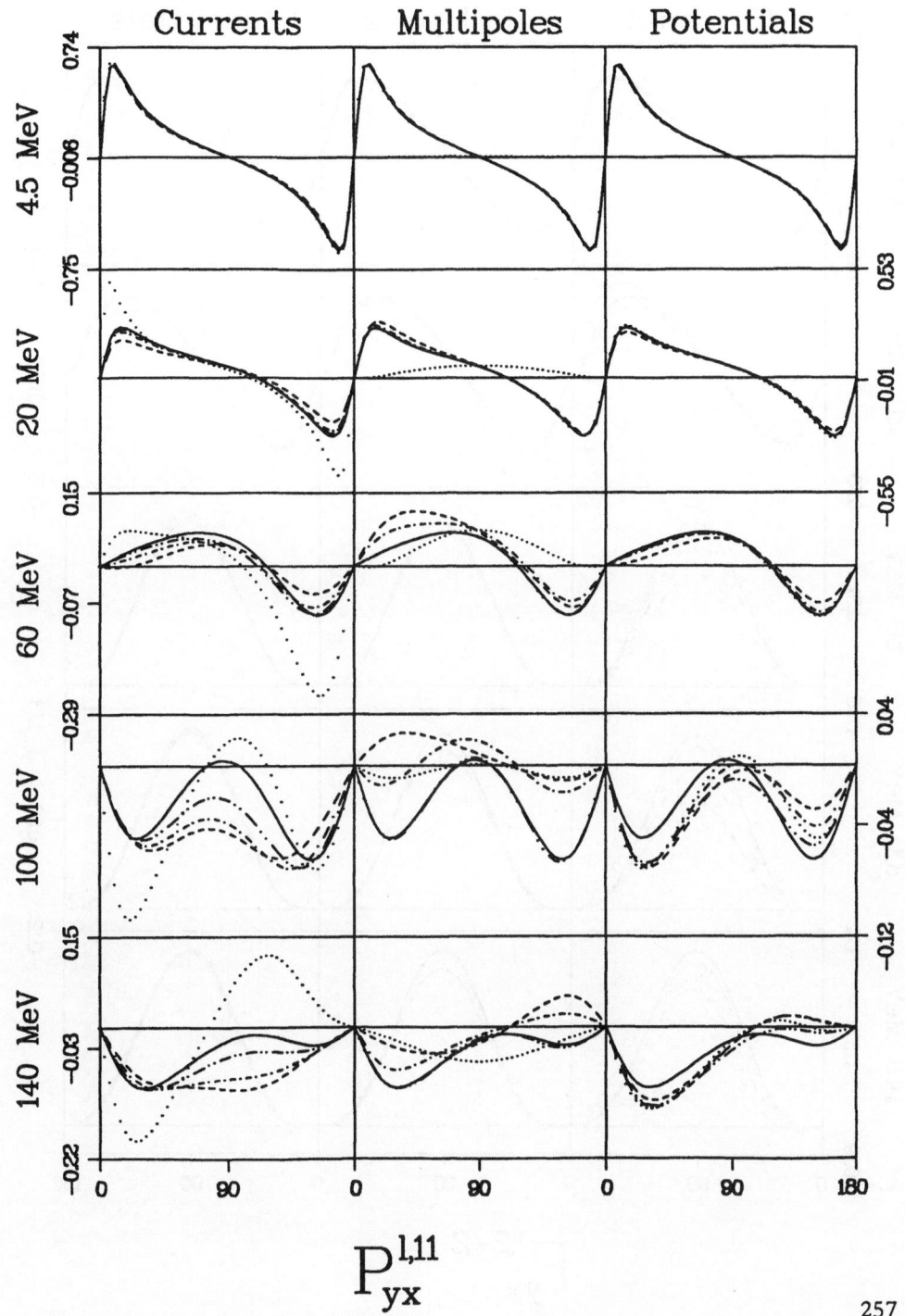

$$P_{yx}^{l,11}$$

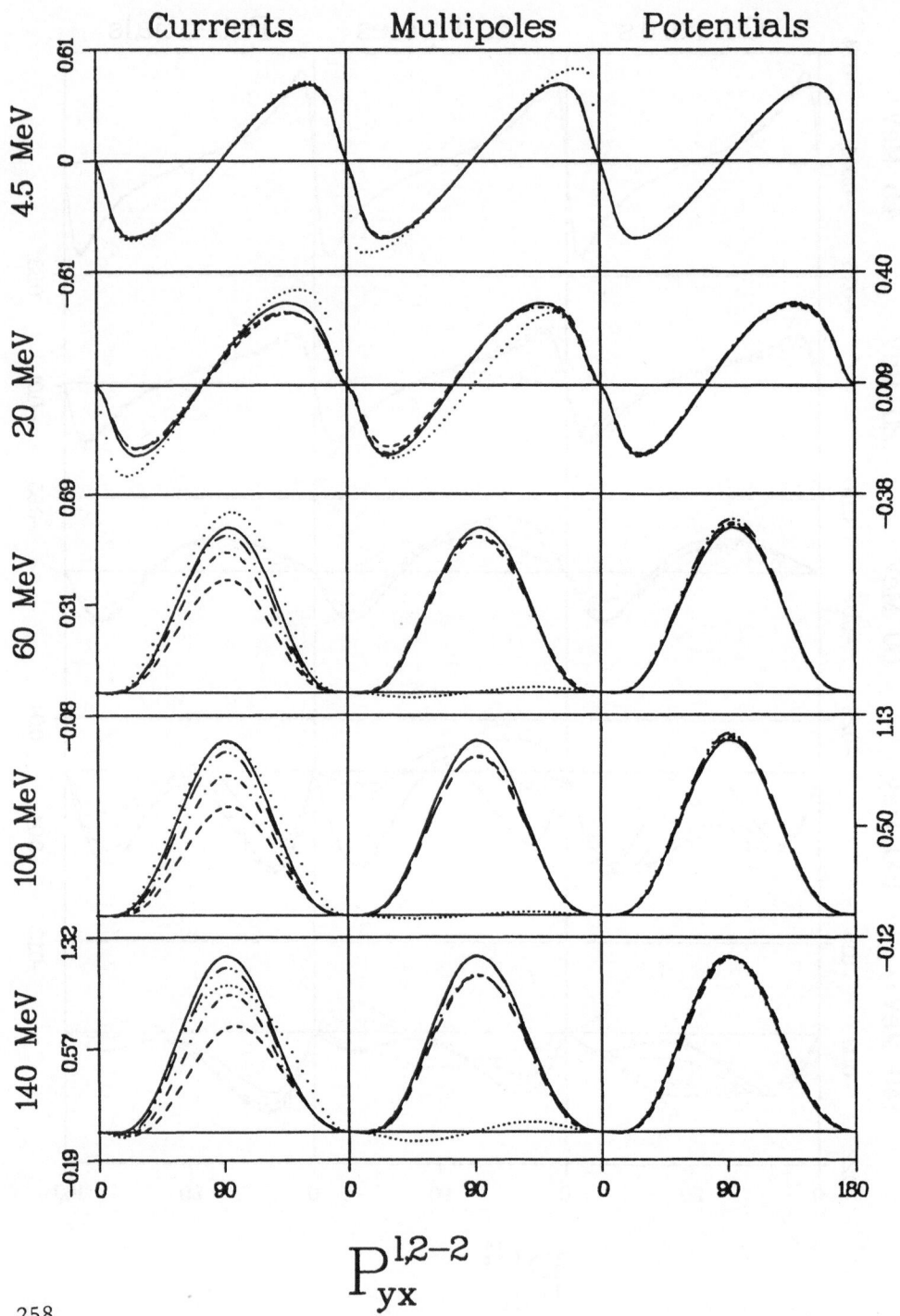

Currents Multipoles Potentials

4.5 MeV

20 MeV

60 MeV

100 MeV

140 MeV

$$P_{yx}^{12-2}$$

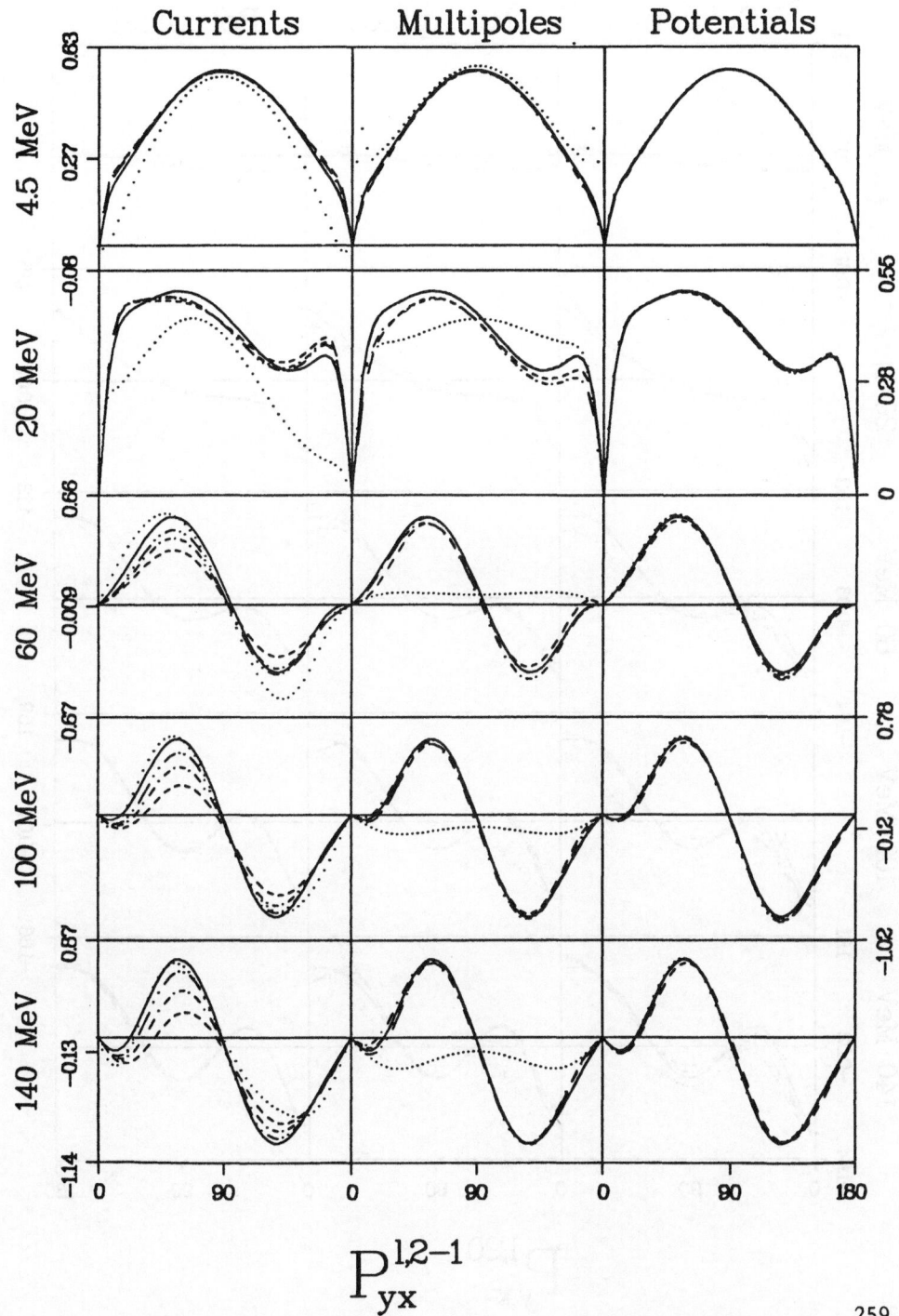

$$P_{yx}^{1,2-1}$$

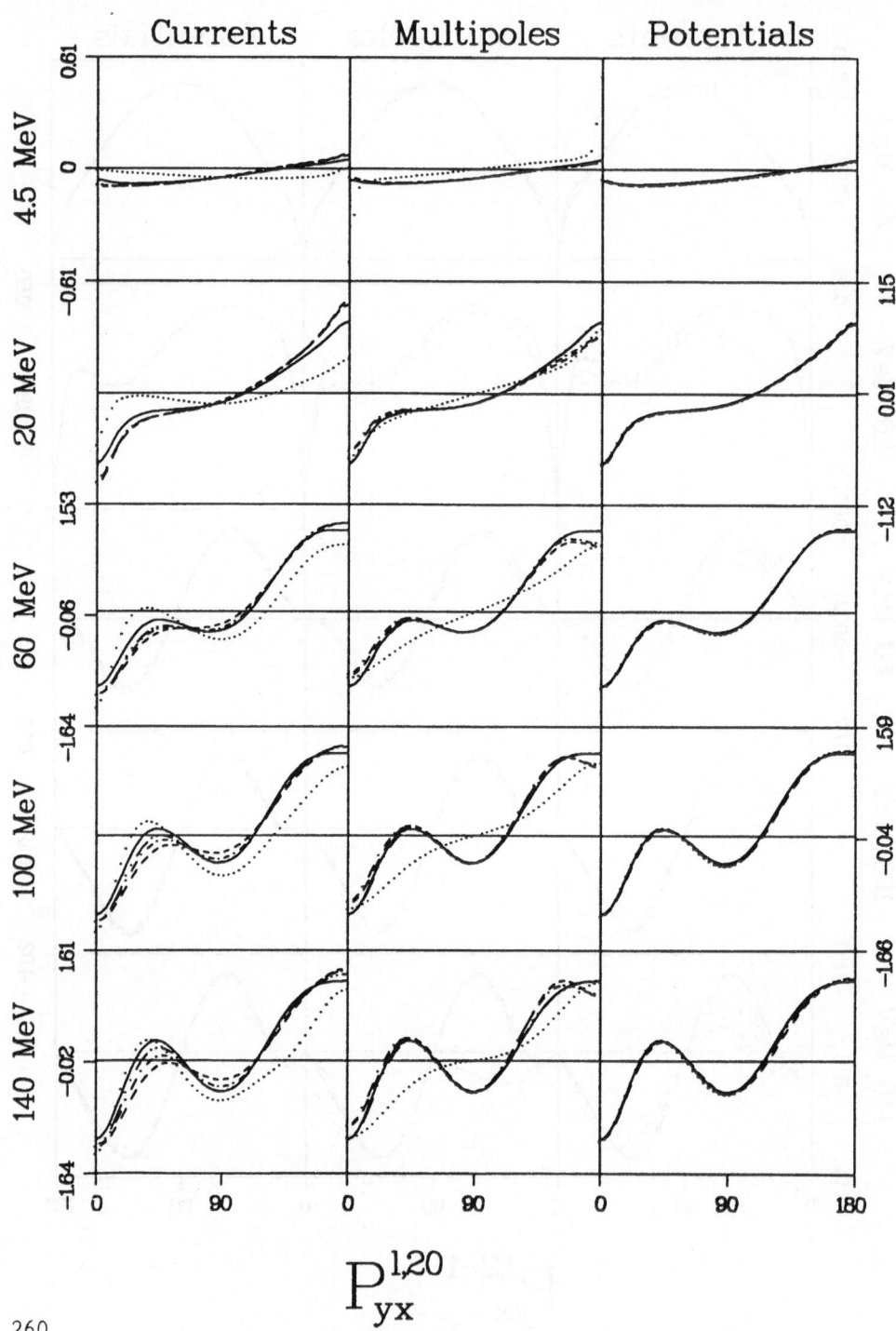

Currents   Multipoles   Potentials

$$P_{yx}^{1,20}$$

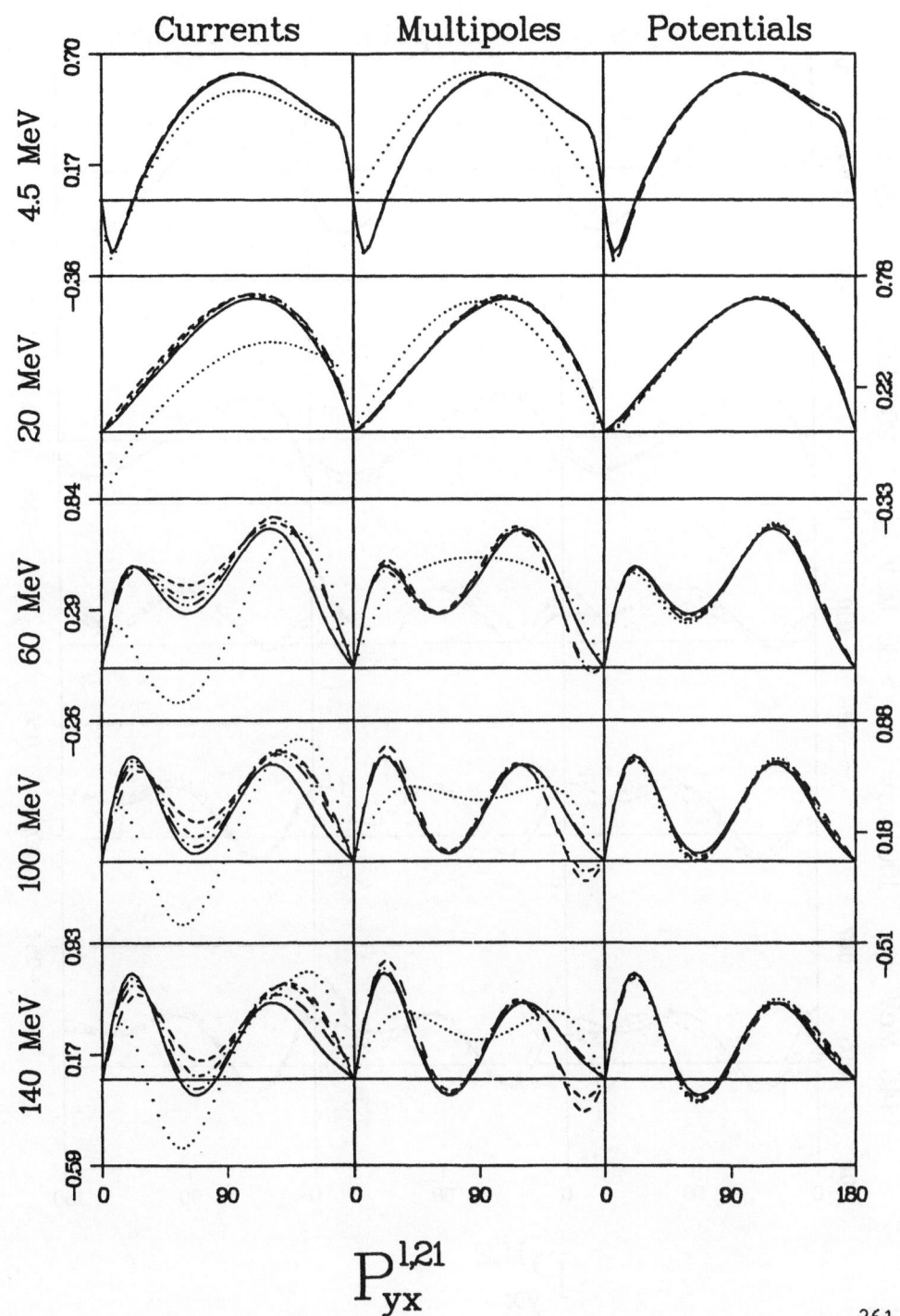

$$P_{yx}^{1,21}$$

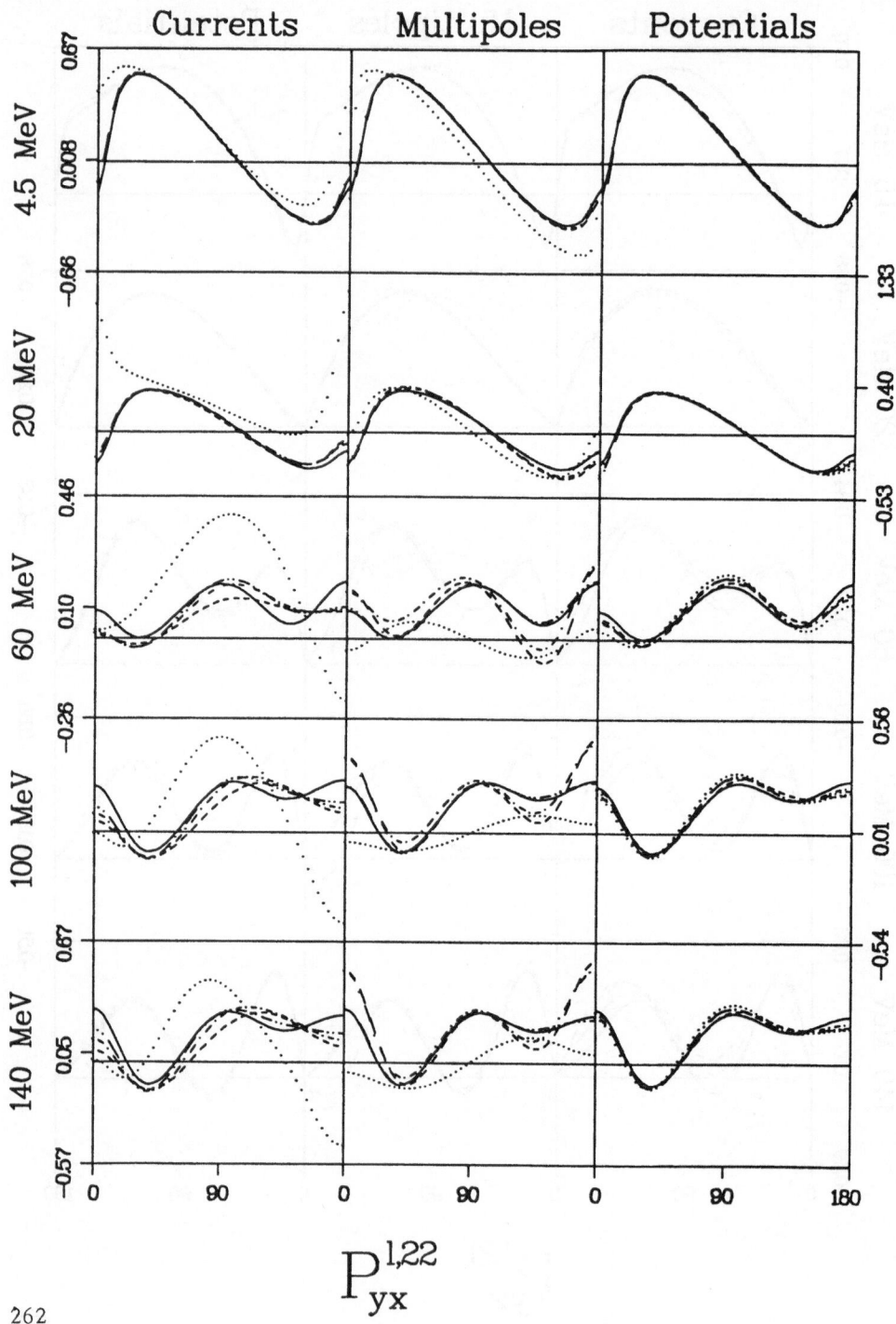

$$P_{yx}^{l,22}$$

5.44 Proton-Neutron Spin Correlation $P_{zy}^{0,IM}$

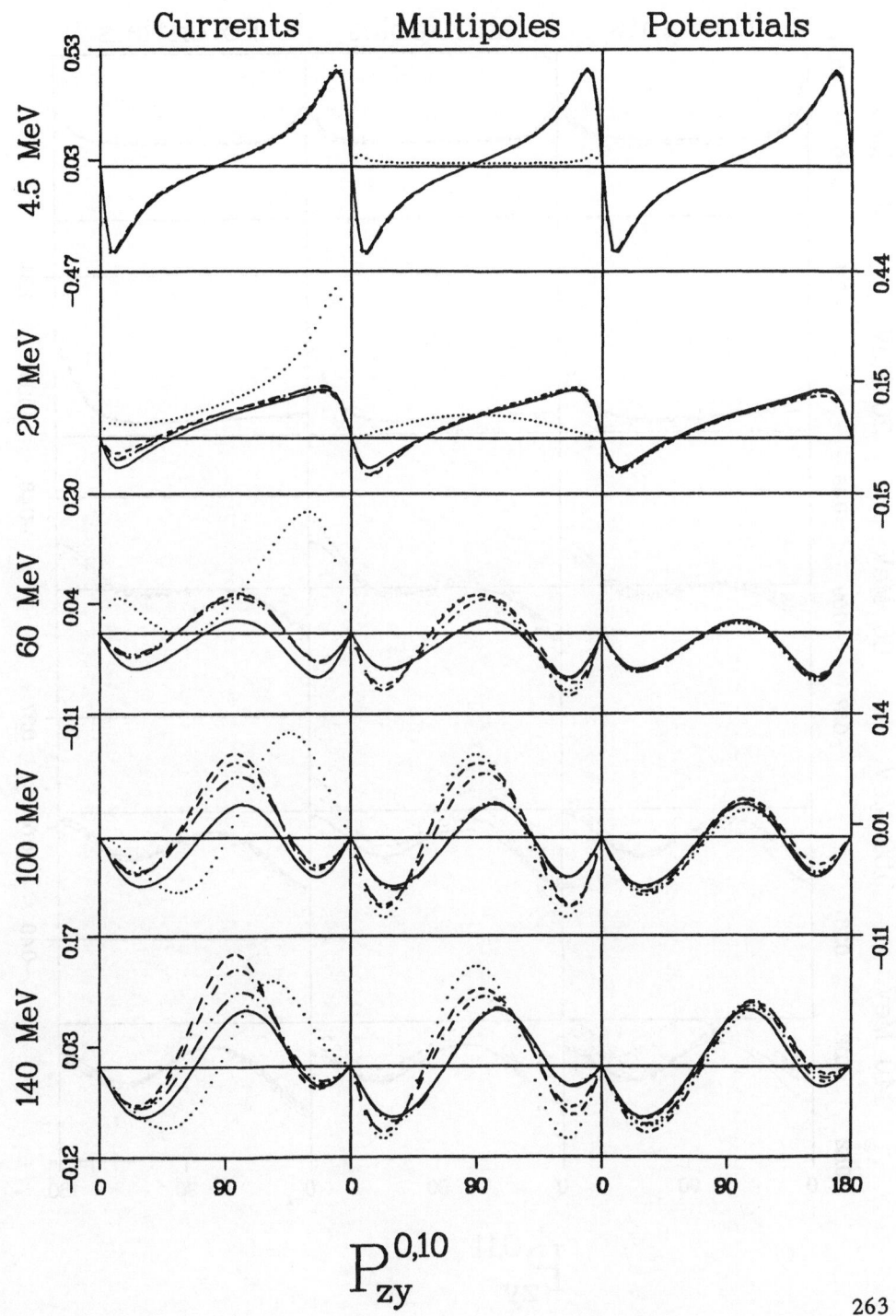

$$P_{zy}^{0,10}$$

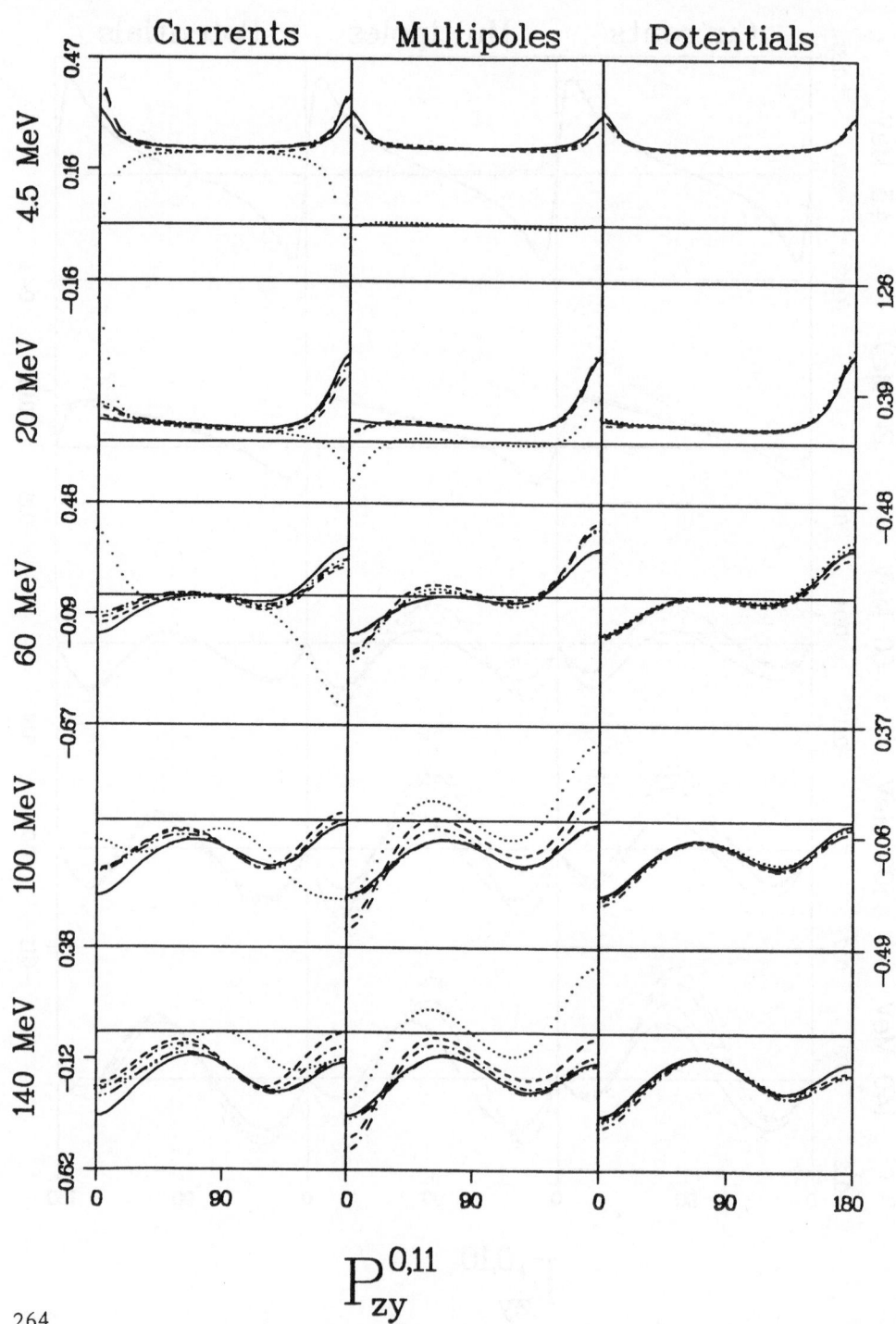

$$P_{zy}^{0,11}$$

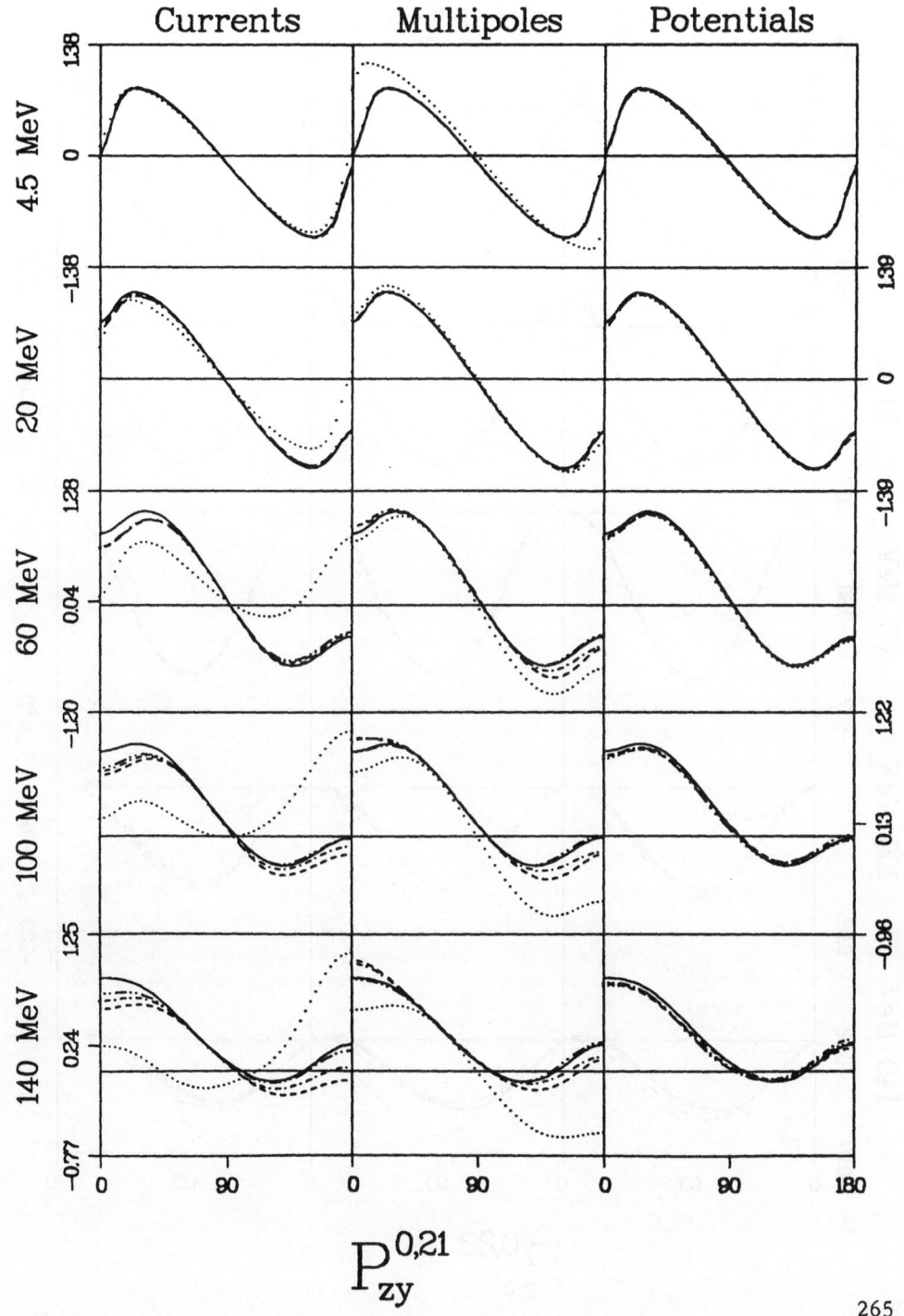

$$P_{zy}^{0,21}$$

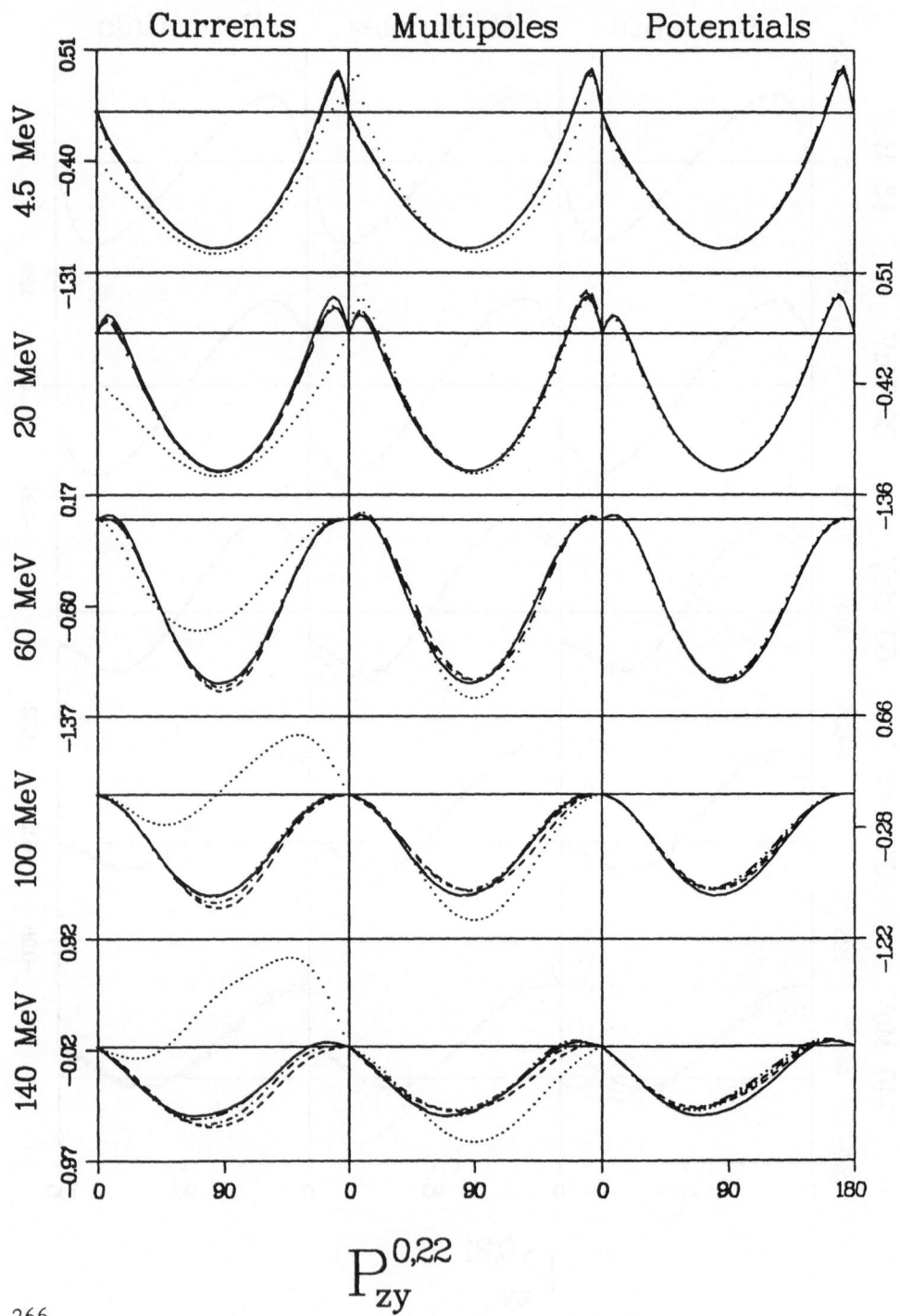

$$P_{zy}^{0,22}$$

5.45 Proton-Neutron Spin Correlation $P_{zy}^{c,IM}$

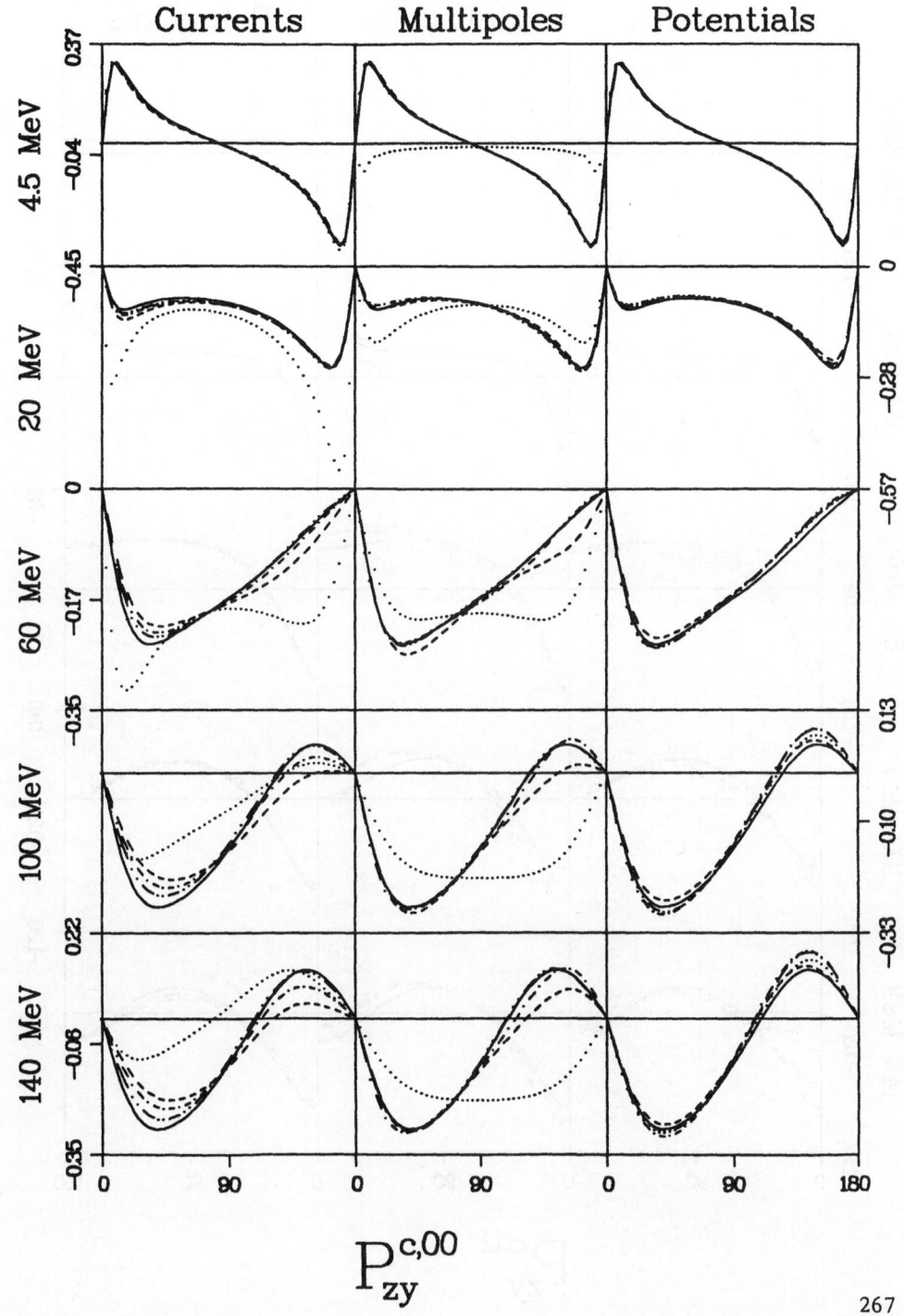

$$P_{zy}^{c,00}$$

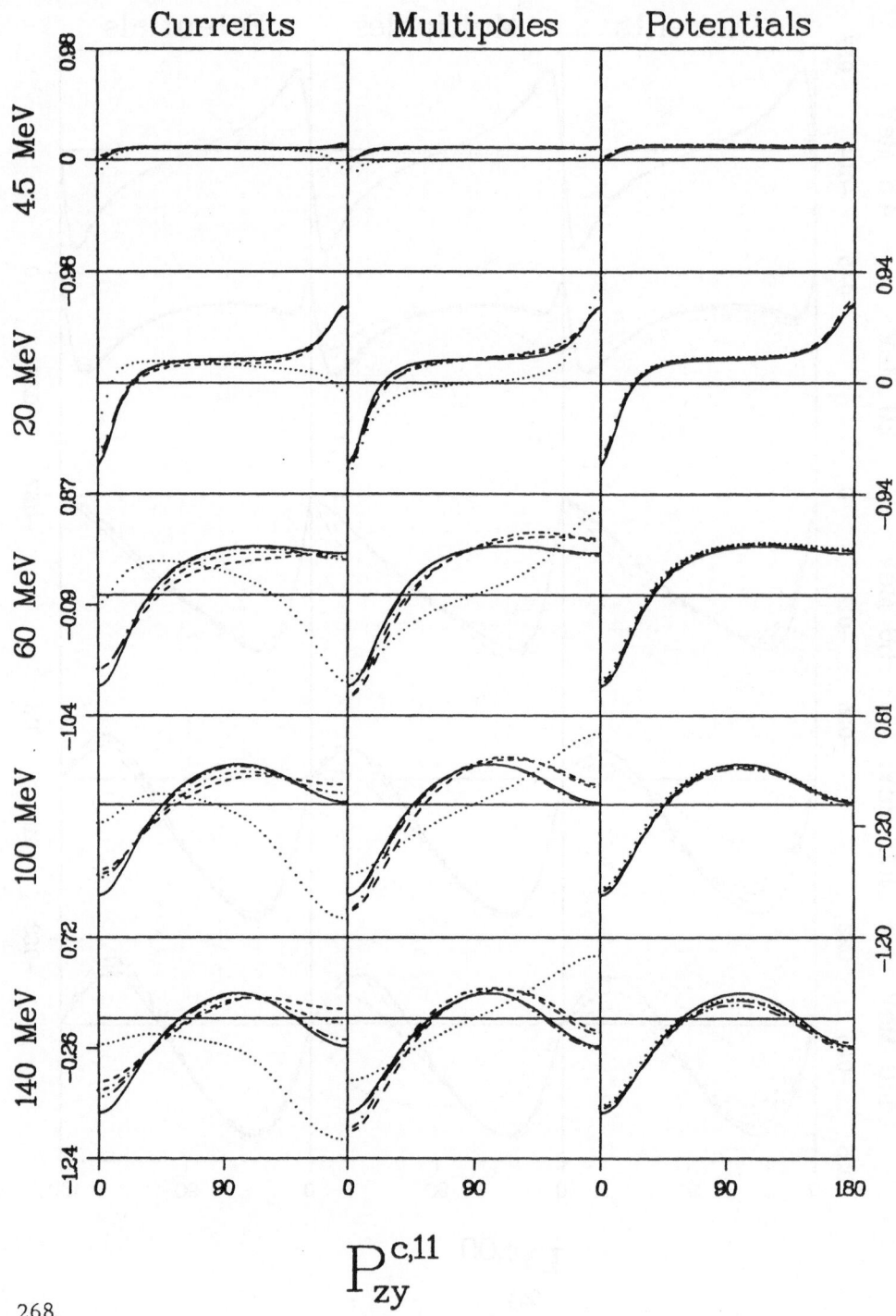

$$P_{zy}^{c,11}$$

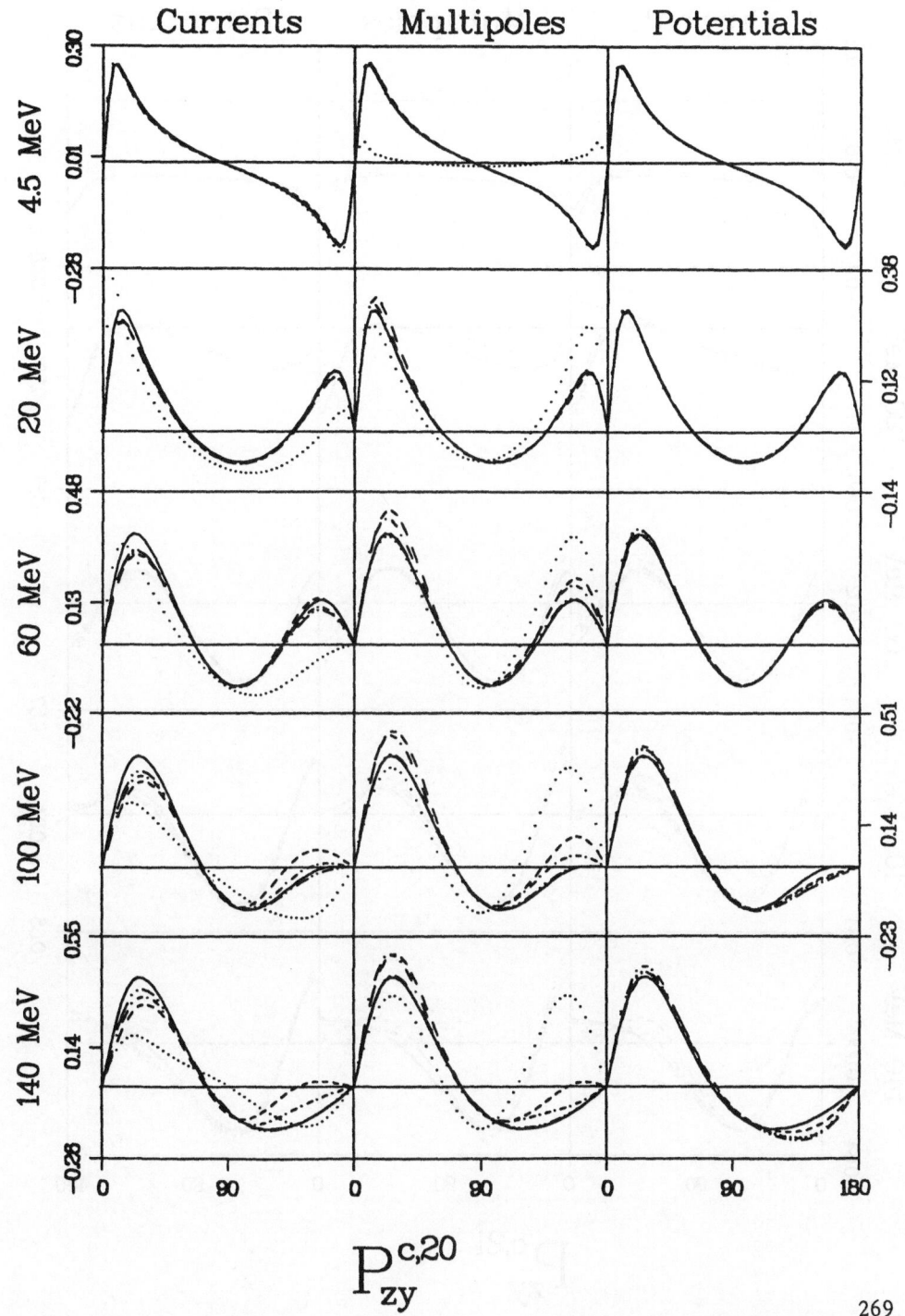

$$P_{zy}^{c,20}$$

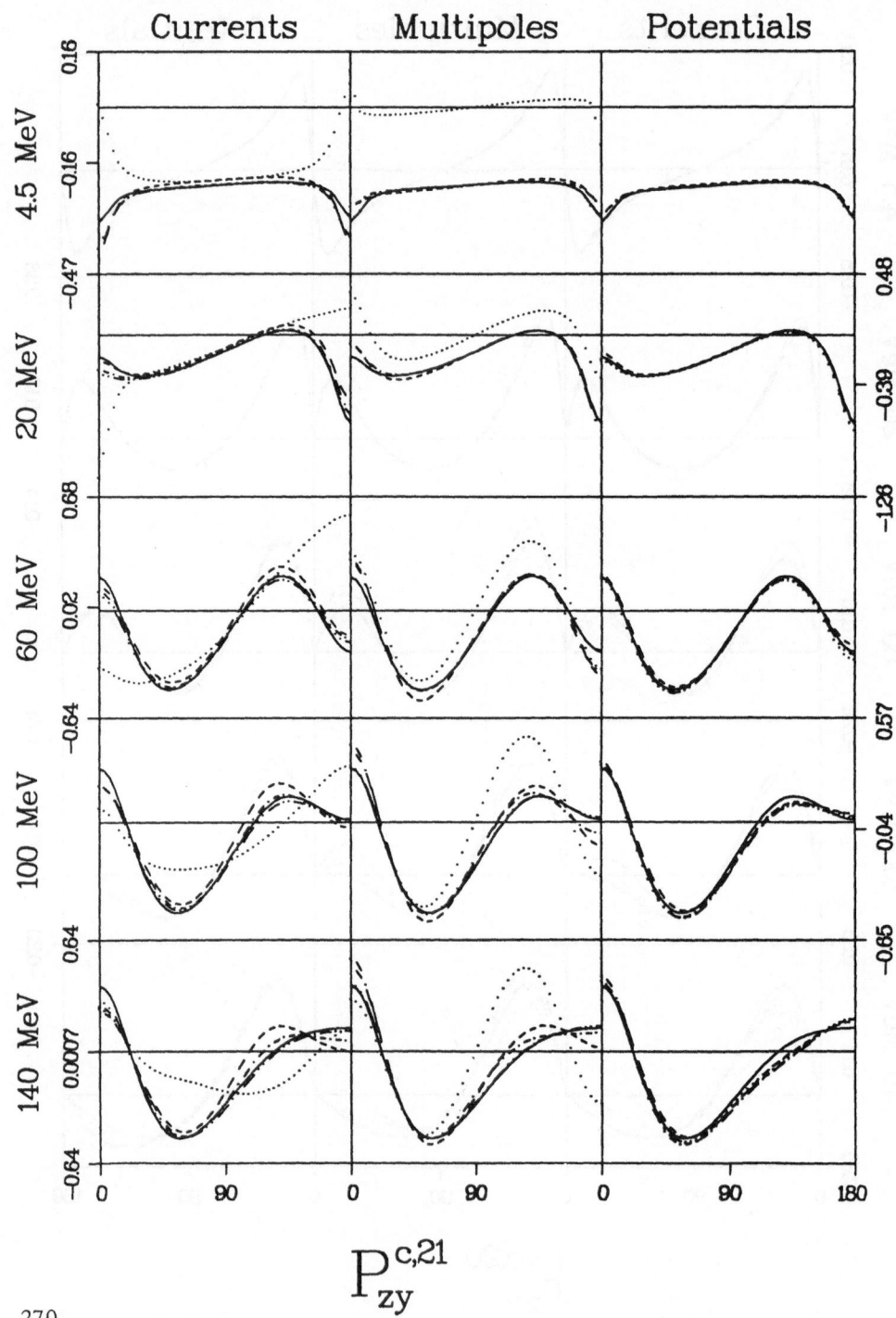

Currents   Multipoles   Potentials

$$P_{zy}^{c,21}$$

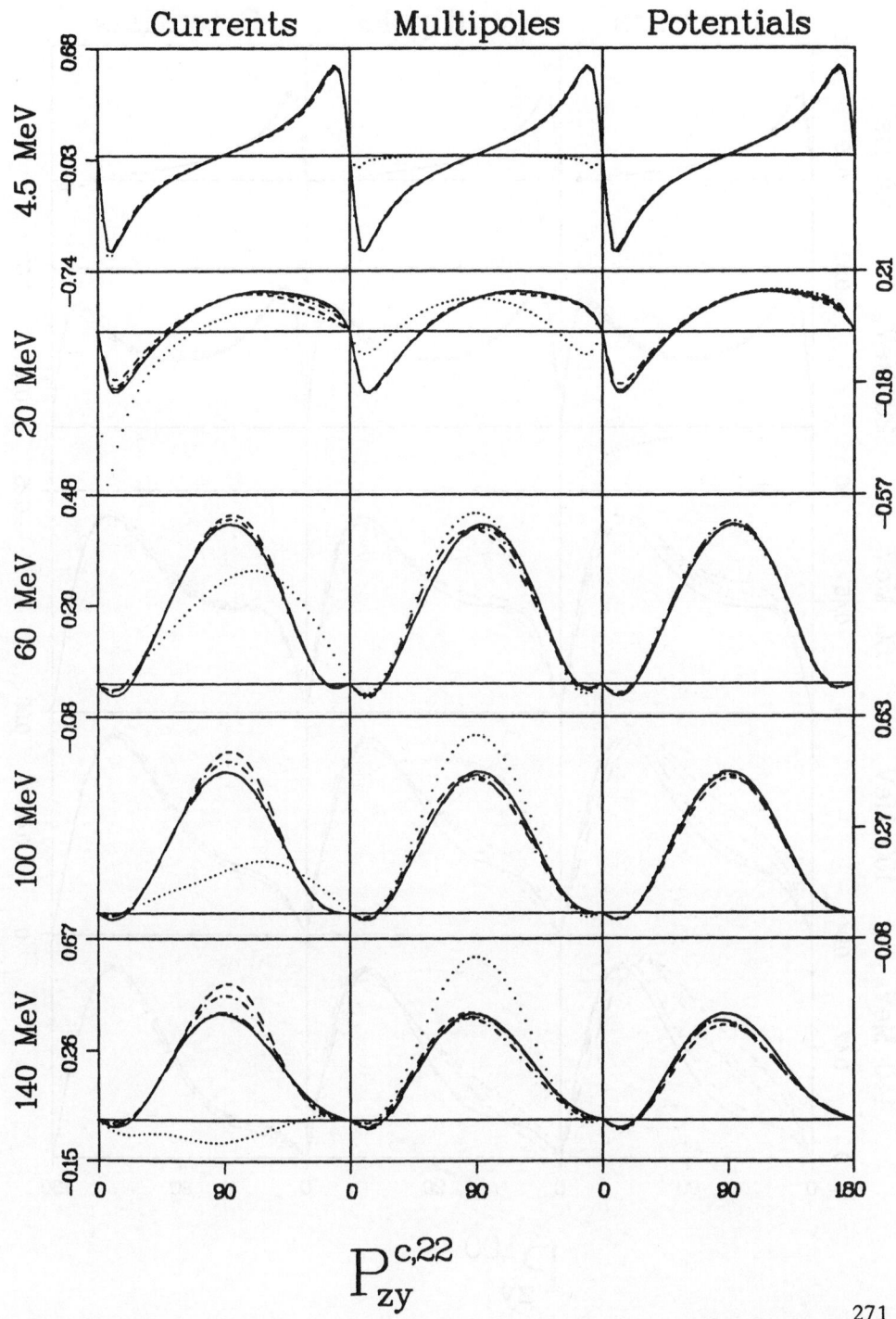

$$P_{zy}^{c,22}$$

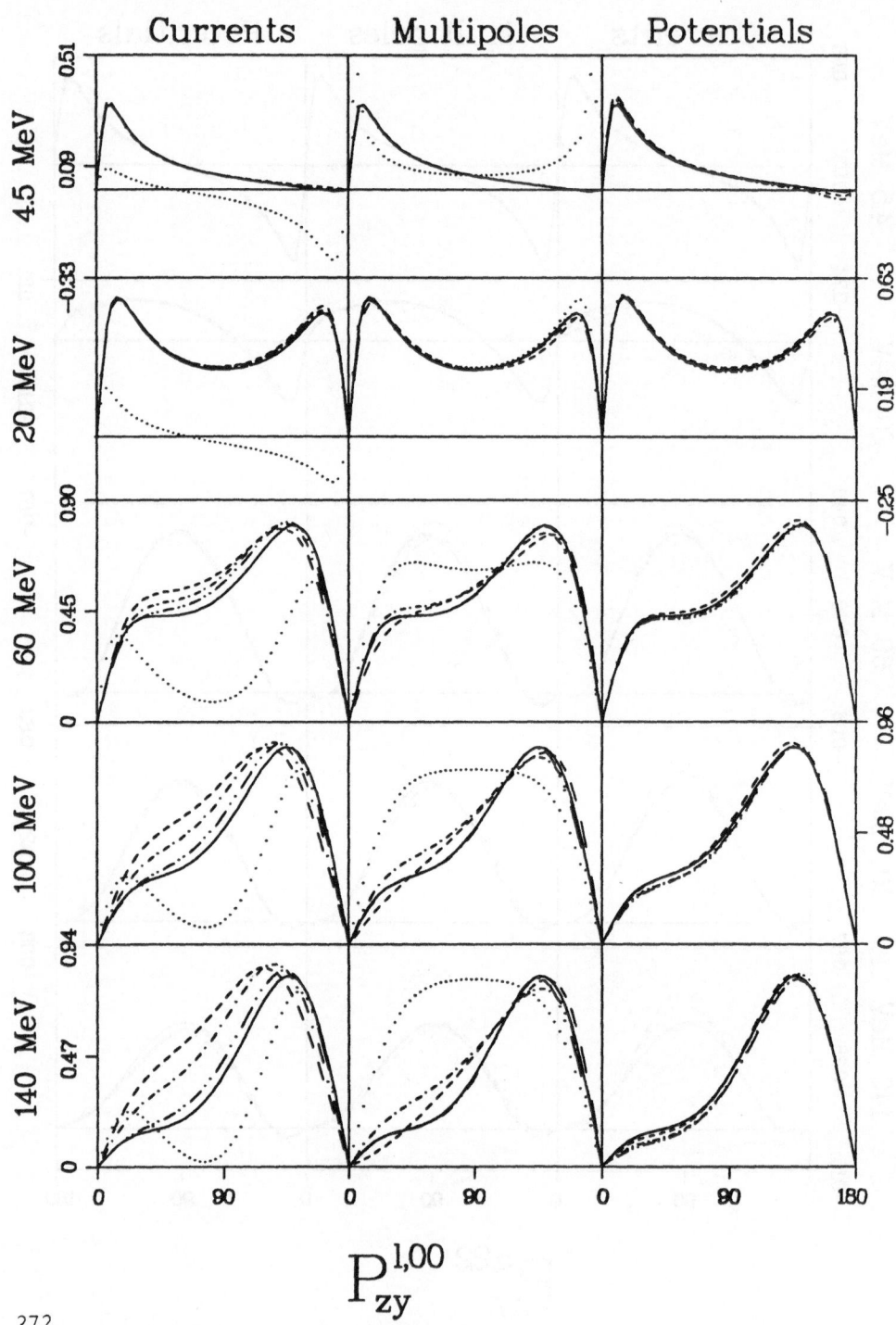

Currents   Multipoles   Potentials

4.5 MeV

20 MeV

60 MeV

100 MeV

140 MeV

$$P_{zy}^{1,00}$$

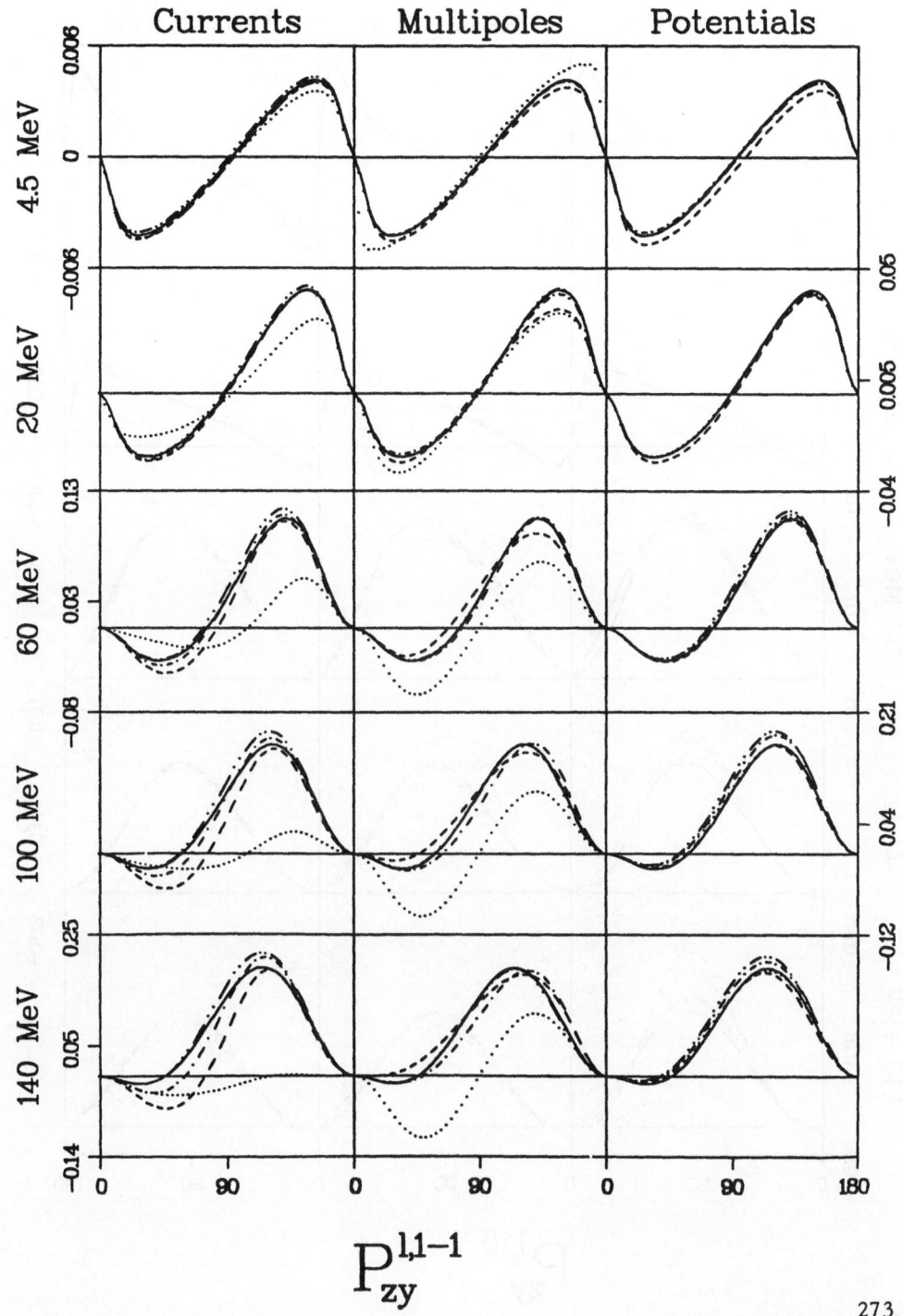

$$P_{zy}^{1,1-1}$$

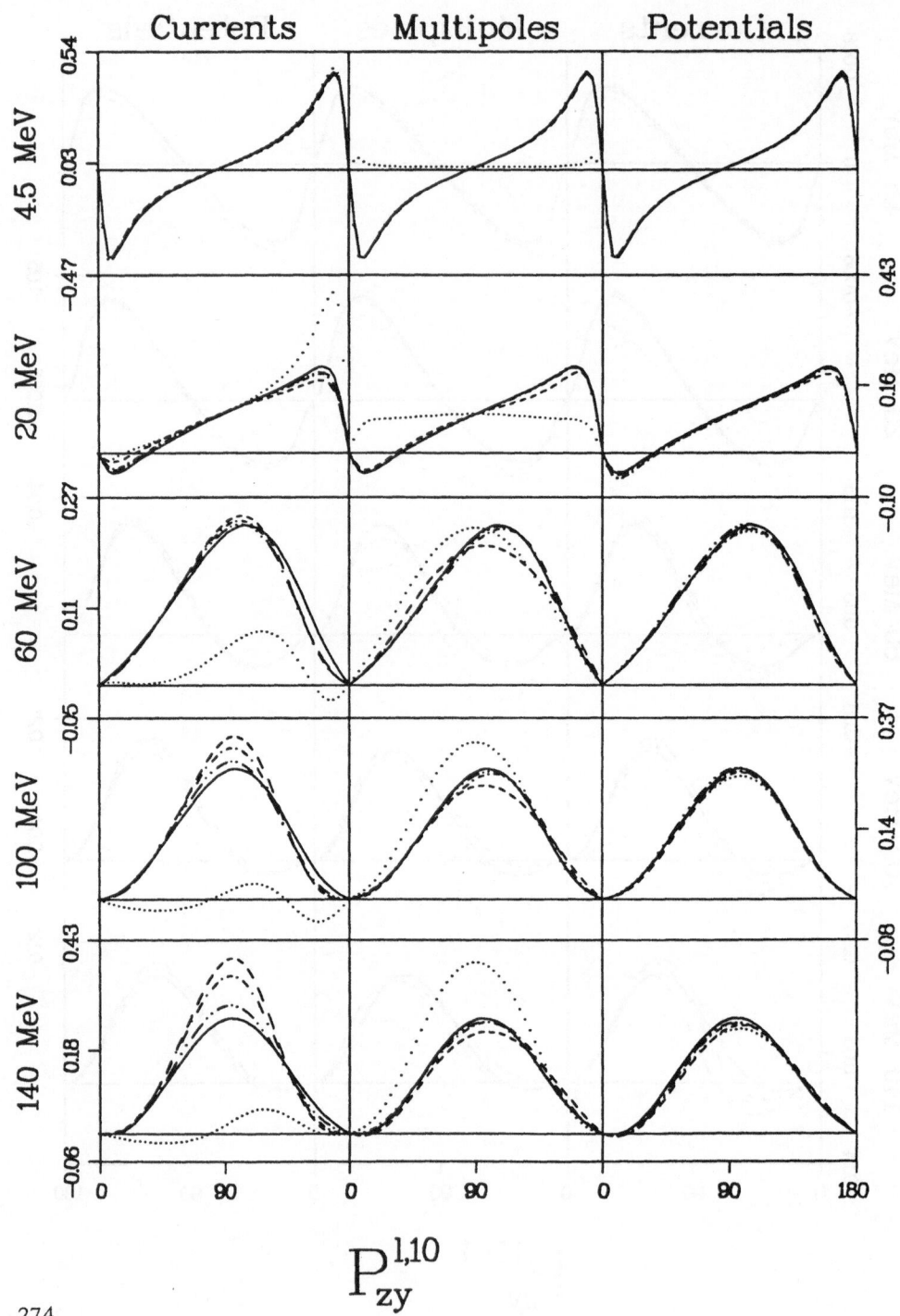

$$P_{zy}^{1,10}$$

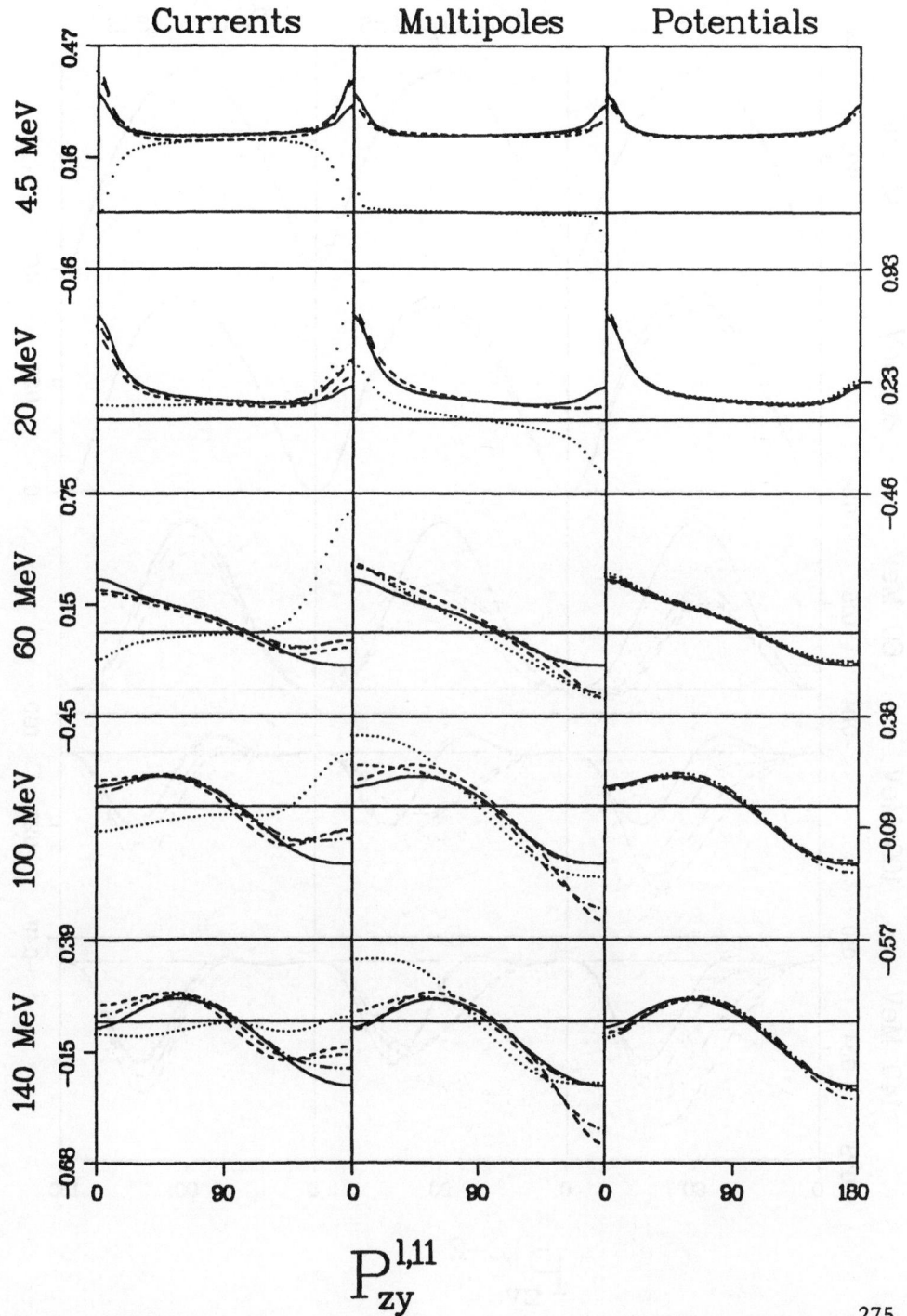

$$P_{zy}^{l,11}$$

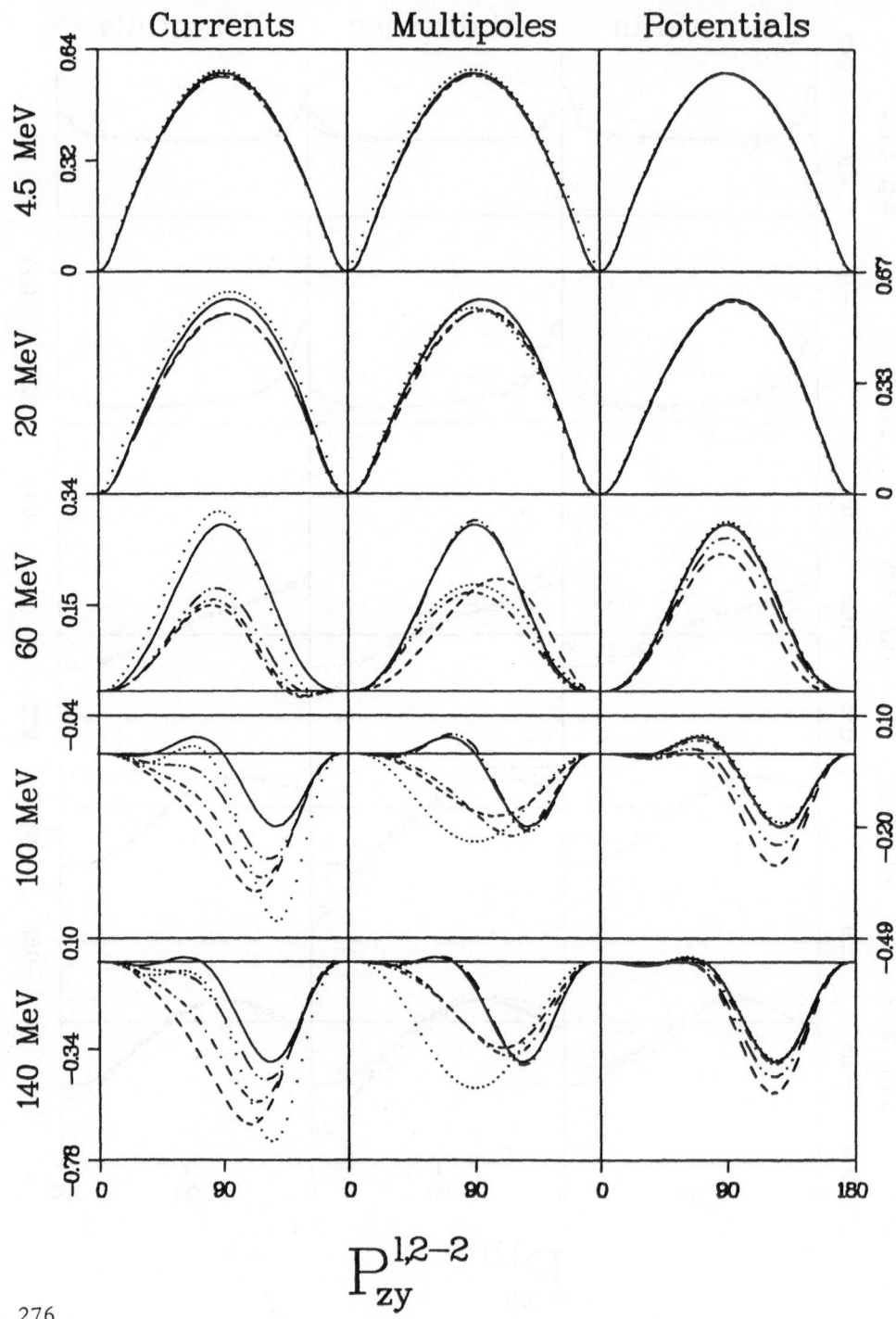

$$P_{zy}^{1,2-2}$$

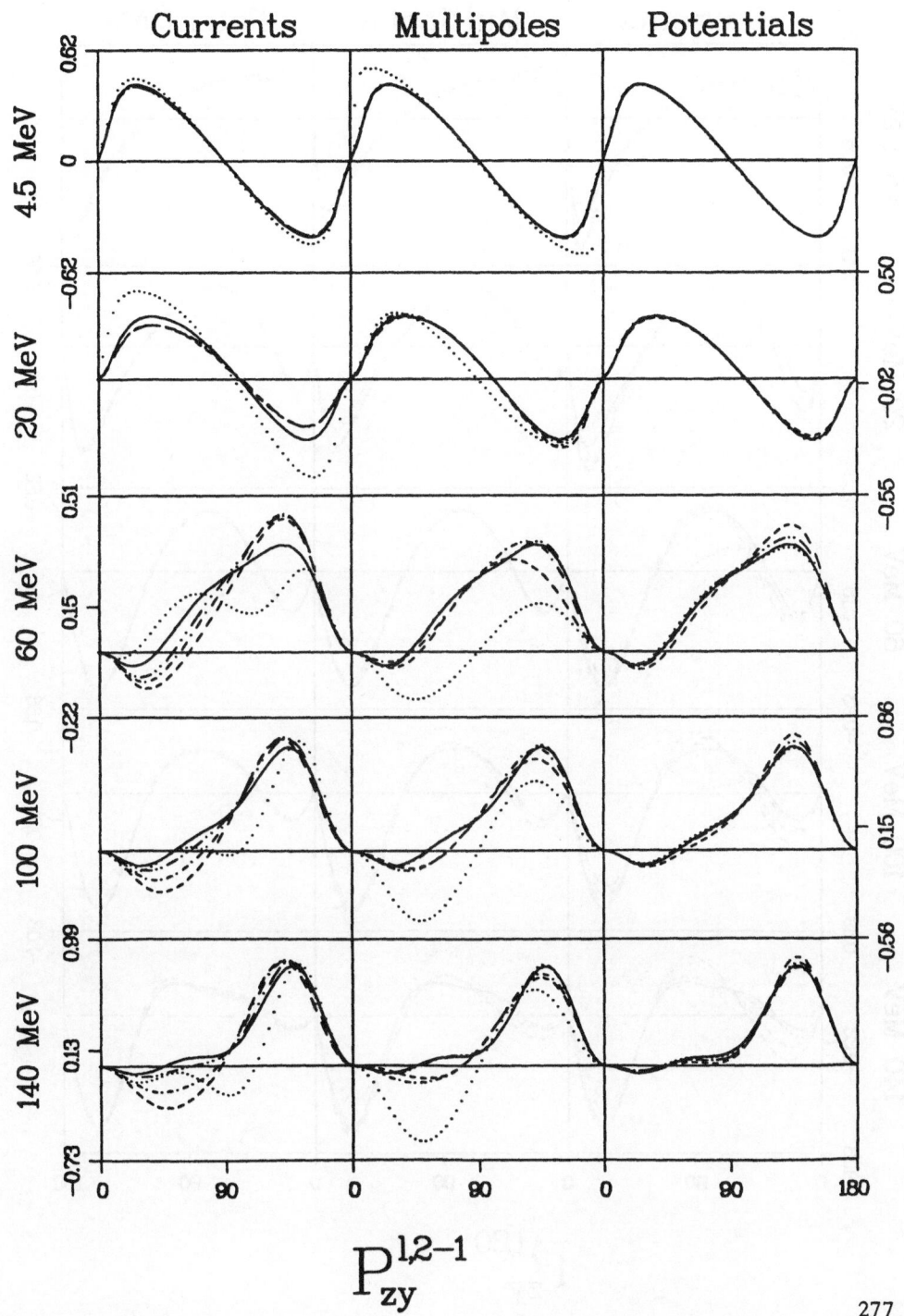

$$P_{zy}^{1,2-1}$$

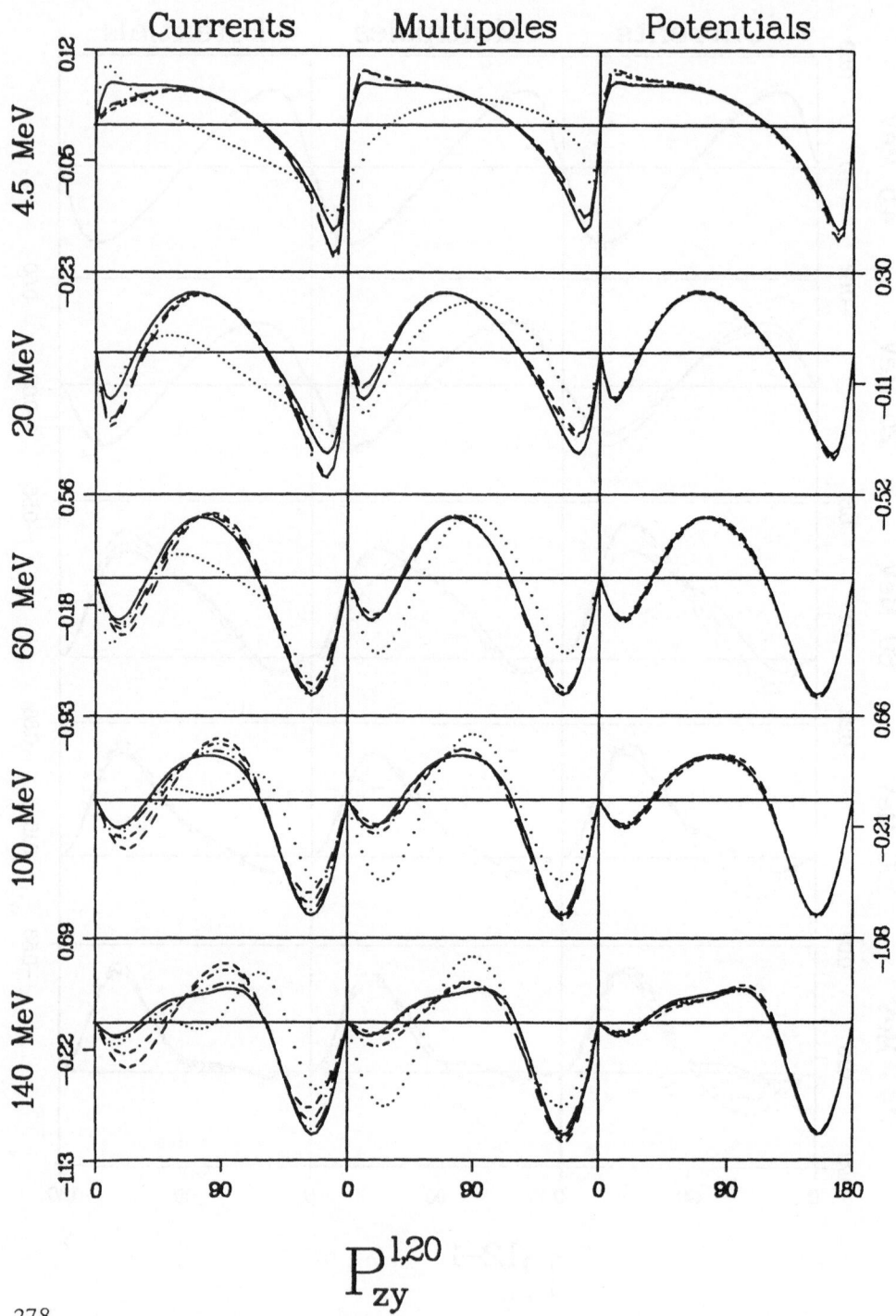

$$P_{zy}^{l,20}$$

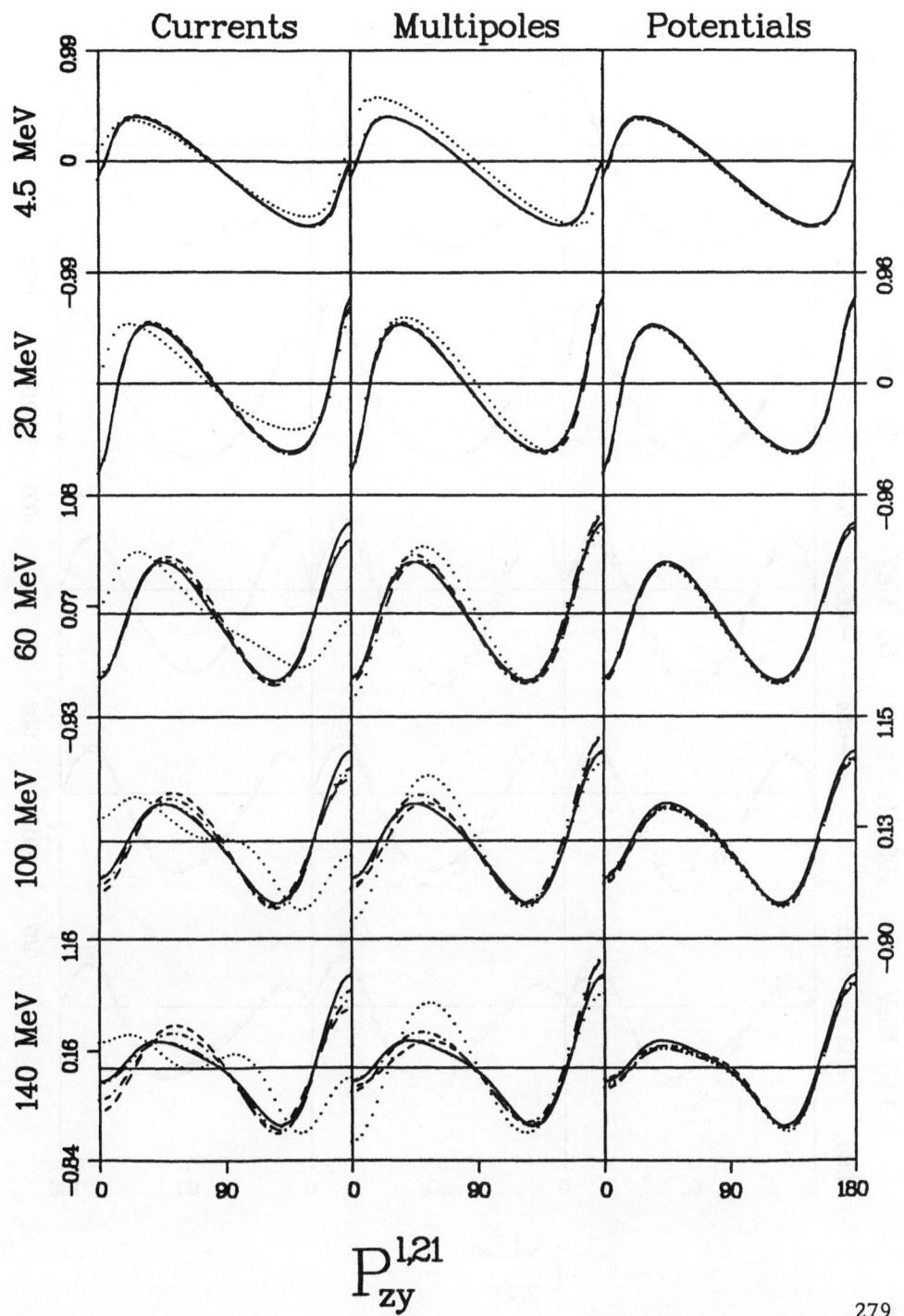

$$P_{zy}^{1,21}$$

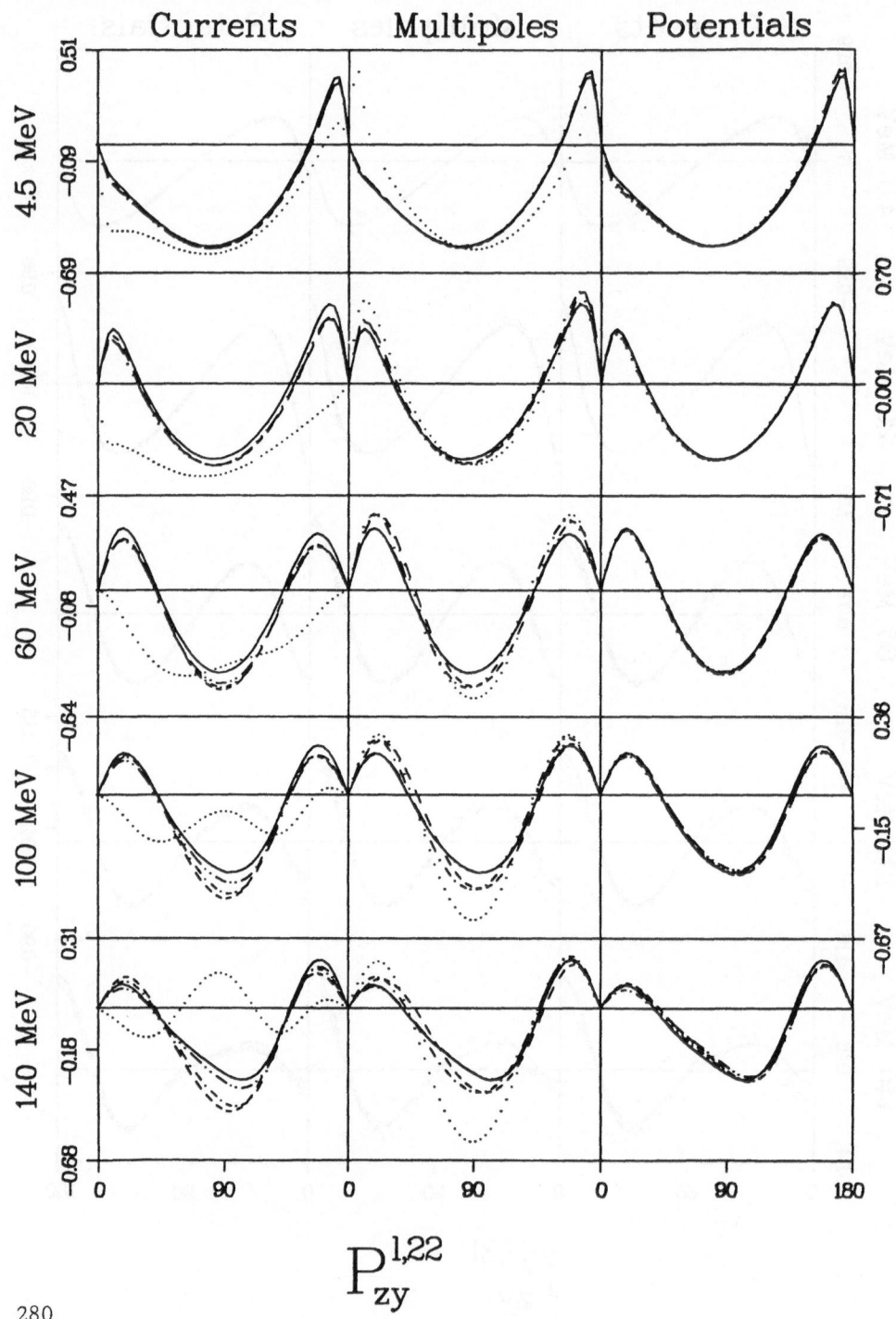

Currents   Multipoles   Potentials

$$P_{zy}^{1,22}$$

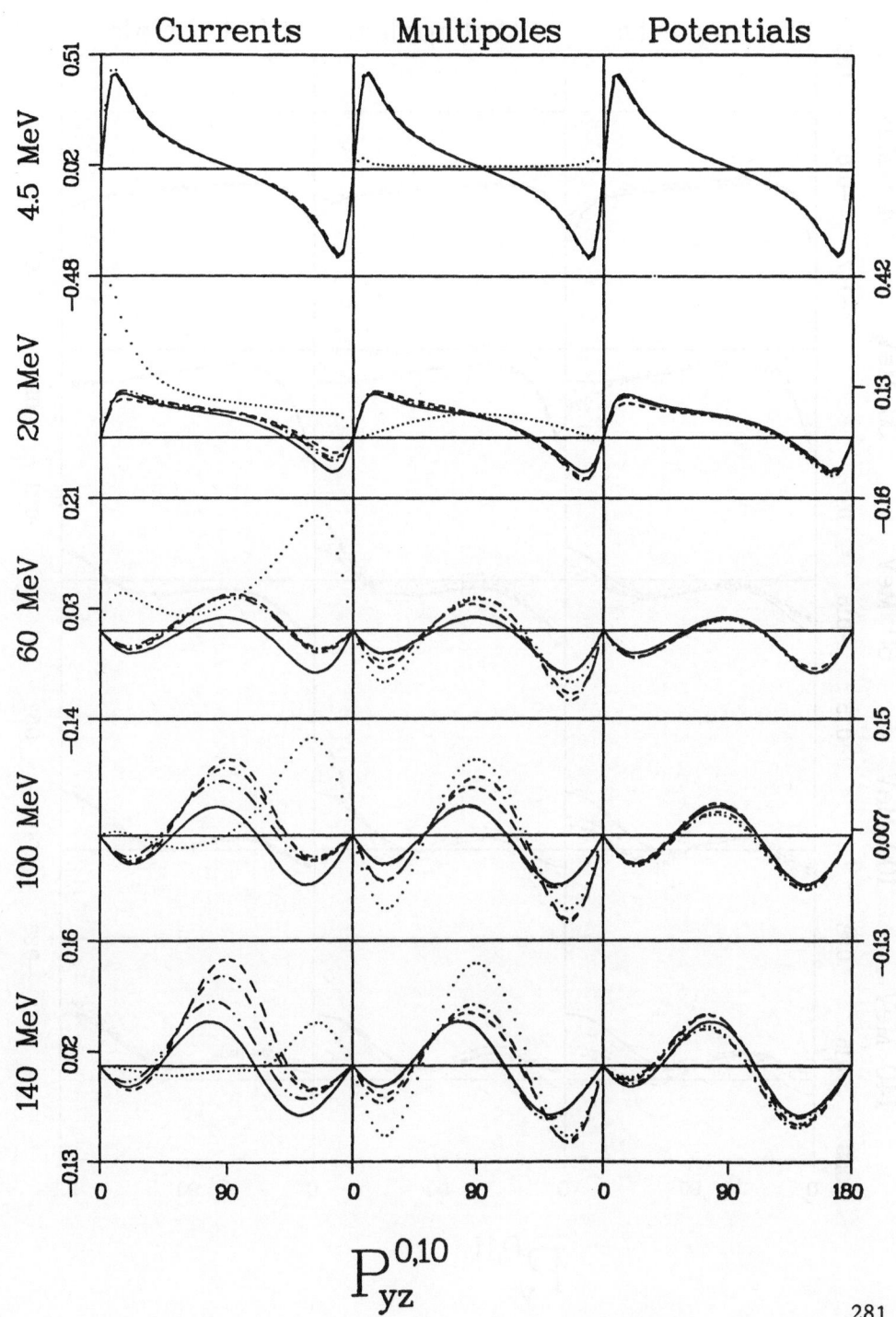

$$P_{yz}^{0,10}$$

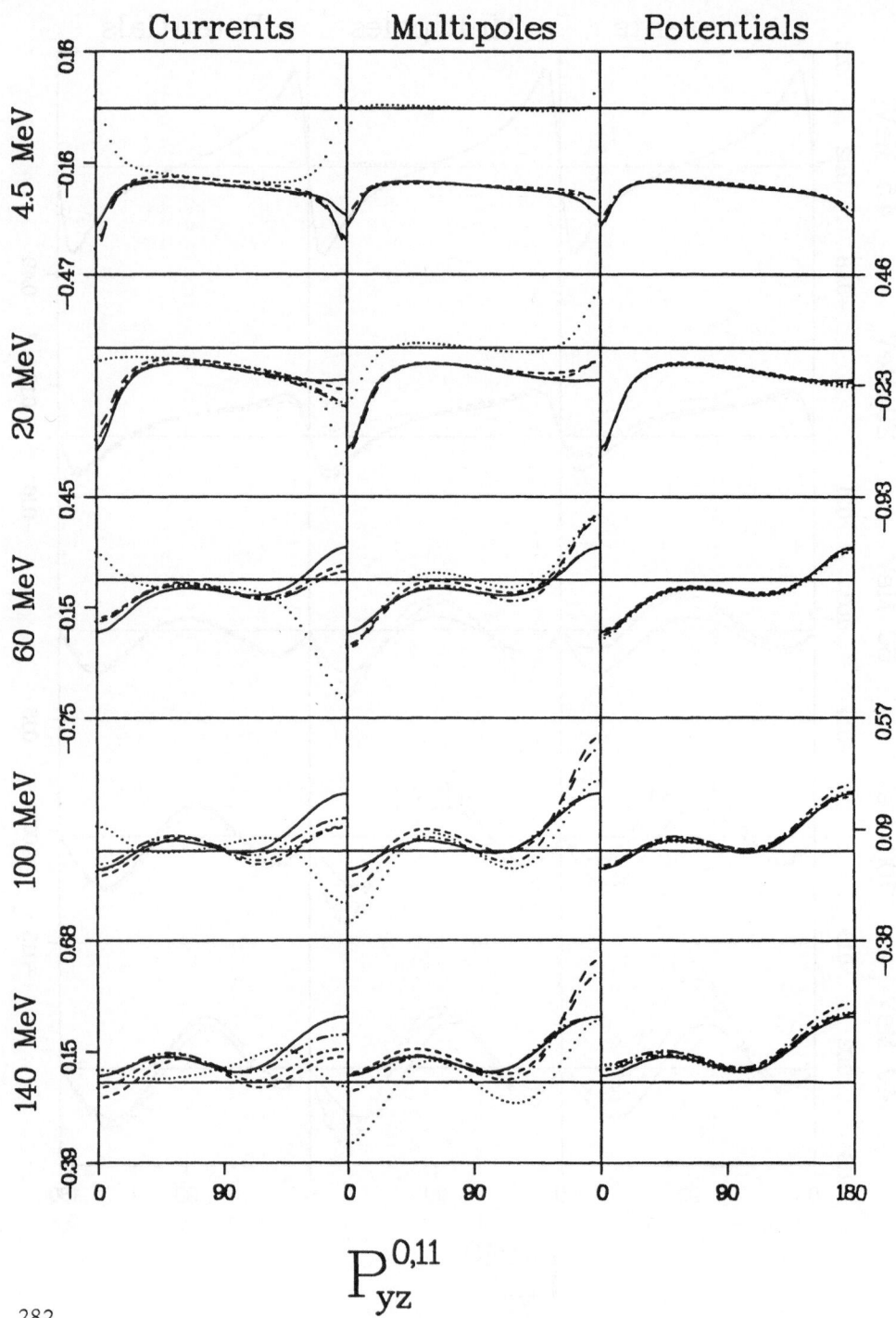

Currents   Multipoles   Potentials

4.5 MeV

20 MeV

60 MeV

100 MeV

140 MeV

$$P_{yz}^{0,11}$$

Proton-Neutron Spin Correlation $P_{yz}^{0,IM}$

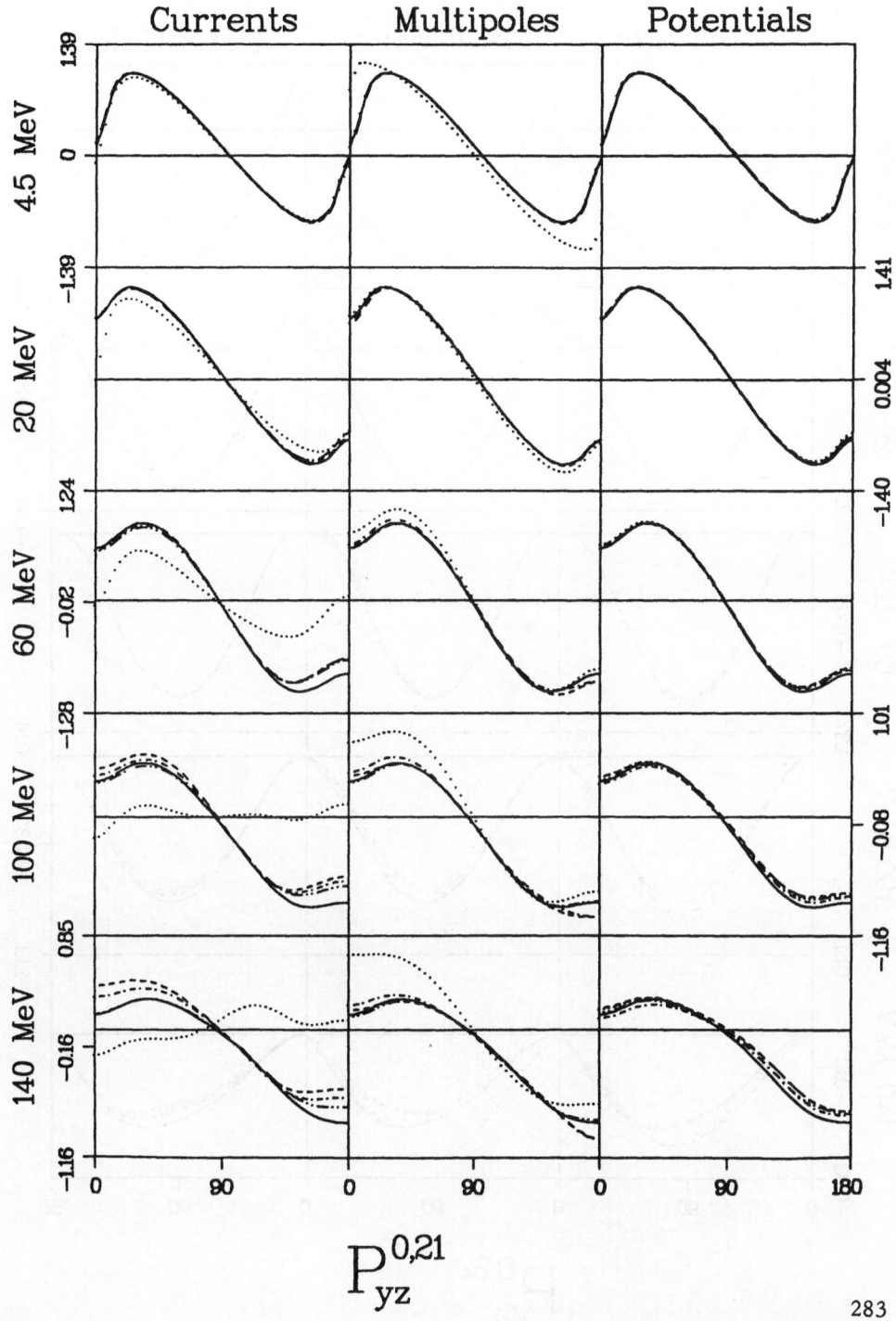

$$P_{yz}^{0,21}$$

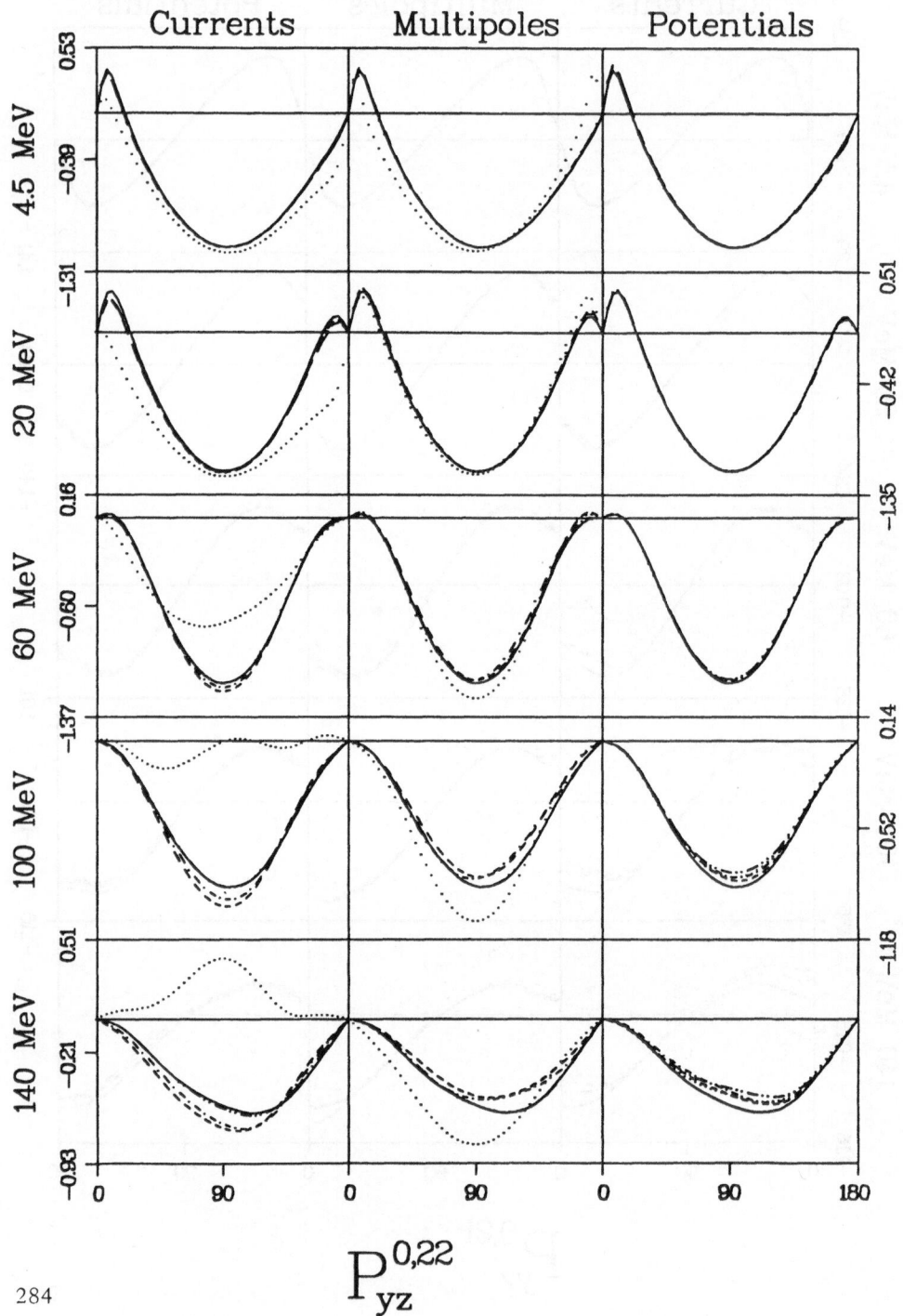

$$P_{yz}^{0,22}$$

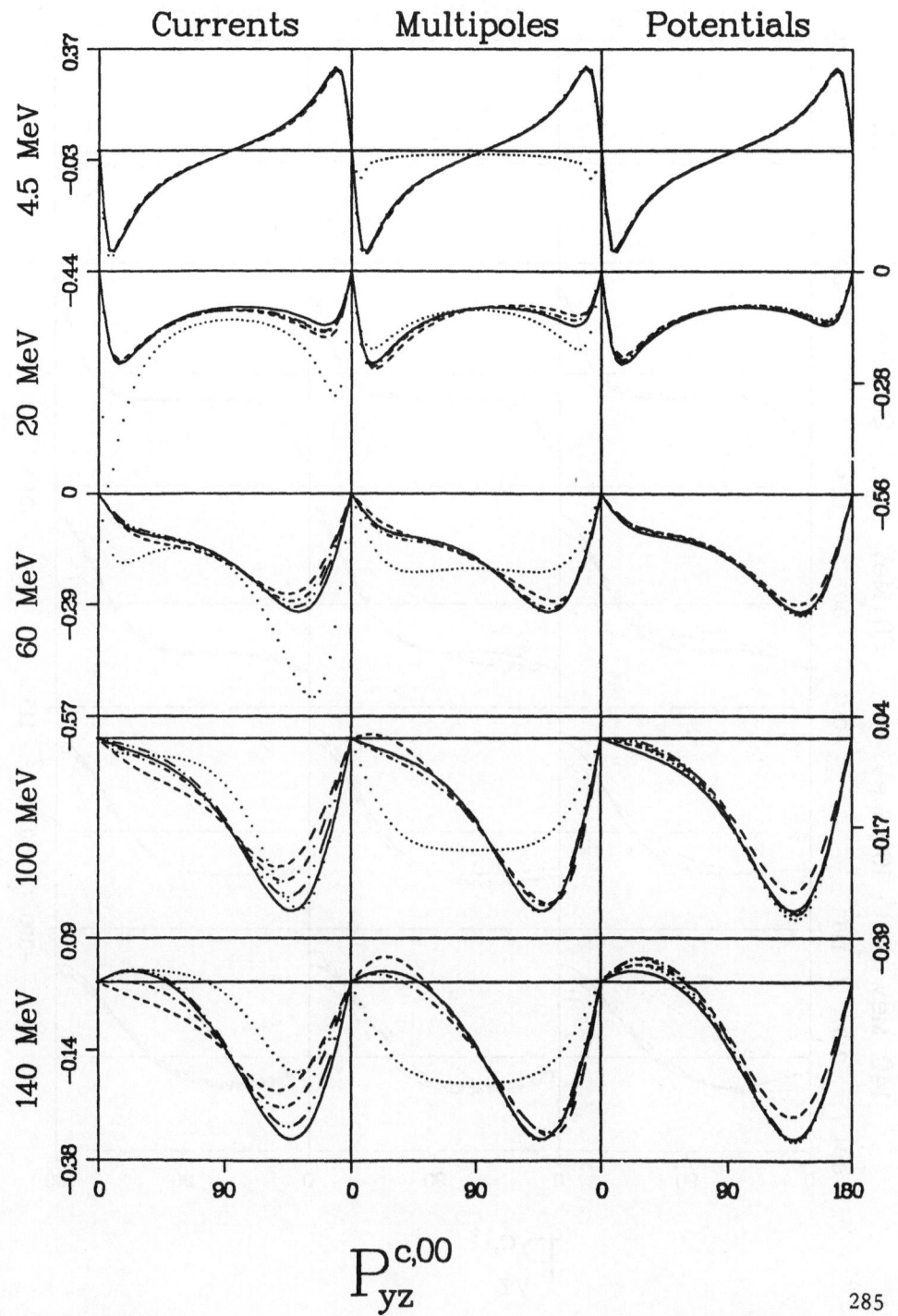

$$P_{yz}^{c,00}$$

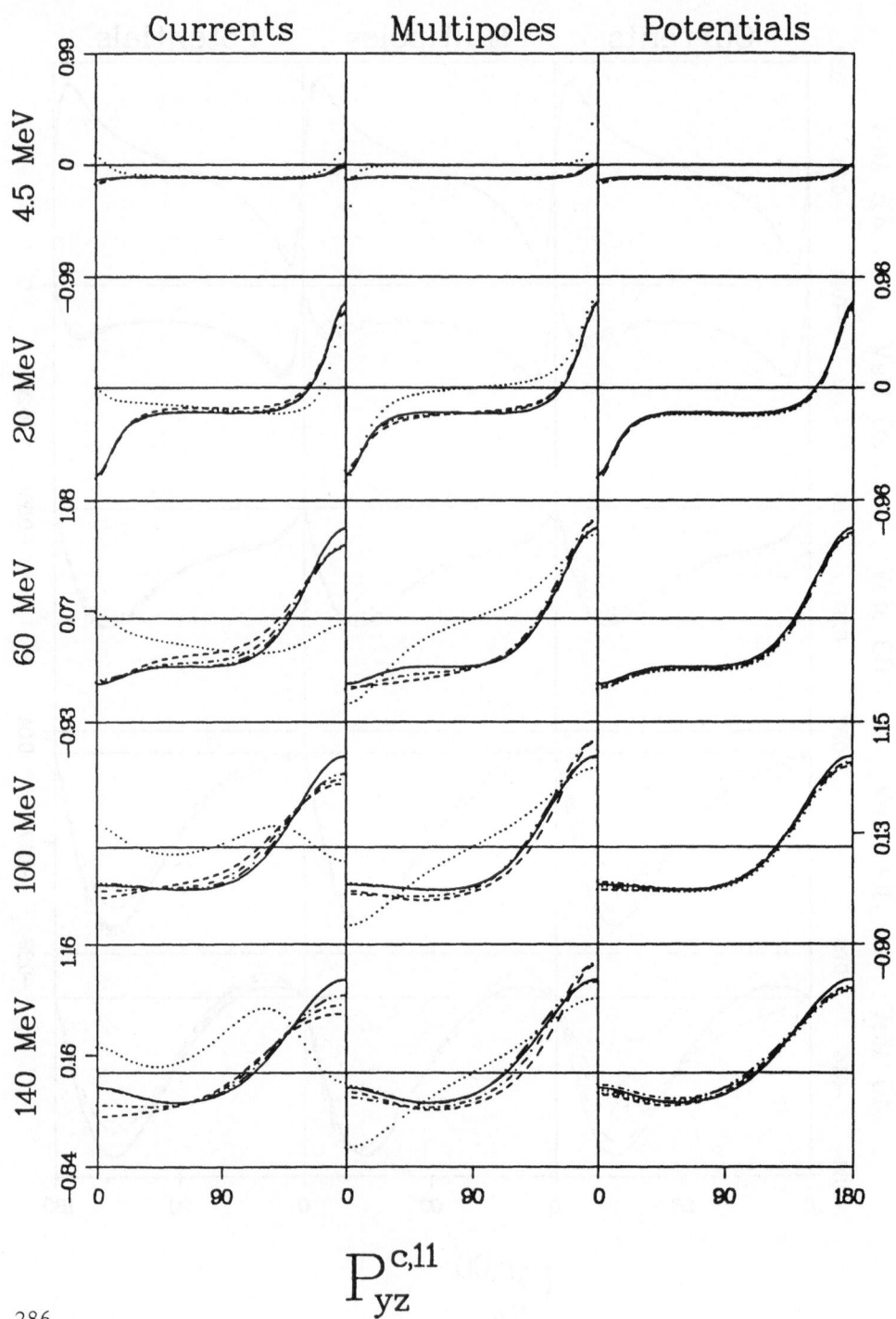

$$P_{yz}^{c,11}$$

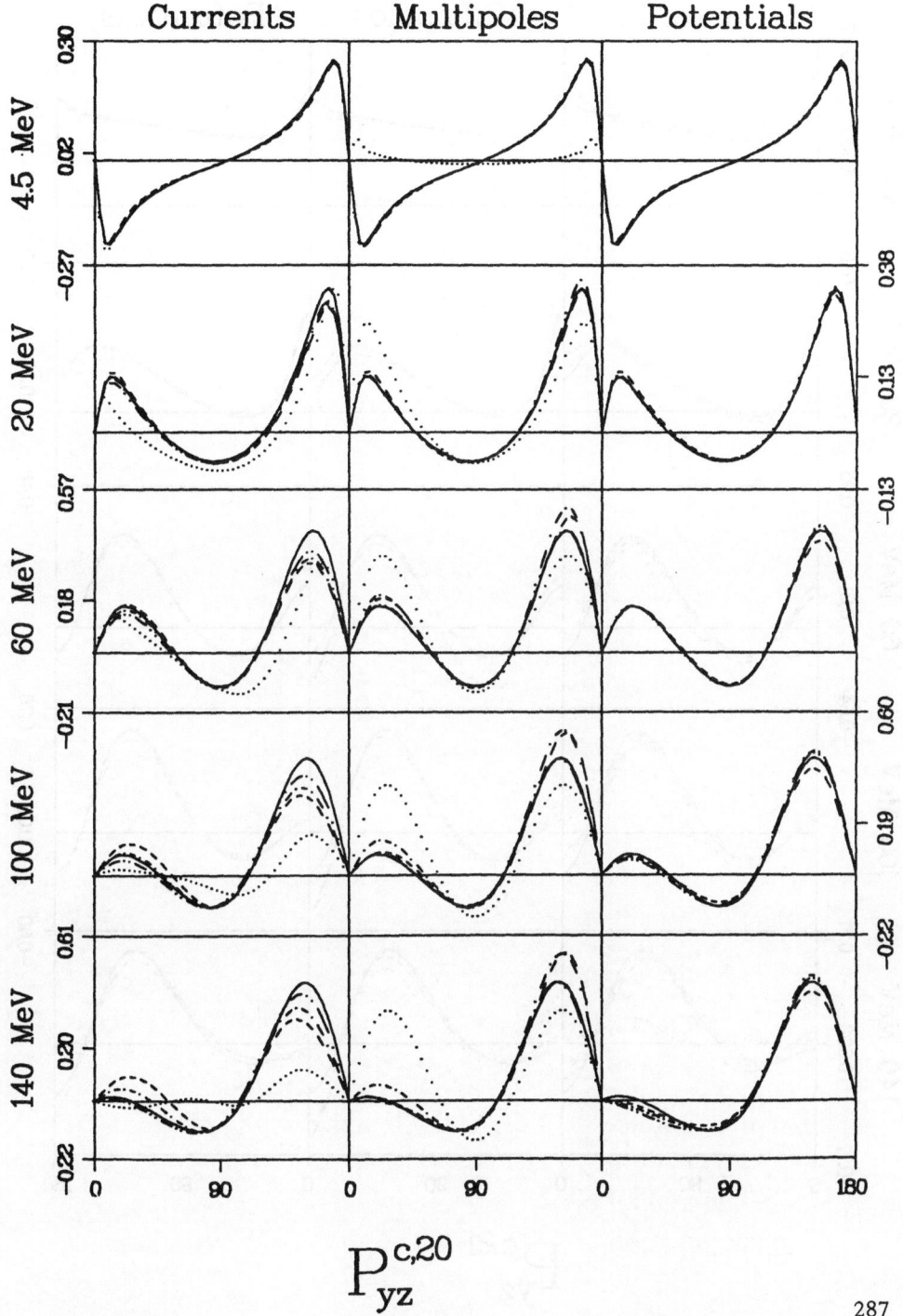

$$P_{yz}^{c,20}$$

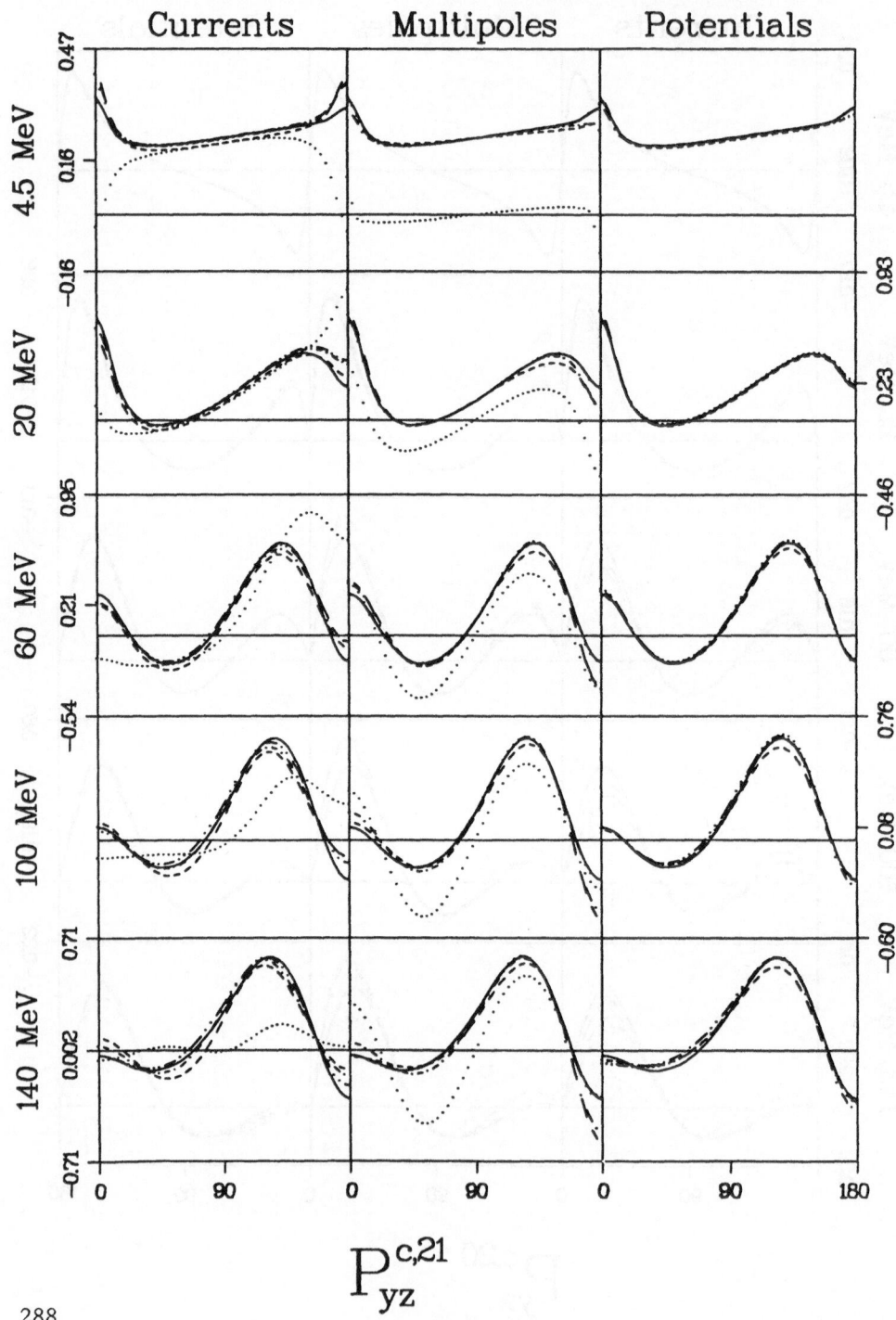

$$P_{yz}^{c,21}$$

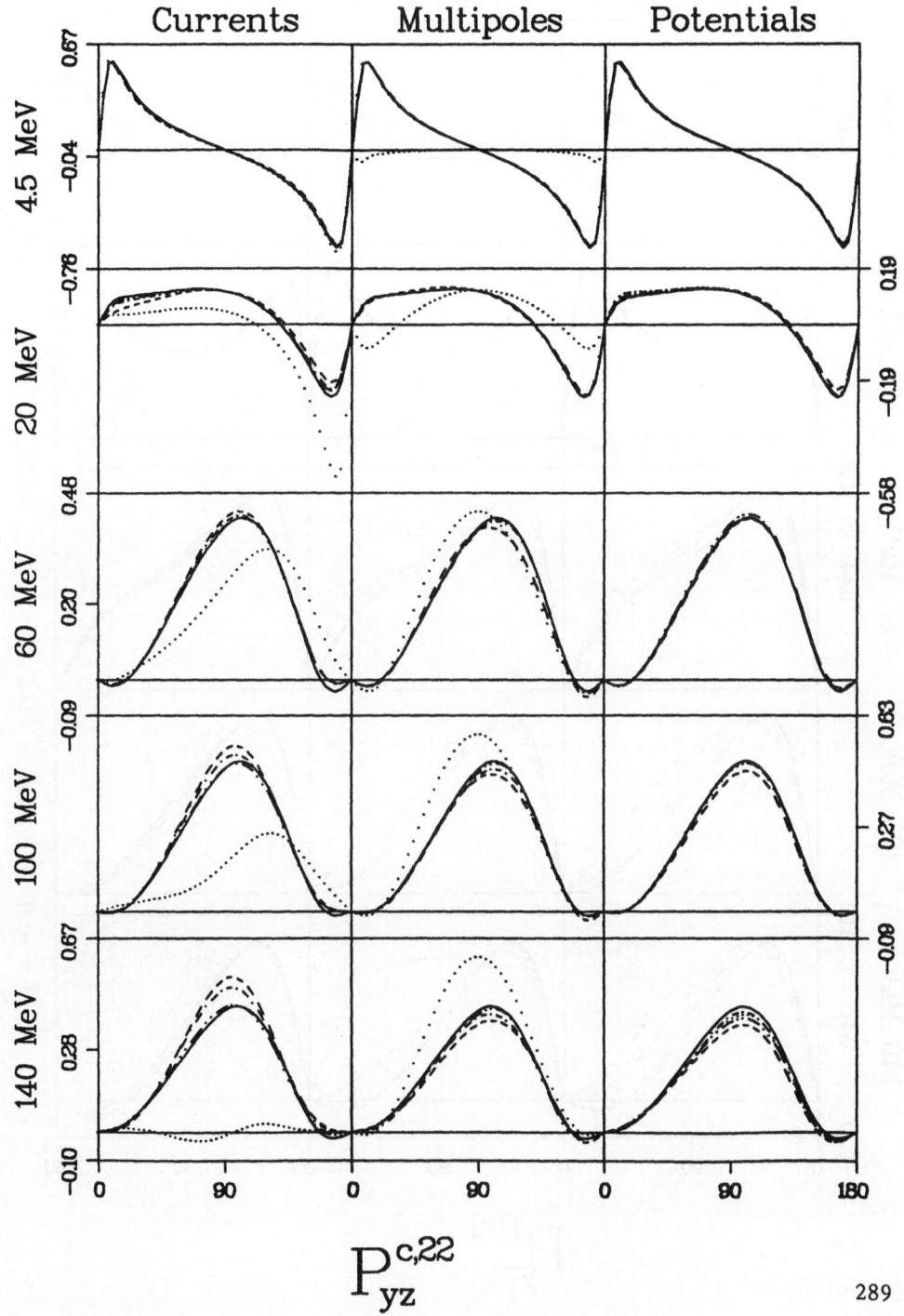

$$P_{yz}^{c,22}$$

5.49 Proton-Neutron Spin Correlation $P_{yz}^{l,IM}$

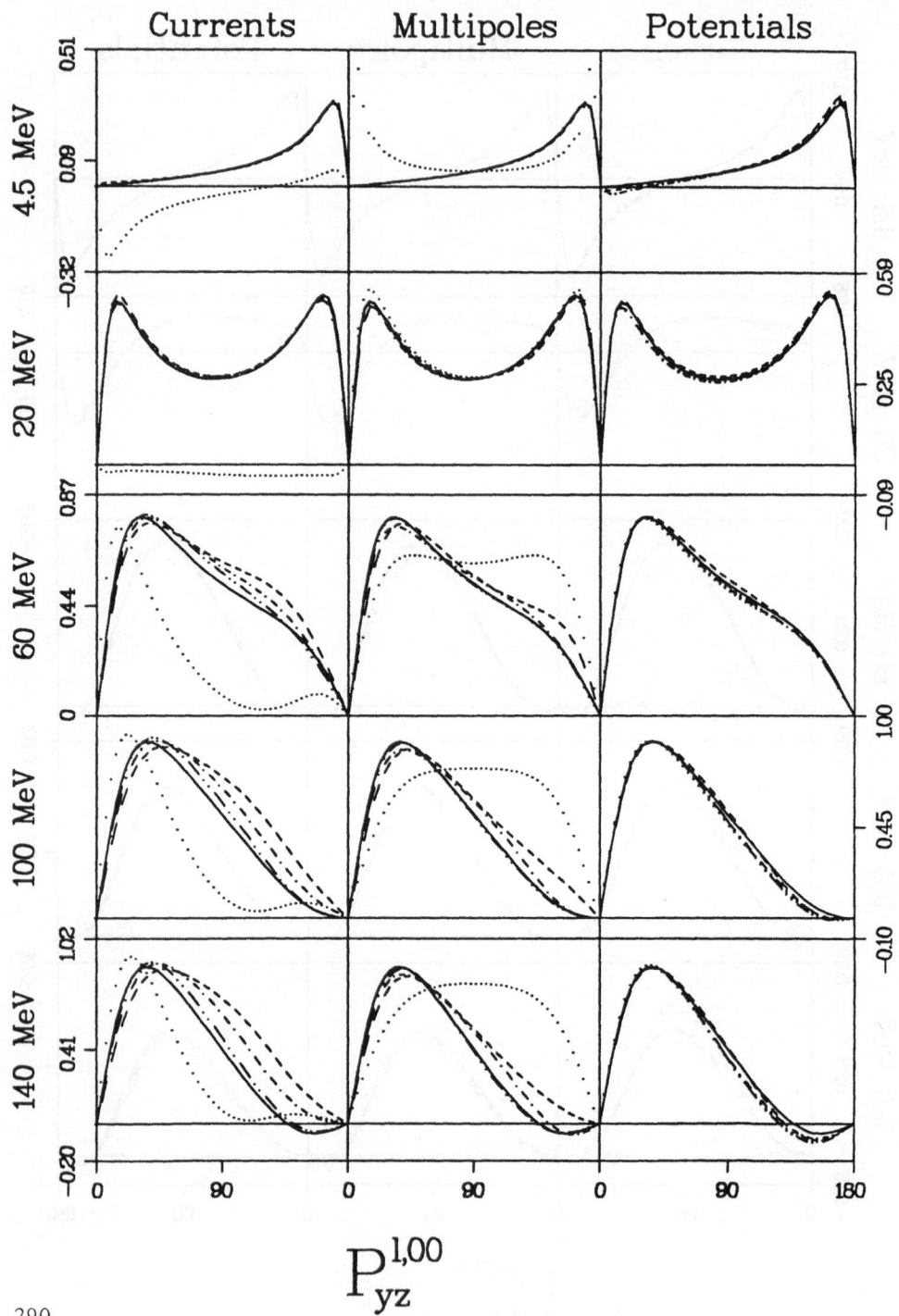

$$P_{yz}^{1,00}$$

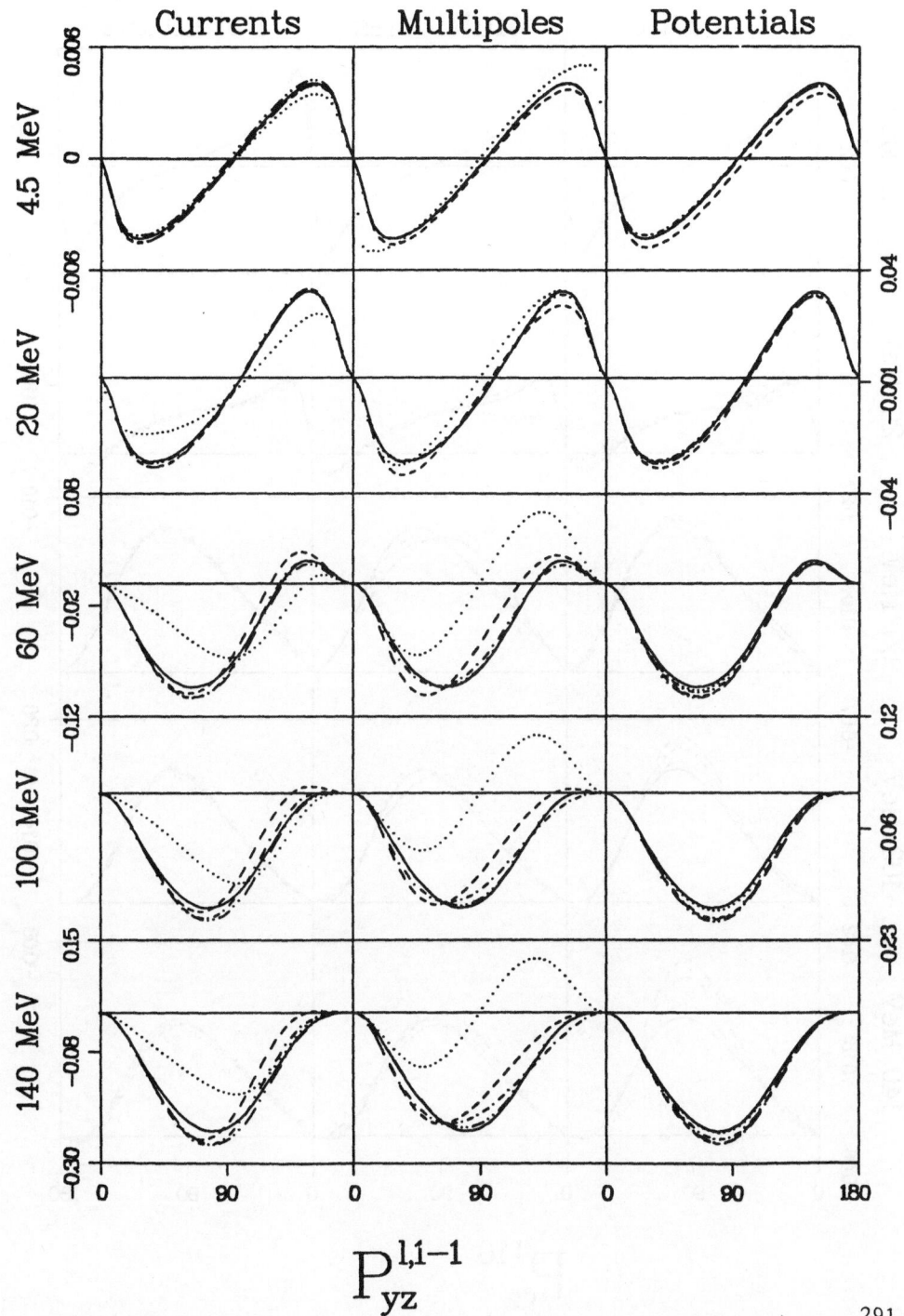

$$P_{yz}^{l,1-1}$$

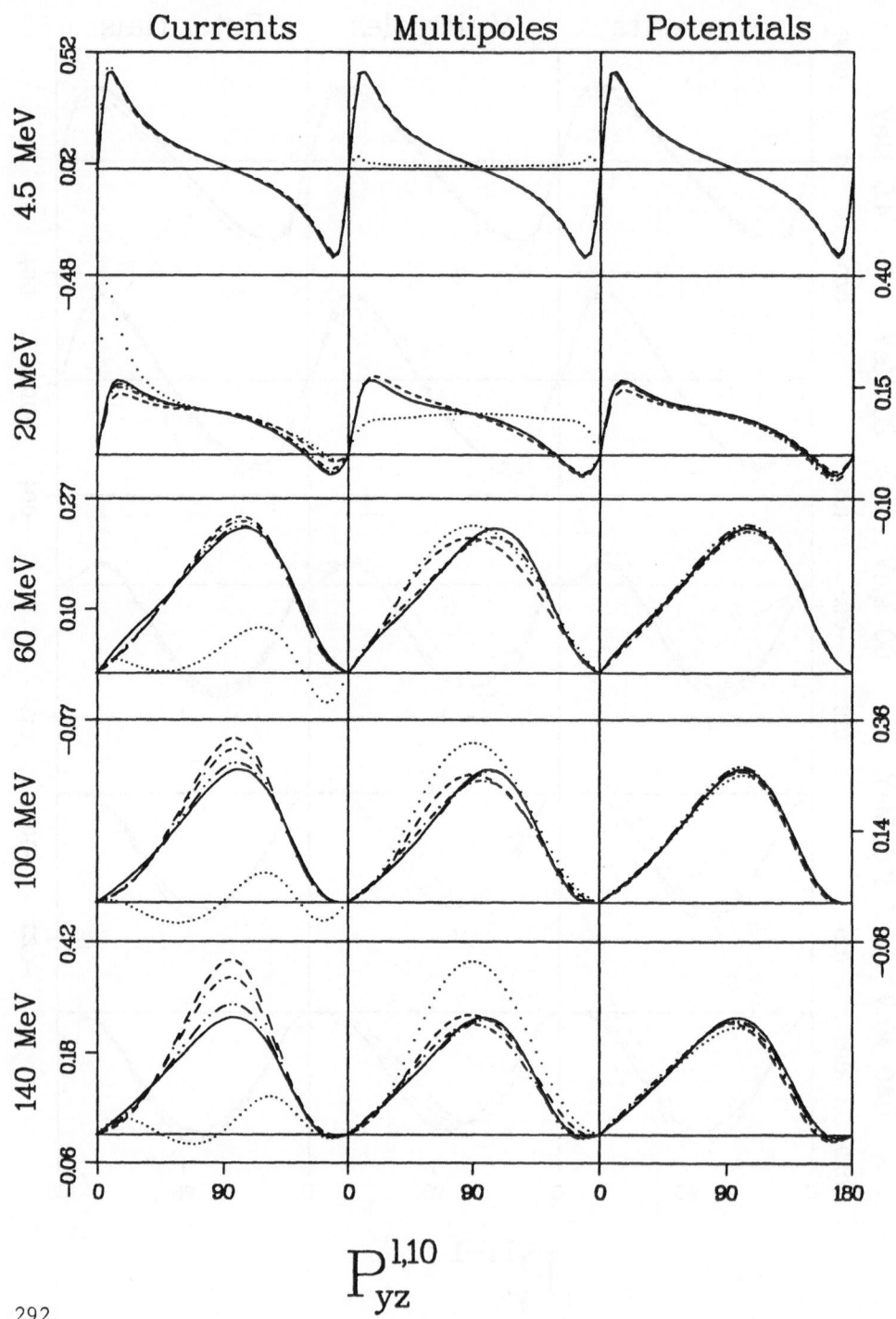

$$P_{yz}^{l,10}$$

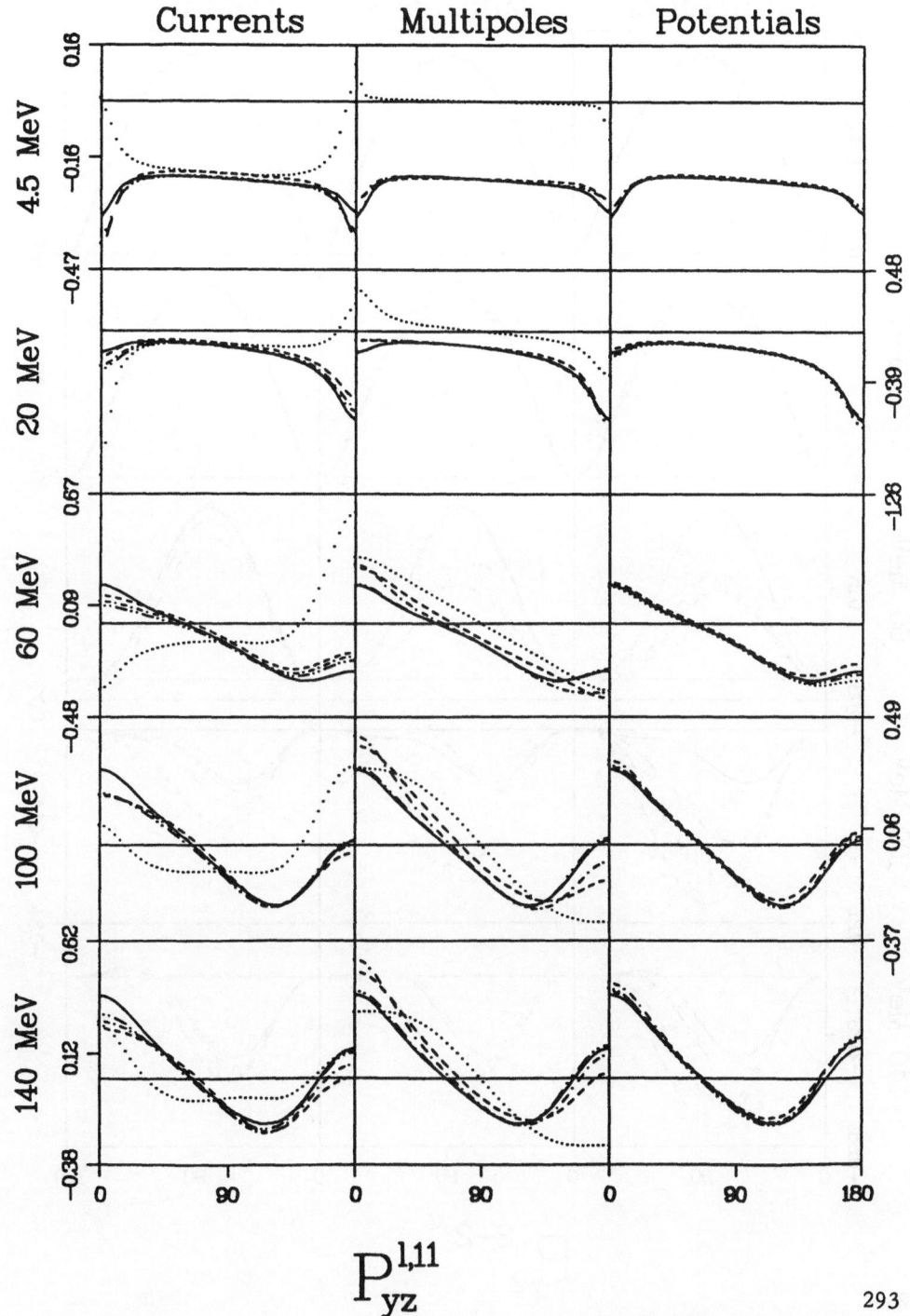

$$P_{yz}^{l,11}$$

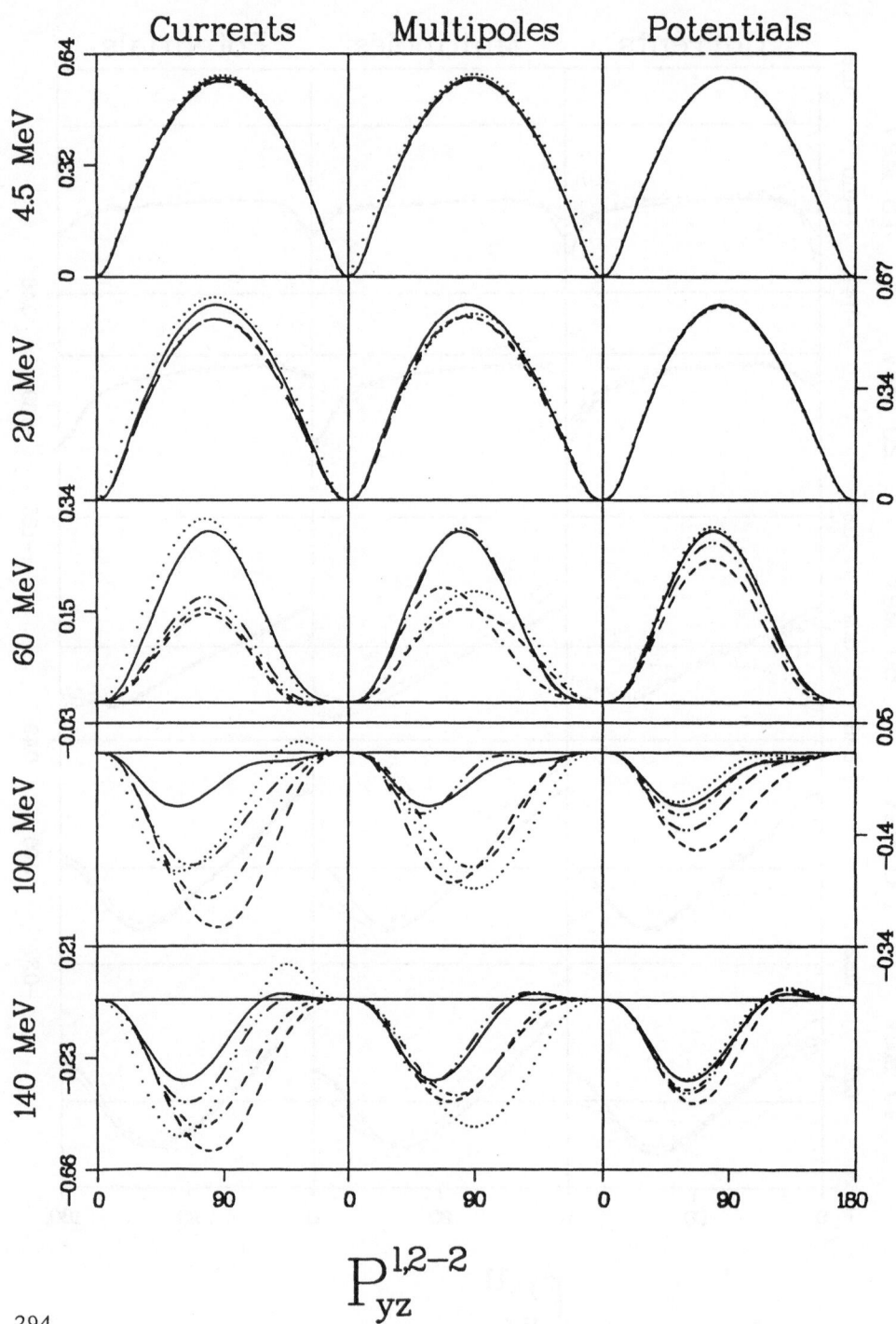

$$P_{yz}^{1,2-2}$$

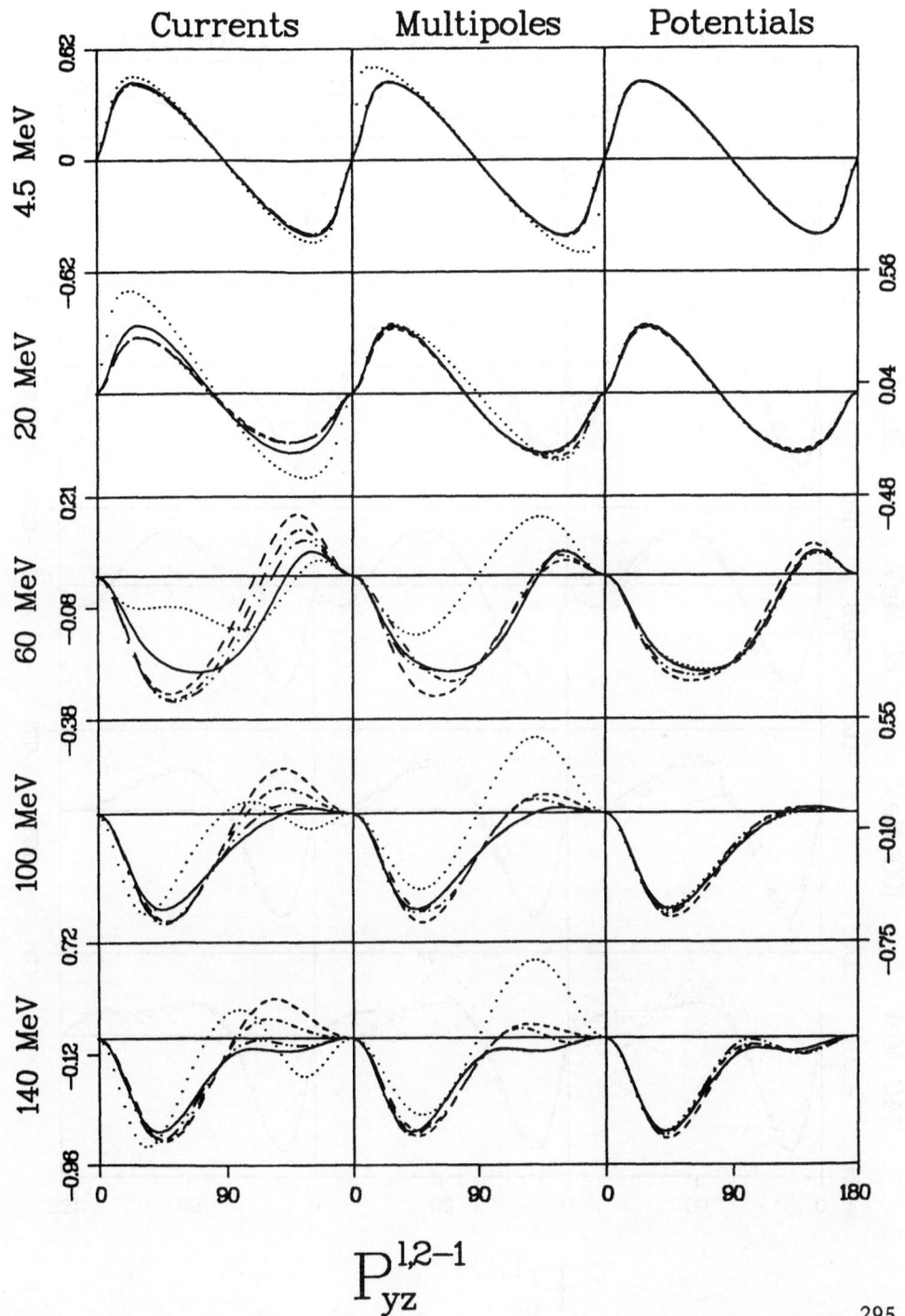

$$P_{yz}^{1,2-1}$$

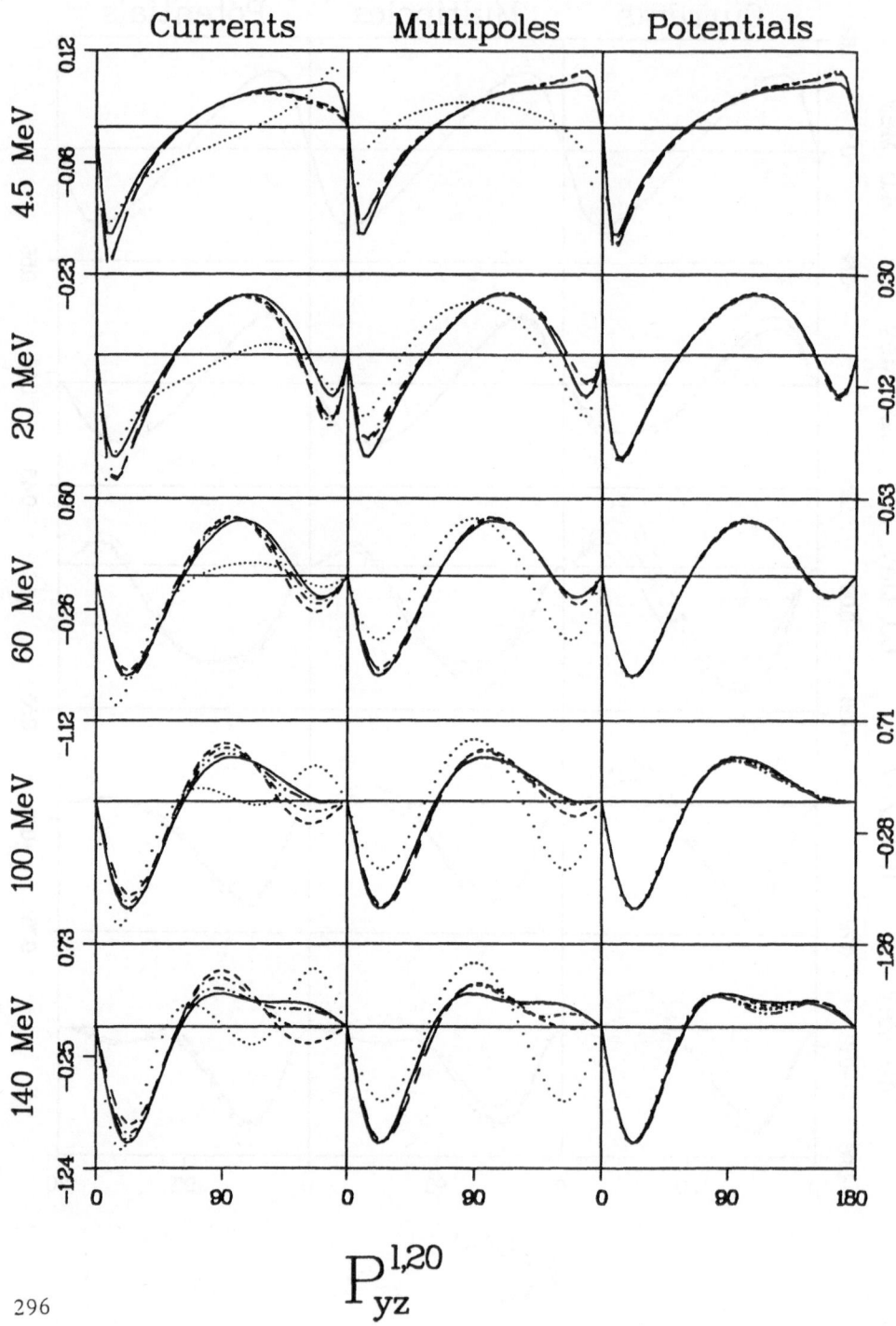

$$P_{yz}^{1,20}$$

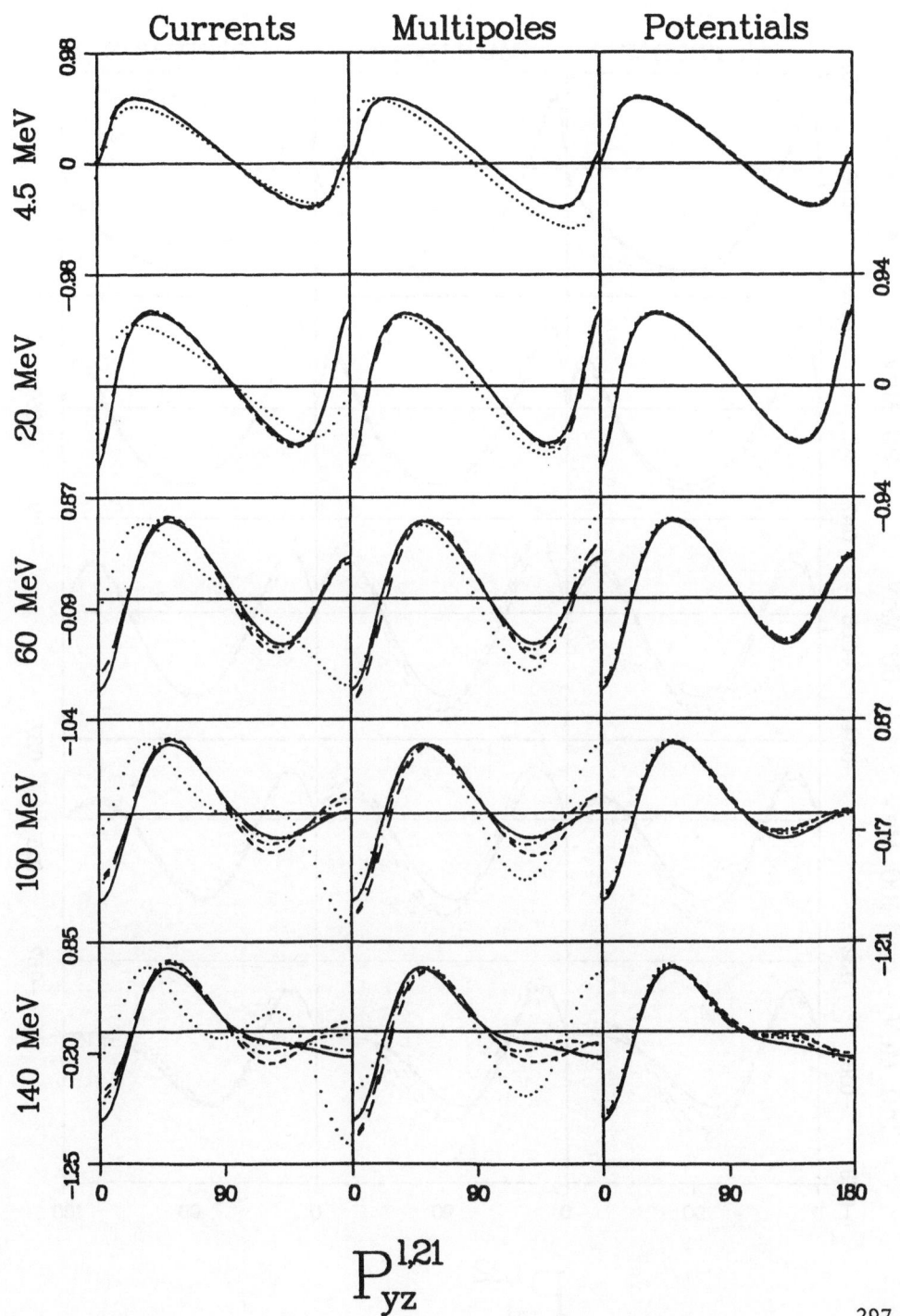

$$P_{yz}^{1,21}$$

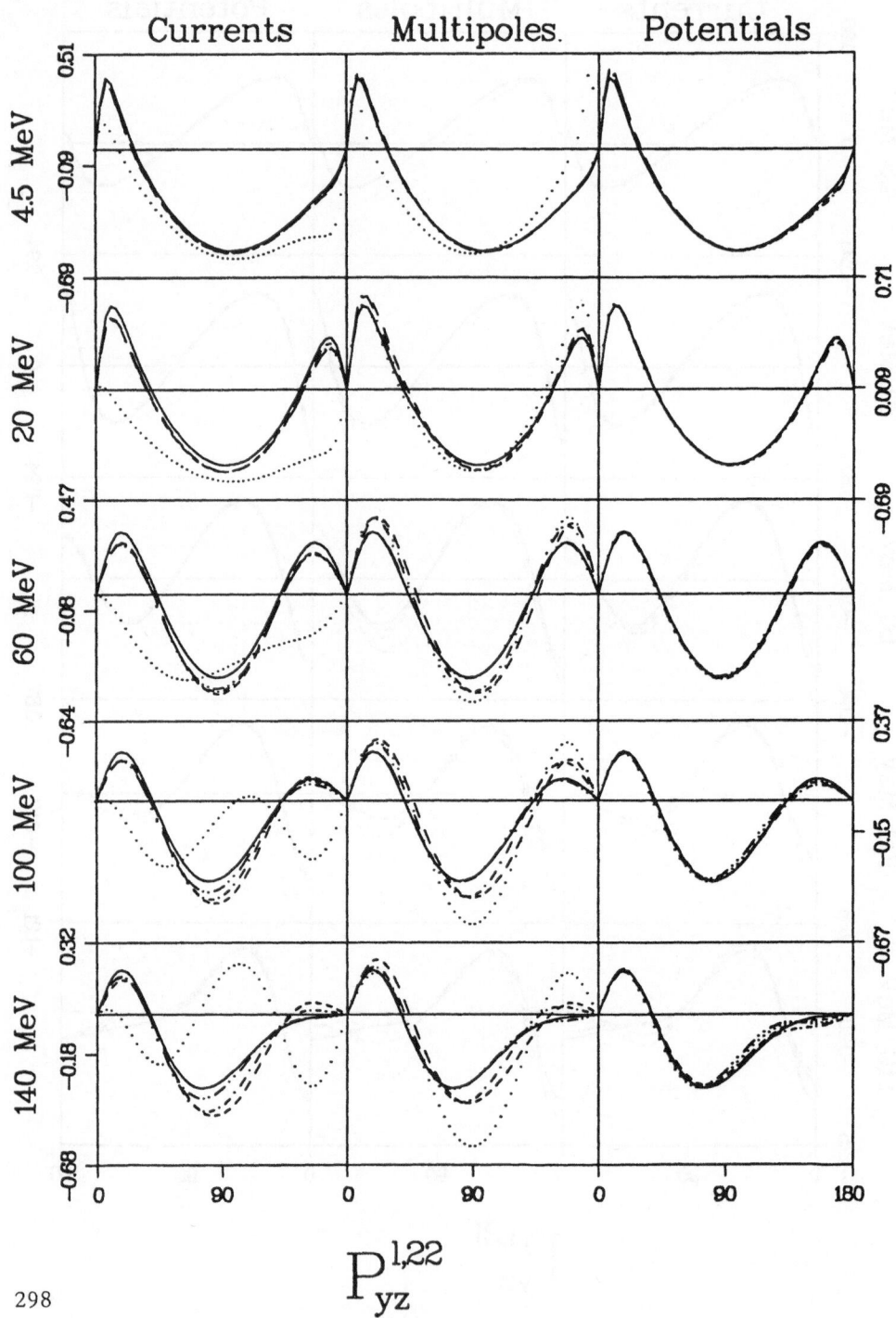

Currents   Multipoles.   Potentials

$$P_{yz}^{1,22}$$

298